D1825898

*l*s

38

Toward a Cognitive Theory of Narrative Acts

Cognitive Approaches to Literature and Culture Series
Edited by Frederick Luis Aldama, Arturo J. Aldama,
and Patrick Colm Hogan

Cognitive Approaches to Literature and Culture includes monographs and edited volumes that incorporate cutting-edge research in cognitive science, neuroscience, psychology, linguistics, narrative theory, and related fields, exploring how this research bears on and illuminates cultural phenomena such as, but not limited to, literature, film, drama, music, dance, visual art, digital media, and comics. The volumes published in this series represent both specialized scholarship and interdisciplinary investigations that are deeply sensitive to cultural specifics and grounded in a cross-cultural understanding of shared emotive and cognitive principles.

Toward a Cognitive Theory of Narrative Acts

EDITED BY FREDERICK LUIS ALDAMA

University of Texas Press ◆ *Austin*

Requests for permission to reproduce material from this work should be sent to:
    Permissions
    University of Texas Press
    P.O. Box 7819
    Austin, TX 78713-7819
    www.utexas.edu/utpress/about/bpermission.html

⊗ The paper used in this book meets the minimum requirements of ANSI/NISO
Z39.48-1992 (R1997) (Permanence of Paper).

Library of Congress Cataloging-in-Publication Data
Toward a cognitive theory of narrative acts / edited by Frederick Luis Aldama.
— 1st ed.
    p.   cm. — (Cognitive approaches to literature and culture series)
    Includes bibliographical references and index.
    ISBN 978-0-292-72157-9 (cloth : alk. paper)
    1. Creation (Literary, artistic, etc.)—Psychological aspects.   2. Creative
ability—Psychological aspects.   3. Cognitive science.   I. Aldama, Frederick
Luis, 1969–
    BF408.T644 2010
    700.1—dc22
                                                                2009050909

For my colleagues and students daring to scratch out lines in the sand betwixt and between the sciences and humanities

A special thanks to Jim Phelan and the journal Narrative *as well as Jeffrey Tibbett for his work compiling the glossary*

Contents

Introduction: The Sciences and Humanities Matter *as* One   1
FREDERICK LUIS ALDAMA

PART I. **General and Theoretical Considerations**   11

1. Arts in the Brain; or, What Might Neuroscience Tell Us?   13
HERBERT LINDENBERGER

2. Narrative as Nourishment   37
ELLEN SPOLSKY

3. Narrative Empathy   61
SUZANNE KEEN

4. The Biolinguistic Turn: Toward a New Semiotics of Film   95
JAVIER GUTIÉRREZ-REXACH

5. Voice and Perception: An Evolutionary Approach to the
Basic Functions of Narrative   119
KATJA MELLMANN

6. Dreaming and Narrative Theory   141
RICHARD WALSH

PART II. **A Selection of New Approaches**   159

7. Cross-Cultural Mind-Reading; or, Coming to Terms with the Ethnic
Mother in Maxine Hong Kingston's *The Woman Warrior*   161
KLARINA PRIBORKIN

8. Theory of Mind and Michael Fried's *Absorption and Theatricality*: Notes toward Cognitive Historicism   **179**
    LISA ZUNSHINE

9. Garden Paths and Ineffable Effects: Abandoning Representation in Literature and Film   **205**
    H. PORTER ABBOTT

10. Consciousness, Ethics, and Narrative: Reading Literature in an Age of Torture   **227**
    PATRICK COLM HOGAN

11. Prophesying with Accents Terrible: Emotion and Appraisal in *Macbeth*   **251**
    LALITA PANDIT HOGAN

    Glossary   **281**

    Bibliography   **285**

    Contributors   **313**

    Index   **317**

Introduction: The Sciences and Humanities Matter *as* One

FREDERICK LUIS ALDAMA

In 1959 when C. P. Snow famously declared the need for building bridges between "the two cultures" of the sciences and the humanities in order to make the world a better place, the impulse wasn't so much ahead of the curve, but simply wrong. It wasn't that in 1959 few knew how and where to build such bridges. The separation itself was a line drawn in sand—specious and artificial.

To understand why our work in the humanities matters, the scholarship presented in *Toward a Cognitive Theory of Narrative Acts* shows just how we can wipe out that erstwhile line drawn in the sand that artificially separates the sciences from the humanities; that separates those disciplines deemed empirical and in pursuit of the universal (the "nomological") from those considered speculative (even if analytical) and in pursuit of the idiosyncratic (the singular or "punctual").

Whether we're talking about mathematics, physics, chemistry, and biology or psychology, literature, linguistics, philosophy, history, and classics, knowledge in each discipline advances along a continuum between the hypothetical and the tested and confirmed. The humanities are part and parcel of the knowledge the human species has acquired about itself over the centuries, together with the social sciences and the natural sciences.

Knowledge is acquired in many ways and may exhibit many degrees of generality, certitude, and power to predict the future. It can be the product of direct observation and a small number of general assumptions, or the result of a very elaborate and long chain of hypotheses and deductions. It can possess a rich factual content or be almost devoid of it, but it must always lead back to factual observations.

Indeed, it is the development of knowledge in the humanities that has allowed many fields to become real scientific disciplines and to become separate scientific endeavors, distinct from the humanities as still embryonic forms of future disciplines. By the end of the Middle Ages and from the Renaissance onward, many fields studied by Aristotle became independent sciences (particularly in the eighteenth and nineteenth centuries). The same thing happened with the domains included in the humanities. The number and the nature (contents) of those domains changed over the centuries and gave birth to new fields of humanities and to the so-called social sciences, along with a certain number of natural sciences.

In our many activities we are reshaping the world that exists independent of our creating; in turn, this transformed world shapes us and our subsequent activities. This takes place in a *one world*—a supersystem composed of interconnected systems of various kinds (physical, cultural, social, biological, and so on) that possess their own peculiar properties and laws. Given that this world is one, as the different tributaries of knowledge gain force and depth they necessarily move with increasing momentum toward one main stream.

In the work of Aristotle, Marx, Einstein, and Chomsky, we see this disciplinary convergence at play. Think of all the fields studied by Aristotle (including the knowledge he himself developed) to understand his one world more completely: from biology (zoology and botany) to medicine and the constitutions of the nation-states in Greece, to law and physics, and, of course, to logic. Chomsky's early and radically innovative formulation of a universal grammar is also a case in point. His research program threw new and important light not only on linguistics, but also on biology (the modularity of the brain), psychology (the mind not as *qualia* independent of the brain, but as a mind/brain composite), and mathematics (recursive function theory). Today, we see modern linguistics in convergence with other research programs such as those advanced in neuro- and evolutionary biology as well as in zoology (animal communication systems). (See Chomsky et al., "The Faculty of Language: What Is It, Who Has It, and How Did It Evolve?") Today, the human genome mapping not only gives us a much sharper understanding of biology (genetics), but proves instrumental in advancing knowledge in geography, archeology, anthropology, and linguistics, among other areas of inquiry.

The scholarship in this collection embodies this impulse toward unification. Each scholar seeks to sidestep subjective opinion and raw speculation in favor of establishing a research program that moves forward through a series of checks and balances (proofs and disproofs) within his or her respective area of scholarly inquiry. While each uses a different theoretical approach, each employs advances in the sciences (especially the biological) to understand better all variety of cultural phenomena that we spin out of ourselves.

Opening Part I of the collection, "General and Theoretical Considerations," Herbert Lindenberger's chapter "Arts in the Brain; or, What Might Neuroscience Tell Us?" shows how research in the neurobiological sciences is indispensable to understanding more deeply our creating and engaging with painting, music, and literature. His examples are many. In one instance, he analyzes why Stravinsky's compositions and not those of Schoenberg went from being rejected to being widely accepted; while the brain evinces a remarkable plasticity, no matter the amount of social and cultural educating, it can only accept a limited amount of chaos before it simply turns off. Ellen Spolsky's "Narrative as Nourishment" uses the conceit of metabolism as a way to develop a cannibalistic approach to understanding innovation in narrative fiction. Her "reality-based" approach applies research in the cognitive sciences to deepen our understanding of how authors strategically create gaps in the information they provide about characters, settings, and events that deliberately lead readers to fail in their inference and mind-reading of characters and authors. For Spolsky such narrative fictions lead to a certain hunger in other authors; they "stimulate others to expand the search for the kind of nourishment that could satisfy." In "Narrative Empathy" Suzanne Keen asks that we reconsider just how a narrative fiction allows us to step into the shoes of and feel for a character. This is not a result of narrative technique, but rather the net result of the narrative fiction as a whole created by the author (the blueprint) and the active participation of the reader. Moreover, the author's creating of the narrative fiction (blueprint) and the reader's following of this blueprint—our filling in all the gaps to solidify in the imagination a coherent and whole storyworld—involves centrally our capacity to empathize and to read mental states of characters and how these mental states relate to the world around them.

Several chapters take an interdisciplinary turn into areas such as film. In "The Biolinguistic Turn" Javier Gutiérrez-Rexach points out just how unlike the human mind is to machines (computers) in order to advance a biolinguistic film studies theory. Gutiérrez-Rexach builds on the work in modern linguistics that formulates our evolution of a universal grammar and "conceptography" and their accompanying specific competencies to deepen our understanding of how film works at the universal and idiosyncratic level. The two chapters that follow offer contrastive complements to one another. The first, Katja Mellman's "Voice and Perception: An Evolutionary Approach to the Basic Function of Narrative," identifies how our evolved cognitive capacity to extract information maps onto the emergence in literature of two narrative fiction devices: "voice" and "perception." Mellmann provides a nuanced understanding of how these devices work (together and apart) to convey information in literary and filmic narratives. The second, "Dreaming and Narrative Theory" by Richard Walsh, asks that we revise narrative theory categories and distinctions—say, between story (the time of the events as occurring within the storyworld) and discourse (the time and ordering of the narrative telling)—in light of our understanding of how lucid dreaming works. Sequences and themes of fictional narratives like novels share much cognitive ground with those created in lucid dreaming—and the psychological effects produced by both are similar. This has deep implications not only in terms of revising many of the tools identified by narratology, but also in determining that it is the medium of telling (discourse) and not the story that really matters when talking about, for instance, literature-to-film adaptations.

Part II of the collection, "A Selection of New Approaches," includes chapters that explore how narrative fictions trigger and even frustrate our Theory of Mind capacity. In "Cross-Cultural Mind-Reading; or, Coming to Terms with the Ethnic Mother in Maxine Hong Kingston's *The Woman Warrior*," Klarina Priborkin teases out the cognitive schemas that prevent real understanding and lead to misreadings of interior states of mind between two generations of Asian American women. That readers also have a Theory of Mind capacity allows Kingston to use certain devices in the creating of her narrative fiction that allow us to step more powerfully into the shoes of her characters. In Chapter Eight, "Theory of Mind and Michael Fried's *Absorption and Theatricality*: Notes toward Cognitive Historicism," Lisa Zunshine broadens the critical scope of a Theory of Mind approach, considering a number of different cultural phenomena within a historical context: how it plays out in the creative work of eighteenth-century authors, playwrights, and artists who were interested in representing unusual states of mind and a suspension of self-consciousness. In an analysis of Chardin's painting "The Soap Bubble," for instance, Zunshine argues that it draws in its audience because of its depiction of a perceived transparency of mind embodied in its painted figures.

In contrast to Priborkin and Zunshine, H. Porter Abbott considers those narrative fictions that engage audiences less through the triggering of our Theory of Mind (our ability to read interior states and intentions from exterior gesture and expression) with characters and their situations and more through the firing up of our puzzle-solving faculty. In "Garden Paths and Ineffable Effects: Abandoning Representation in Literature and Film," Abbott explores a number of resolvable and irresolvable "garden-path" narrative fictions (literature and film) that challenge our cognitive impulse to link mimetically the fictional with the real world. Those that contain too little information (or in the case of Joyce's *Finnegans Wake*, too much) and too large a gap in their narrative detail resist our meaning-making process, and we shut off. However, authors and directors can find just the right balance between too much and too little, allowing us to parse the information and make meaning, but often in ways that lead us to titillatingly ambiguous results.

Considerations of Theory of Mind also play an important role in Patrick Colm Hogan's "Consciousness, Ethics, and Narrative: Reading Literature in an Age of Torture." Here, however, Hogan sets his sights on how appraisal theory, emotion (guilt, for instance), and Theory of Mind work in our conflict (and resolving of conflict) between right and wrong. In a discussion of *King Lear*, Hogan shows how it is narrative prototypes—and not rules of behavior—that centrally inform our ethical choices. Lalita Pandit Hogan uses appraisal theory to offer a nuanced analysis also of a Shakespeare play. In "Prophesying with Accents Terrible: Emotion and Appraisal in *Macbeth*," rather than follow a moral-oriented analysis of the play as is the tradition, appraisal theory reveals *Macbeth* to be a play that evinces a complex variety of emotions in its characters (and audience) that arise out of multiply layered conflicts in realizing goals and ambitions.

While each of the scholars included in this collection uses many different tools to analyze an array of cultural objects, they all seek to follow an interdisciplinary research program that formulates hypotheses, tests them, and revises them with the aim of moving toward a unified knowledge. This does not mean, as Lindenberger astutely reminds us, "that science will rob us of our traditional task of conserving the past and placing artworks within an appropriate historical context, or, for that matter, of subjecting them to the interpretation and evaluation that their consumers have long depended upon." Rather, as with other like-minded scholars, we can learn from science, and this may very well affect, as Lindenberger concludes, "the methods that we choose to exercise our interpretive and evaluative skills."

Toward a Cognitive Theory of Narrative Acts represents but a small sample of the growing body of scholarship out there that shares this serious interdisciplinary impulse. Here in my backyard at the Ohio State University we have David Huron, who turns to cognitive science and neurobiology to explore how music stimulates the imagination, triggers our appraisal faculty, and moves us emotionally (*Sweet Anticipation*); Timothy Schroeder in philosophy turns to neuroscience to shed light on what motivates us to act, to feel pleasure and displeasure—to desire and imagine (*Three Faces of Desire*); Jim Phelan turns to the field of ethics in his study of how short stories and novels move us to feel and act (*Experiencing Fiction*). In my own work I seek to have a firmer grasp on how novels, films, comic books, and the like, *tick*. (*Your Brain on Latino Comics* and *Why the Humanities Matters* offer recent examples.)

Beyond Ohio State University there are scholars such as Robert Storey, Joseph Carroll, Jonathan Gottschall, and Nancy Easterlin, among many others, who have turned to advances in evolutionary biology to explore the repetition and rupture of human behavior as expressed in the arts. Further solidifying the importance of art and music in our everyday activities, Steven Mithen (*Singing Neanderthals*) turns to research in archeology and evolutionary biology. Advances in brain science research allow neuroscientist Isabelle Peretz to mine the field of musicology (*The Cognitive Neuroscience of Music*) to understand the differences and similarities in how we process music and language. In film and drama studies, scholars such as Greg Smith (*Film Structure and the Emotion System*), Per Persson (*Understanding Cinema*), Bruce McConachie, and F. Elizabeth Hart (see their co-edited *Performance and Cognition*) turn to advances in cognitive psychology and cognitive linguistics to understand better how movie directors and dramatists employ and re-deploy certain cognitive and emotive mechanisms and processes to innovatively engage the minds of audiences across time and space.

Others have used new technologies such as fMRI, EEG imaging, and PET scans to compare networks of neural response in fiction and nonfiction activities (see Paul Mathews and Jeffrey McQuain's *The Bard on the Brain*). John G. Nichols (*From Neuron to Brain*) and Joseph LeDoux (*The Emotional Brain*) investigate how the biochemical and biophysical transmission of nerve signals creates in humans a higher-minded self that has evolved the capacity for metarepresentation—a key ingredient in our evolved faculty for communication, including the visual, plastic, and written arts. Michael Gazzaniga explores those neural mechanisms that underlie ethics, empathy, conceptual thinking, and self-awareness that allow us to make and engage with the arts (*Human: The Science be-*

hind What Makes Us Unique). V. S. Ramachandran's neuroscientific research on perception and the senses generally aims to throw light on aesthetics (*A Brief Tour of Human Consciousness*). Antonio Damasio's neurobiological studies extend our knowledge concerning the functioning and role of emotions (*Looking for Spinoza*). And others such as Marc D. Hauser (*Moral Minds*), as well as Jorge Moll and Joshua Greene along with their respective research teams, explore the neurological basis for ethical behavior and attitudes.

In our one world, we all tell stories based on actual or made-up events. Our narrative-making and -engaging capacity is universal. And, while biology constrains what we can do with oral narrative fiction (the limits of our memory, for example), we have developed ways to throw off such restraints in a variety of metarepresentational media, including the form of written narrative fiction.

In short stories and novels, for instance, there can be deep temporal layers as well as subtle forms of filtering a story through the perspective of a character, such as the use of free indirect discourse and psycho-narratology. The work done in narratology, coupled with research developed in the cognitive sciences and neurobiology, can deepen further the use of specific narrative techniques, as well as shed light more generally on the processes involved in our making and consuming of narrative fictional works. As Lindenberger discusses in Chapter One, the recent discovery of the mirror neuron system—a set of neurons that become active in the brain's superior temporal sulcus and the Broca region (or language center) when we perform *and* observe an action—reveals more precisely how we produce language, feel empathy, interpret and understand other people's emotional states, and know the difference between fact and fiction.

I mentioned already Lindenberger's tempering of the interdisciplinary move in the humanities. Before wrapping up this introduction, let me offer several additional points of caution.

The use of science to advance an understanding of how, say, narrative fiction works—or opera, or comic books, or visual arts, and so on—is not an end-all and be-all to productive humanistic inquiry. The turn to advances made in cognitive psychology, evolutionary biology, and the brain sciences generally is not meant to be some sort of life jacket that will save the humanities. To consider it thus necessarily risks confusing and even subordinating the study of music, novels, films, comic books, opera, painting, and the like, with the methods and goals that characterize science.

As the work here attests, we are entitled to use any knowledge coming from any discipline that will shed light on our object of study; that line drawn in sand between the humanities and sciences needs smoothing over. Hence we would do well to consider how information in the sciences helps further deepen our knowledge of how, say, music, literature, painting, drama, and so on, work cross-culturally to engage audiences (readers, listeners, viewers). Likewise, it is a good idea to formulate a hypothesis, speculate in as hardy a way as possible without any limitations (no fear of what others think), and develop one's thought as far as possible concerning the notion of, say, music, literature, drama, and the like. For instance, one might speculate or hypothesize that recent studies in neuroscience on the localization within the brain of ethical dilemmas—and solutions to these dilemmas—can deepen a theory of narrative acts, allowing one to propose predictive hypotheses that concern cognitive and emotive capacities shared by all human beings. And, one might turn to advances in neurobiology to see what kinds of explanations there are on our everyday faculty for empathy and Theory of Mind to understand better how this works in narrative fiction. However, it would be a mistake to abandon partially or totally a specific study of empathy or ethics in literature under the pretext that whatever there is to know is to be taken from neurobiology.

The humanities are an integral part of the knowledge the human species has acquired about itself over the centuries, together with the social sciences and the natural sciences. And it is important for us to cherish the knowledge we have acquired and to continue developing it, because the practical and theoretical human activity called knowing (learning, experimenting, verifying, and teaching), together with the theoretical and practical human activity called technology and technological invention, are the specifically human activities that allowed us to survive and to radiate across the planet. Knowledge is quite literally essential for our survival. In whatever shape you want to slice global reality (social, natural, and so on), and to whatever degree of certainty that we may possess knowledge at any given moment, such knowledge (however limited and provisional) is a necessary stepping-stone for the acquisition of more precise knowledge today and tomorrow.

Some scholars currently worry about the future of the humanities. The scholarship presented in *Toward a Cognitive Theory of Narrative Acts* refuses to worry. To the extent that their contents become more and more precise knowledge, the humanities will give birth to (or transform themselves into) new scientific fields, new sciences, new scientific research programs. Imagine what a science of written narrative fiction will be when we study more deeply phenomena such as free indirect discourse with the help of developmental psychology, neurobiology, and linguistics. The more we learn about certain subjects and problems (techniques such as free indirect discourse, for instance), the more this work will become a solidly established discipline with its own set of problems studied, understood, and subsumed under a series of scientific research programs, approaches, and methods.

So rather than bemoan our own obsolescence as scholars in the humanities, as the scholars presented here attest, we would do better to focus our energies in developing concepts as tools sufficiently clear with sufficiently precise boundaries for us to dig for answers to our different sets of questions and problems. The research presented by the scholars in this volume considers the ways in which our work in the humanities to know our world better converges especially with the research programs in the neuro- and cognitive sciences, ethics, biology, psychology, philosophy, and linguistics, among others.

Each of the scholars represented in these pages knows well that any individual phenomenon studied requires a cognitive methodology of its own; we will not be able to study John Cage's music with the cognitive methodology of, say, particle physics. While each one requires its own tools of study, each of the scholars presents a "research program" that moves toward a knowledge of the whole—a unified theory.

In each of the chapters, then, what we see is scholarship that seeks the means to make its approach more systematic, clear, and precise, so that we can all see with better clarity how music, art, literature, film, and so on, work in and of themselves, and also how they are interconnected. Hence in each we see the identification of a method and taxonomy specific to the particular phenomenon (art, literature, and so on) as well as a reach toward an understanding of those universals that connect these particulars with the whole.

Given that the world is made up of "interconnected systems" (25), as Mario Bunge aptly identifies, the scholars here follow methodologies that aim to deepen our knowledge of emergence and convergence of knowledge. *Toward a Cognitive Theory of Narrative Acts* reminds us that the real, deep reason that our work in the humanities matters, yesterday, today, and tomorrow, is that it is knowledge.

PART I

GENERAL AND THEORETICAL CONSIDERATIONS

Arts in the Brain; or,
What Might Neuroscience Tell Us?

HERBERT LINDENBERGER

Let's start by revisiting what may well be the most debated passage in the history of literary criticism, Aristotle's theory of catharsis as at once the goal and the pleasure of tragedy. The key lines appear early in the *Poetics:* "Tragedy, then, is an imitation of an action that is serious, complete, and of a certain magnitude, . . . in the form of action, not of narrative; through pity and fear effecting the proper purgation of these emotions" (22).

I shall not rehearse the long-standing dispute as to what Aristotle meant by "purgation" or "catharsis"—are we to take him figuratively, or is he being literal?—in which case we would be responding to tragedy with our guts. Rather, I should like to see how the knowledge about the brain that has been accumulating in recent years might create a context within which to view Aristotle's view of tragedy and, as this chapter will develop further, to look at some influential discussions that have taken place in two other art forms, painting and music.

First, I state a key postulate within contemporary neuroscience: wherever we may think we perceive our reactions to events, whether in real life or in art, these reactions are processed not in the gut or in the chest but rather in the brain and, in particular, in those parts of the brain that are dedicated to specific purposes.[1] Take, for example, the two emotions, pity and fear, whose arousal and expulsion Aristotle sees as central to achieving the peculiar effect of tragedy. A key area in the brain generating fear is the amygdala.[2] Empathy for others, which seems close to what Aristotle meant by pity,[3] is also associated with particular regions of the brain, notably the ventromedial prefrontal cortex. And these areas do not act in isolation but are part of a complex network of interchanges.

The emotional disruptions that we experience from the arousal of pity and fear are of course felt within our body. The processes that neuroscience describes involve a network starting with signals sent by our senses to specific locations within the brain, from which new signals are eventually sent back to the body. Aristotle's goal in the *Poetics* is not, like that of a modern brain researcher, to map out our mental and emotional processes, but rather to describe the ideal way of composing a tragedy. Since he sees the end of tragedy as the achievement of catharsis in its spectators, only certain plots, and certain types of hero giving shape to these plots, are able to attain this end. For instance, a tragedy centered around the fall of an evil ruler is unable, according to Aristotle, to excite either pity or fear (26). Nor would a plot that enacts the crushing of an innocent hero be able to tap these emotions. As Aristotle puts it, "Pity is aroused by unmerited misfortune and fear by the misfortune of a man like ourselves" (27). And in order for the plot to set off the ideal mixture of pity and fear, Aristotle arrives at his celebrated description of the tragic hero, a figure caught between two extremes, someone neither wholly good nor wholly bad but one whose fall is brought about by "some error or frailty" (27), the so-called *hamartia*, sometimes rendered today as "missing the mark."

Indeed, all of Aristotle's further descriptions of the parts of a tragedy are directed to setting up the effect of catharsis. For instance, the plot, as he presents it, must be complex rather than simple (25–26, 30–31), one in which the turning point, or peripety, should ideally be coordinated with the hero's recognition of his situation (26, 31). And the play needs as well to give the audience a sense that its incidents follow one another inevitably, and above all, they must convey the illusion that they could probably happen in real life (25).

And yet is it possible to explain in contemporary terms how the confluence of pity and fear in an audience can result in feelings of satisfaction and well-being? And how is it that feelings of pity or fear alone, from Aristotle's point of view, fail to do the trick? If I may cite anecdotal evidence from my own play-going, performances I have witnessed of works such as *Macbeth*, Ibsen's *Ghosts*, and O'Neill's *Long Day's Journey into Night* have resulted in something of the experience that Aristotle described—yet I also recognize that these experiences were likely mediated by my long-standing acquaintance with the *Poetics* and its legacy in later criticism.

On the basis of a recent study of subjects with lesions in that part of the prefrontal cortex responsible for feelings of empathy (Koenigs et al.), one could speculate that these lesions might well prevent such subjects from feeling the pity for the tragic hero that Aristotle stipulates as necessary for the experience of catharsis. Similarly, sociopaths, whose prefrontal cortices show a reduction in grey matter, could not be expected to respond empathetically to the hero's plight (Moll et al. 801). Moreover, a case study of a patient with calcified deposits in the amygdala revealed an inability to recognize fear either in herself or in others (Damasio, *Feeling of What Happens*, 62–67); damage of this sort would likely not allow her to experience the fear that a tragedy within the Aristotelian mode seeks to awaken in the audience.

It is conceivable that researchers will be able to locate what happens within the brains of audience members at the climactic moments of a tragedy. As I shall indicate, subjects watching a film have already been "wired" to gauge what was going on in their brains. Can one imagine that Aristotle's long-controversial theory may be clarified by means of a similar experiment in the theater? An imaging study comparing subjects' experience of empathy with the experiences of those undergoing actual pain demonstrated that degrees of empathy can be measured in intensity just as pain can be measured (Singer et al.): does this mean that we may be able to test a play's success in engaging empathetic feelings toward its characters?

There is another aspect of Aristotle's theory of tragedy on which recent brain research may shed some light. I refer to the work done on so-called "mirror neurons" and "resonance behaviors" by a group of Italian neuroscientists in the late 1990s (Rizzolatti et al.). The first of these, as observed in monkeys and infants, consists of a subject imitating specific movements that it observes in another. The second, and more complex, form involves a subject internalizing actions that it observes in order to understand them. More recent research on mirror neurons suggests that the imitative capacities that human beings display when they mimic the behavior of others may well account for the development of empathy and the evolution of culture.[4]

What this recent research further suggests is the mechanism by means of which an audience comes to empathize with a dramatic character that it watches onstage. One might note that Aristotle, two chapters before he presented his theory of catharsis, spoke of "the instinct of imitation . . . implanted in man from childhood, one difference between him and other animals being that he is the most imitative of living creatures" (21). Although this sentence seems consonant with the directions that the work on mirror neurons is taking, Aristotle's main concern in the rest of this chapter is not so much the reactions of the audience, as it is in the section on catharsis, but rather the way that writers organize their observations with literary devices to create a viable dramatic work. Thus, he uses the term *imitation* primarily to speak of the dramatist's final product as the "imitation of an action" (22).

And yet his discussion of the human instinct for imitation, together with the catharsis theory and his concept of the ideal tragic hero, also suggests the identification that takes place between the audience and the dramatic character, and this mode of identification is also central to what research in mirror neurons is uncovering.[5] Remember that Aristotle defines the tragic hero as "a man like ourselves" (27)—in short, somebody with whom we can identify and whose movements and talk trigger imitative reactions within ourselves. If, say, we have empathetically followed the hero's development in the course of *Oedipus Tyrannus* or some later tragedy, imagine the strong emotions we would share with him at the moment of his tragic self-recognition. At the same time, our consciousness of the as-if nature of our imitative reactions, our understanding that we are only simulating what we observe in others, allows us to set a limit to our identification—with the result that, shaken up though we may be, we feel relieved (purged, shall we say?) to be able to return to our normal worlds.

To find similar correspondences between art and the brain, let us turn to painting, and to an example drawn from a treatise written almost two millennia after Aristotle's. I refer to a passage from Leonardo da Vinci's *Treatise on Painting*, in which the painter is advised to create the illusion of relief, that is, depth or contrast, even if it is achieved at the expense of beauty of color:

> Which is of greater importance: that the form should abound in beautiful colors, or display high relief? Only painting presents a marvel to those who contemplate it, because it makes that which is not so seem to be in relief [parere rilevato] and to project from the walls; but colors honor only those who manufacture them, for in them there is no cause for wonder except their beauty. . . . A subject can be dressed in ugly colors and still astound those who contemplate it, because of the illusion of relief. (1: 63)

And in another passage, an implicit reference to earlier painters who sacrificed the realistic representation of forms for the sensuousness of color, Leonardo warns,

What is beautiful is not always good. I say this in reference to those painters who so love the beauty of colors that, not without great regret, they give their paintings very weak and almost imperceptible shadows, not esteeming the relief. (1: 63–64)

Despite Leonardo's knowledge of writings on optics, his observations on the difference between seeing objects in color and seeing them in relief would have been influenced less by earlier theories than by his own experience as a practitioner of painting—not to speak of his role observing the practices of his predecessors. Yet this is a conflict that contemporary neuroscience can explain. The visual system, as Livingstone (49–52) and Marmor ("Eye and Art" 4, 7–8) have shown, has two distinct subdivisions, each of them deriving from a different stage of evolution. Our ability to view relief in a painting is dependent on that subdivision which we share with other mammals and which was developed at a relatively early stage to aid in depth perception and presumably to help discern other animals nearby; though sensitive to differences in brightness, it is also color-blind. The other, more recently evolved, subdivision is common largely among primates and allows color to be recognized—that is, as an additional way of seeing the world beyond the perception of depth (Livingstone 24–45).

When Leonardo objects to the lack of relief in many late medieval and early Renaissance paintings, he refers to the inability of their painters to create sufficient difference in shading to allow viewers to note the depth in, say, the folds of garments, a subject to which he devoted a whole section of the *Treatise* (1: 203–208). Both in his practice and in his theoretical writings Leonardo recognized that gradations of luminance enable us to make distinctions between objects and between parts of objects.

Livingstone illustrates the differences between these two modes of seeing, the one centered on luminance, the other on color, by reproducing a series of pictures ranging from an early Christian mosaic to Post-Impressionist work: each picture is printed both in color and in a black-and-white reduction (112–137). When seeing the picture in black and white, the viewer can more readily discern the degree of luminance contrast; and the greater the contrast, the greater the opportunity to recognize depth. To achieve the latter effect, a painter must find a compromise between the two subdivisions within our visual system—sometimes, as Leonardo recommends, sacrificing beauty of color in favor of depth.

Since the relative brightness of colors can be manipulated through the choice of pigments and the admixture of white, painters who, like Leonardo, seek to portray depth on a two-dimensional surface design their palette to emphasize contrasts in luminance. At the opposite extreme, as in some of Monet's misty scenes, luminance contrast is at a minimum, with the intended result that his surfaces look flat. At least since the Renaissance, and centuries before the two subdivisions of the visual system were explained by neuroscience, painters figured out how to mix colors to achieve whatever degree of depth or flatness they sought.

If the distinction I have shown between color and depth illustrates how painting is processed by distinct parts of the brain, a look at still another art form, music, can show how the same areas within the brain can be activated by a variety of experiences, aesthetic and otherwise. My example of music's effects on the brain dates from precisely three centuries after the lines I quoted from Leonardo's treatise. I quote from the review that the novelist and composer E. T. A. Hoffmann wrote of Beethoven's Fifth Symphony in 1810, two years after the completion of this piece:

> . . . Beethoven's instrumental music unveils before us the realm of the mighty and the immeasurable. Here shining rays of light shoot through the darkness of night, and we become aware of giant shadows swaying back and forth, moving ever closer around us and destroying everything within us except for the pain of infinite yearning, in which every desire, leaping up in sounds of exultation, sinks back and disappears. Only in this pain, in which love, hope, and joy are consumed without being destroyed, which threatens to burst our hearts with a full-chorused cry of all the passions, do we live on as ecstatic visionaries. (238)

Anybody familiar with the history of aesthetics will recognize here the characteristic vocabulary associated with the concept of the sublime—"mighty" [des Ungeheueren] and "immeasurable," words that conjure up an overwhelming and limitless power; negative words such as "pain," "destroyed," and "threatening" that counterbalance positives such as "exultation" and "ecstatic." Note also how imprecise Hoffmann's description is in comparison, say, to the statements by Aristotle and Leonardo quoted above; indeed, to make his point he uses images such as "rays of light" and "giant shadows swaying" in place of simple discursive prose.

Hoffmann's experience hearing this symphony is clearly in the tradition established in late antiquity by Longinus, who drew his examples of sublimity largely from Homer, and revived in the eighteenth century by such influential treatises as Edmund Burke's *Enquiry into the Sublime and Beautiful* (1756) and Immanuel Kant's *Critique of Judgment* (1790). By the beginning of the nineteenth century sublimity could be located in a wide variety of areas—landscape (in particular, the Alps or the English Lake District), epic poetry (above all, in Milton), music (in Handel's oratorios), drugs (as in De Quincey's *Confessions of an English Opium Eater*), and religious experience (in, for example, Chateaubriand's and Schleiermacher's apologies for religion).

But in the course of the nineteenth century it was in music above all that those who sought to experience the sublime found spiritual excitement. Beginning with Beethoven, in particular with the *Eroica*, whose length, loudness, and expanded orchestra raised the stakes for sublimity as no earlier music had done, music could perform the trick more powerfully than other artistic genres, its most serious rivals being perhaps opium and the ascent of challenging mountain peaks.

Just as Burke and Kant had defined sublimity by opposing it to that far milder experience they called "beauty," so Hoffmann, in the paragraphs preceding and following the above quote, sets Beethoven, whom he associates with the traditional sublime vocabulary of "awe," "fear," "terror," and "pain," against his two predecessors in the classical style: Haydn, whose realm he characterizes as "a world of love, of bliss, of eternal youth, as though before the Fall; no suffering, no pain"; and Mozart, whose symphonic writing, though it shows "dread lying all about us . . . , withholds its torments and becomes more an intimation of infinity" (237–238). Whatever magical qualities Mozart may reveal, he merely "intimates," but, unlike Beethoven, does not unleash, the sublime.

In view of the vagueness and the need for metaphor with which sublime experience has traditionally been described, one might wonder how it could ever lend itself to scientific investigation. And yet a goodly amount of research on the effects of music has appeared in recent years. In 1980, even before functional magnetic resonance imaging (fMRI) had started, my Stanford colleague in pharmacology Avram Goldstein conducted a study of what he called the phenomenon of "thrills" that people experienced when listening to classical music. Goldstein defines a thrill as "a chill, shudder, tingling, or tickling," with "hair standing on end" or "goose bumps" (127), most of these terms being common in traditional descriptions of the sublime. His subjects, who included Stanford students in medicine and music and also employees of the Addiction

Research Foundation, reported physical effects most often in the upper spine and back of the neck, from where these effects spread to neighboring parts of the body (needless to say, nobody "felt" the effects in the brain, from which they obviously originated). Subjects chose their own musical selections; Goldstein, in fact, found that these same subjects felt far fewer thrills when hearing the pieces selected by others. Subjects' reactions were self-reported, though Goldstein listened to pieces together with his subjects and recorded their indications of the intensity, frequency, and duration of thrill. In addition Goldstein administered naloxone, an opiate receptor antagonist, to his subjects and discovered that thrills were attenuated for some subjects taking this substance; as a result, he speculated that their emotional responses to music "may be mediated in some manner by endorphins" (126).

A subsequent paper, by a cognitive psychologist, Jaak Panksepp, expands on Goldstein's findings by measuring what he calls the "chills" (a term he prefers to Goldstein's "thrills"), and at one point he labels the phenomenon he is investigating a "skin orgasm" (203). Whereas Goldstein had used classical music, Panksepp asked his subjects, all of them undergraduates in an experimental-psychology class, to bring favorite examples of pop music, to be divided between those that the subjects judged "sad" and "happy" pieces. The maximum chills, as reported by his subjects, occurred at moments of high musical intensity—and also, it turned out, in the sad rather than the happy songs. Whereas Goldstein did not find significant gender differences among his subjects' reactions, female subjects in Panksepp's study felt stronger emotional reactions than males while listening to sad songs. As in Goldstein's study, subjects reacted more strongly to their own selections than to those selected by others. But Panksepp goes beyond Goldstein in speculating about evolutionary origins to the chills set off by sad music in "the neural circuits for separation distress that lead young animals to cry out when they are lonely and lost" ("Emotional Sources of Chills," 198–199).

With the development of brain imaging, the chills that had earlier been described by means of self-reporting could be located in specific regions of the brain. In a positron emission tomography (PET) study published in 2001, Anne J. Blood and Robert J. Zatorre examined ten McGill University students, all of whom had had considerable earlier musical training. The authors, like Panksepp, preferred to use the word *chills* over *thrills*. As with the studies described above, the students chose their own selections—though in this experiment all the examples were classical music. Each subject's selections were also measured in another student, who served as a control; as in the preceding study, chills proved stronger for those who had selected the music. The areas in the brain that showed activation were those normally associated with expectation and achievement of rewards. These areas included, among others, the ventral striatum, the amygdala, the hippocampus, and the dorsomedial midbrain. The authors conclude with a statement linking music-induced chills with activities normally seen as quite distant from music consumption: "We have shown here that music recruits neural systems of reward or emotion similar to those known to respond specifically to biologically relevant stimuli, such as food and sex, and those that are artificially activated by drugs of abuse" (11823).

The reward processing that this paper describes by means of a PET scan is corroborated in the higher-resolution imaging made possible by an fMRI in a paper of 2005 by Vinod Menon and Daniel J. Levitin.[6] Unlike the preceding paper, which used musicians as subjects, this one used college-age non-musicians—the purpose being to prevent expert bias from influencing the results. And again, unlike the two preceding papers, subjects did not choose their own musical selections. Controls were established not, as in the earlier papers, by playing selections for other members of the group, but rather by having the subjects listen to scrambled versions of the music: although pitch, loudness, and timbre remained the same, the order of notes was changed so that the music lost its temporal structure. The scrambled selections, it turned out, did not display the same degree of brain activation as the original pieces.[7] And in contrast to the Panksepp study, this experiment did not seek to distinguish the effects of happy and sad music.

The study demonstrated not only that its musical selections activated the regions of the brain associated with reward and affect, as the Blood and Zatorre paper had done, but also that the higher resolution possible in fMRI scanning located an additional network of activations in widely separated areas—in this instance, the nucleus accumbens, the hypothalamus, the insula, and the orbitofrontal cortex. On the basis of these findings, the authors speculate that listening to music leads to increased dopamine levels analogous to the effects of a number of addictive drugs. I might add that the selections that Menon and Levitin used included Beethoven's Fifth Symphony and Mozart's *Eine kleine Nachtmusik*. One remembers that in Hoffmann's passage this Beethoven symphony is the prime manifestation of the sublime, while Mozart would likely count as an example of the beautiful, or of something in between. But the authors of the study did not test for differences in affect between these two composers. Would brain imaging be able to distinguish between the sublime and the beautiful? Or if both composers can induce similar chills, is the old dichotomy still useful at all?

Thus far I have sought contemporary scientific commentary to elucidate some celebrated commentators on several art forms from widely separated periods of time. When I chose the passage from Leonardo I was already aware of the research on how the brain processes color. But I deliberately chose the selections from Aristotle and Hoffmann before I had looked for recent accounts that might help explain the phenomena they were describing.

What conclusions might one draw from this juxtaposition of discourses—first, the language traditional to understanding the various arts, and, second, that of contemporary science? The most obvious answer is that the great artists and critics of the past knew what they were doing, that science simply helps corroborate what they (and we) knew all along. (Aristotle would not, by contemporary definition, be classified as an artist, while Leonardo was both an artist and a critic, and Hoffmann was at once distinguished as a composer, conductor, fiction writer, and music critic.) If, say, Sophocles structured *Oedipus Tyrannus* to create what Aristotle called a "complex" plot in which peripety and recognition coincide, he did so knowing that he would thus overwhelm his audience with pity and terror (and with their consequent catharsis). And if Beethoven manipulated rhythms and volume in the Fifth, as well as delaying closure at the end in a thoroughly unaccustomed way, he knew he could awaken certain emotions that earlier music had left untapped. Do we need the confirmations of science to tell us what artists and critics of the past already understood, albeit in the languages of their own time? It may well be that the high prestige of science in today's culture, especially in comparison with the declining prestige of the humanities, enables these confirmations to validate the importance of art in our lives.

But there is more we can learn about the arts than simply the fact that the effects of art can be observed emanating from specific networks in the brain and that these effects can be measured. Once it has been demonstrated that the emotions triggered by music are processed in the same areas as those elicited by, say, drugs, food, and sex, or that drama may tap regions that also play a role in the empathy we exercise in everyday life, art may well lose some of that magical aura it has often claimed, especially during the past two centuries. Art thus becomes the creation not so much of those lofty figures whom we seek to honor with the term *genius* but simply of those skilled enough in their various crafts to manipulate their consumers' brains to produce particular effects. Producing and consuming art, in short, becomes an activity as "natural" as, say, having sex or cooking and enjoying food.

And if art is seen as natural, one would ask next what role it has played in human evolution. The notion that music, for instance, possesses survival value for the species goes back to Darwin himself:

> I conclude that musical notes and rhythm were first acquired by the male or female progenitors of mankind for the sake of charming the opposite sex. Thus musical tones became firmly associated with some of the strongest passions an animal is capable of feeling, and are consequently used instinctively, or through association, when strong emotions are expressed in speech. (*Descent of Man*, 2:336n.)

But the insight that Darwin briefly played with here has recently become a matter of lively debate. On one side one finds figures like Daniel Levitin, who extends Darwin's argument about the role of music in stimulating sexuality and, on the other side, figures in cognitive psychology and linguistics like Steven Pinker and Dan Sperber, who find no evolutionary basis for music;[8] Pinker, in fact, goaded others into taking stands on the issue by once referring to music as "auditory cheesecake" (quoted in Levitin, *This Is Your Brain*, 242). Steven Mithen, a specialist in prehistoric archaeology, has devoted a whole book, *The Singing Neanderthals*, to arguing music's place in evolution, above all through its relation to language learning. Walter Freeman, a neurobiologist, has assigned music, together with dance, a central evolutionary role in enabling the bonding necessary to create human societies.

If one accepts the view that music helped shape human evolution, one may ask if similar roles can be assigned to other arts. But that would not be a proper way to frame the question, for the division of the arts, as we know them, is a relatively recent phenomenon. As Mithen points out, music cannot be separated from language in the songs that mothers sang to their young (69–84);[9] nor can it be separated from dance in courtship rituals or, for that matter, from the visual art that decorated dancers' bodies and whatever they were wearing to help attract the opposite sex. The idea of art that stems, say, from a visit to the Louvre or from a performance of the Mahler Third in Disney Hall is quite a different matter—even if our neural circuits still light up similarly to the way they did for our ancestors—from what it must have been in prehistoric days.

In view of the changes that our notions of what constitutes art have undergone—not to speak of the differences in the arts at any one time among diverse cultures—the passages by Aristotle, Leonardo, and Hoffmann that I have linked to our present knowledge of the brain may seem quite limited in scope. After all, they refer to forms of art drawn from Western culture and dominant only during certain eras. Aristotle's favored form of drama, based on a coherent, compact plot that drives to a single powerful moment of catharsis, is wholly different from the various dramatic modes developed, say, in Japan, China, and India, and it is also quite foreign to the sprawling dramatic cycles performed in Western Europe during the Middle Ages. Leonardo's treatise is devoted to the proper execution of representational painting, while Hoffmann's music criticism, like that of all his contemporaries, can take for granted the naturalness of the tonal system—yet representation and tonality, as we look back historically, flourished only during a limited number of centuries within Western culture.

Note, for example, the profound change of style (not to speak of content, if the two can even be separated) that marked the beginning of the twentieth century in all the arts. The rupture that we have come to call Modernism signaled the undoing of the aesthetic principles upon which the quotations I used earlier were based. The linear plot celebrated by Aristotle gave way, both in fiction and drama, to texts that moved in unpredictable directions and that often questioned their ability to render any real world. Poems abandoned easy coherence, nor did they make it easy for readers who treasured strong emotional responses; as H. Porter Abbott argues in his chapter within this volume, a text by Gertrude Stein or Samuel Beckett is not relevant for "what it is *about*" but rather for "what it cognitively *is*." Paintings often refused to represent any discernible world, and, even when they pretended to be representational, gave up on perspective and the various tricks developed over centuries to create the illusion of reality. Music often abandoned tonality altogether, at times adopting chance operations; and even when employing the tonal system, it often held back from the sublime effects that had dominated during the preceding century.

Consider these celebrated lines near the end of *Adonais* (1821), Shelley's elegy for Keats:

The One remains, the many change and pass;
Heaven's light forever shines, Earth's shadows fly;
Life, like a dome of many-coloured glass,
Stains the white radiance of Eternity. (426)

Note the acid comment on this passage by F. R. Leavis as part of his program to rethink the history of English poetry from a Modernist perspective: "The famous imagery is happily conscious of being impressive, but the impressiveness is for the spell-bound, for those sharing the simple happiness of intoxication" (232). Leavis establishes a link between the effects of art and those of other stimulants that the various researchers on "thrills" and "chills" had made in studying reactions to music. And he firmly rejects these effects.

Contrast these lines with the opening of Ezra Pound's Modernist suite, *Hugh Selwin Mauberley* (1919):

For three years, out of key with his time,
He strove to resuscitate the dead art
Of poetry; to maintain "the sublime"
In the old sense. Wrong from the start— (61)

Not only does Pound explicitly distance himself (or his surrogate speaker) from the style of "the sublime," but his lines are designed to refuse any of the intoxicating effects that Leavis decries in Romantic poetry. Pound's prosy, ironic manner in this passage would likely affect different areas of the brain than Shelley's.

The empathy for the plight of tragic heroes that Aristotle saw as central to Greek drama and that persisted over the ages within diverse dramatic styles was resoundingly rejected by that avowedly anti-Aristotelian Modernist dramatist Bertolt Brecht. Brecht advocated a style of acting that sought to prevent his actors from identifying with the characters they impersonated, and he worked assiduously—often rewriting plays such as *Mother Courage* and *Life of Galileo*—to keep his audiences from identifying or empathizing with his heroes. Audiences were expected not to allow their emotions to be engaged, but rather to think rationally about the issues his plays were raising. Brecht was clearly aiming his effects at different parts of the brain than earlier dramatists—intent on tapping the audience's emotions—had done. Yet as subsequent stage directors have found, people generally go to the theater for an emotional experience, and despite the alienation effect (to use Brecht's famous term) that supposedly operates in these plays to prevent the viewer's identification with characters, audiences (and actors as well) have discovered ways to bring the proscribed emotions back.

Although I am unaware of any studies—whether by imaging or self-reporting—on the differences in effect between Modernism and earlier eras of literature, the brain's processing of abstract painting has actively stimulated researchers. Certain forms of non-representational painting—Russian Constructivism, the Dutch De Stijl movement, Op Art, and kinetic art, for example—have proved amenable to examining how the mind processes what the eye sees. As the neurobiologist Semir Zeki puts it, the "emphasis on lines in many of the more modern and abstract works of art . . . [derives] from the experimentation of artists to reduce the complex of forms into their essentials or, to put it in neurological terms, to try and find out what the essence of form as represented in the brain may be" (111). As it turns out, painters such as Malevich and Mondrian have provided a fertile field for brain imaging because of the fact that any particular cell in the visual cortex is sensitive to only a particular orientation—horizontal, say, or vertical or diagonal at a certain angle—of a line. Paintings composed of such lines, as many key works of the past century are, create a series of active brain responses that show up clearly when imaged.[10] Without the distraction of easily identifiable representational content, which might awaken empathy (as in an early Picasso portrait of poor people), or emotions of awe (as in a Turner seascape), the viewer experiences a play of forms that elicits a potentially wide range of responses. Representational painting can evoke many of the same emotions—pity, fear, disgust, for instance—that literary and narrative texts tap. By contrast, non- and partially representational painting creates pleasures, sometimes of a strongly emotional sort, not necessarily shared by other art forms.

Although neuroscientists cannot (and ordinarily choose not to) explain why a particular abstraction is "great," or even "better" than others, they can show with considerable precision how the visual system creates certain illusions for viewers. For example, Mondrian achieves the illusion of motion in his famous painting *Broadway Boogie Woogie* by juxtaposing colors—a yellow bar containing grey squares against an off-white background—of equal luminance, with the result that he plays on the differences between the two subdivisions of our visual system mentioned earlier in this paper (Livingstone 154–155, 157). Two Op Art painters, Isia Leviant and Bridget Riley, figure prominently in neuroscientists' writings on modern art because some of their paintings achieve an even more pronounced sense of motion than the Mondrian mentioned above (Livingstone 151, 160, 162–163; Zeki 162–163; Marmor, "Illusion and Optical Art," 150–151, 162–165). Zeki, for instance, explains the extraordinary motion we experience in Leviant's *Enigma*, a painting that juxtaposes circles and spokes, on the basis of how different cortical areas become activated (163).

Certain representational movements, above all Pointillism and Fauvism, have also proved fertile sources for commentary on how the brain processes visual art. Like the Op Art painters, both the Pointillists and the Fauvists had employed radically new techniques to play with their viewers' vision. Standing as they did at the threshold of Modernism, these painters give us recognizable objects from everyday life at the same time that they challenge earlier notions of what it means to represent the real world. A Seurat painting plays on the differing ways we view it from distinct distances—the Pointillist dots dominating at close distance, yet disappearing at far distance, while in between these extremes we feel a special vibrancy, a virtual motion. As Livingstone puts it, "You can see simultaneously both the separateness of the dots and a blending together to form a single larger surface" (176). Marmor explains the peculiar power of Pointillist painting from the fact that whereas one subdivision of our visual system "recognizes . . . tiny dots but does not respond to color," the other subdivision "recognizes color [but] cannot discriminate very small lines or objects" ("Eye and Art," 15).[11]

The differences between these two subdivisions are central to our experience of Fauvist paintings as well. A Fauve painting, we remember, gives us deliberately "false" colors that fill the shapes of familiar objects and forms. Yet despite being "misled" by the colors that artists such as Matisse or Derain applied during their Fauvist period, we recognize their representations with relative ease. The reason we do so, as Livingstone explains, is that the wrong colors they give us display the same relative luminance as the right colors would (133–137). Thus, the colorblind system that reads for depth and shape is able to function along with the system that interprets color. Moreover, as Zeki discovered in imaging subjects who viewed different types of art, colors in abstract painting are activated in areas additional to those activated in representational painting; and in Fauvist painting the areas that display activation are different from those in both abstract and representational painting (197–204). As a result of his experiments showing the difference in neural activity between a Corot and a Mondrian, on the one hand, and a Corot and a Fauvist painting, on the other, he concludes that "artists are unknowingly exploiting the organization of the brain" (204).

Despite the shock effects that Modernism initially created in the visual arts, in the course of the twentieth century viewers gradually assimilated the experiments both in abstraction and in such distortions of representation as the Fauvists and the Cubists created. Not so with the more radical experiments in musical Modernism, above all in non-tonal and chance-organized music. Despite the initial shock effects created by a number of certain early-twentieth-century works such as Stravinsky's *Le Sacre du printemps*, audiences have learned to accept these works as long as they remained within the tonal system. By contrast, the non-tonal works of Arnold Schoenberg, whether during his earlier atonal period or after he had developed his serial method, have had to struggle to find an audience, except for hardcore aficionados (of which I happen to count myself one). A possible explanation is that non-tonal works do not resolve their dissonances as tonal works do; Schoenberg himself boasted of his "emancipation of the dissonance" (193). Theodor Adorno, as part of his defense of Schoenberg against Stravinsky, praisingly described the former's musical language as "undisguised, corporeal impulses of the unconscious, shocks, and traumas" (35). A cognitive psychologist, David Huron, agreeing with Adorno's depiction of the disturbing element central to Schoenberg's music, has recently analyzed the composer's way of thwarting the expectations of traditional listeners by means of tone rows that are designed to prevent these listeners from finding comfort in any vestiges of tonality; as a result of this active thwarting, Schoenberg's work, as Huron puts it, should be called "contratonal" rather than "atonal" (339–344, 352).[12]

A brain-imaging experiment by Isabelle Peretz examined the reactions of subjects to consonant and dissonant pitch combinations in order to determine how and where in the brain dissonance is processed. The subjects included a group of normal persons without musical training plus one subject, I. R., a woman with brain damage whom Peretz has used in a number of studies on music.[13] Passages of classical music from various periods (though not atonal pieces) were played in different versions, each with an increasing degree of dissonance. Whereas the normal subjects declared the dissonances they heard "unpleasant," I. R. was unable to hear anything unpleasant in these dissonant chords. As a result of the images she recorded in I. R.'s damaged brain, as well as in her control group's brains, Peretz concluded that musical perceptions are first processed bilaterally in the superior temporal gyri, after which they are relayed to emotional systems in the paralimbic and frontal areas. Our judgment of the relative pleasantness of consonance and dissonance is thus built into our brains, or, as Peretz puts it, "The brain . . . is pre-wired for processing consonant pitch levels" ("Cortical Deafness," 939). But she also warns that this would still need to be demonstrated across cultures.

Dissonance obviously functions differently in actual musical performance than in this experiment, in which the dissonant chords came not from composed pieces but rather from distortions of classical passages. Still, the fact that the brain judges dissonance to be inherently unpleasant can help explain the difficulty that listeners have experienced assimilating much twentieth-century music. I for one can testify to feeling the chills associated with musical sublimity when I hear certain nontonal works such as Berg's *Three Pieces for Orchestra* or Schoenberg's *Die Jakobsleiter*. But I also suspect that I do not speak for a large band of listeners.[14]

The difficulties that listeners feel with the Second Viennese school are likely less than those often reported with the music of John Cage, above all the pieces organized by chance operations, a procedure he initiated about 1950. For example, Cage's one foray into opera, *Europeras 1 & 2*, combines actual arias from the standard operatic repertory with orchestral accompaniments that have been scrambled by chance operations governed by means of a computer program. Thus, while hearing familiar arias (with sometimes two going on at once), the audience can often identify instrumental phrases that, though drawn from the original scores, have been displaced and assigned to instruments different from those for which they were intended. The result is not simple dissonance but cacophony of the most striking sort—compounded as well with deliberately absurd stage antics. Just as the Menon-Levitin paper ("Rewards"), described earlier, scrambled familiar musical pieces (with the result that the subjects' brains were not activated as they had been in the unscrambled pieces), Cage's work did not award the musical pleasures ordinarily associated with music—though, according to reviews, it struck many of its audience members at the original Frankfurt production of 1987 as a wildly comic romp.[15]

By contrast, the lithographs that Cage composed by chance operations have, like Modernist art in general, proved much more accessible than his music. These works, done during his last two decades, consist of lines and forms whose size and positioning on the sheet were determined by his chance-oriented computer program. Like the Constructivist art that obviously influenced Cage's print-making style and whose effects on the brain Zeki analyzed (109–142 passim), these works, chance-organized though they may be, do not, like his music, affect those parts of the brain that react with feelings of discomfort or indifference.

The varying responses that viewers and audiences have shown to the stylistic ruptures that marked Modernism in the various arts during the preceding century cannot hide the possibility, perhaps even likelihood, that any one work, whether Modernist or not, will tap the same areas of the brain in most of those experiencing it. This is evident, for example, in Op Art, whose effects can be explained quite precisely by citing matters such as the varying degrees of luminance and the geometrical relations of lines to one another.

An experiment imaging a group of subjects who watched a half-hour segment of a 1966 movie, *The Good, the Bad, and the Ugly*, demonstrated—"unexpectedly," as the authors put it—that "brains of different individuals show a highly significant tendency to act in unison" (Hasson et al. 1638). This movie was chosen because its high-action nature allowed subjects to react to a number of emotional peaks. Whatever cultural differences there may have been among these subjects, the same areas of the brain were activated at particular moments: for example, the fusiform gyrus, the area in the ventral occipito-temporal cortex responsible for facial recognition, reacted in face close-ups, while other areas were activated for buildings and inanimate objects. The researchers express surprise about another activated area, the middle postcentral sulcus, for which they could not at first find a correlation with the film images; only at closer inspection did they realize that this region reacted to images of delicate hand movements (1636–1637).

Whether or not audiences at musical events act similarly in unison may be put to the test in a concert hall currently being constructed at McGill University, where a segment of the audience will be assigned sensors to monitor heart rate, skin electrical responses, and facial musculature, while another segment will register their reactions on hand-held devices (Balter). Does music, one wonders, create the same degree of uniformity as film? Are the ears alone less engaged in a performance than that powerful combination of eyes and ears demanded by cinema? And what if it proved possible to monitor the audience in a large opera house for a performance, say, of a Wagnerian *Gesamtkunstwerk*, in which, at least according to the composer's intentions, all the arts—drama, music, visual design, dance—are supposedly working on the spectator at once?

Whatever uniformity of response neuroscientists may find among viewers and consumers of the arts, certain individuals will sometimes fail to hear, see, or understand artistic media in the usual way. Take, for example, that phenomenon called amusia, which refers to difficulties in perceiving and processing music. Congenital amusia is not yet known to have a single underlying cause, and it manifests itself in several ways, most commonly in the failure to recognize differences in pitch (Ayotte et al.). A study at the University of Newcastle upon Tyne compared sensitivity to pitch difference in ten tone-deaf subjects with that in ten who perceived music normally (Foxton et al.). The former proved unable, for instance, to distinguish between two contiguous notes of different pitch, nor could they tell if a gliding tone was moving upward or downward. But amusical individuals also display varying musical deficiencies: some hear pitch but not rhythm or timbre, others rhythm and timbre but not pitch (Levitin, *This Is Your Brain*, 184).

On the other hand, musicality is one of the strong suits of patients afflicted by Williams syndrome, a serious genetic disorder whose developmental basis has only recently become well understood. Williams sufferers, despite some cognitive as well as physical impairment, are quite emotionally responsive to music and can sometimes even perform it well.[16] To cite still another variant in musical perception, patients with autism spectrum disorder, though they may be technically proficient in playing music, show no signs of emotional involvement (Levitin, *This Is Your Brain*, 253).

It is clear that even if we can chart our customary responses to works of art, persons with brain impairments, whether congenital or caused by later damage, are unable to perceive many of the crucial signals that writers, composers, and visual artists embed in their works to communicate with their audiences. To take examples from the two recently mentioned disorders, autistic persons are unable to read the emotions of others and, in particular, they lack feelings of empathy. As a result, one could scarcely expect them to identify with the protagonist of a novel or to grant a tragic hero the pity that Aristotle prescribed. Williams sufferers are eminently social and empathetic (Meyer-Lindenberg 386); when cognitively up to the task of following a play, they might well feel moved by a character's plight.

The problems that amusical people encounter in processing music find a parallel in the difficulties of the color-blind in viewing paintings. Whereas about 4 percent of the general population is thought to suffer from some form of amusia (Foxton et al. 802), as many as 10 percent of American men are afflicted by the most common form of color blindness, namely, a deficiency in distinguishing red and green (Marmor, "Eye and Art," 19). The preponderance of men among the color-blind is due to the fact that the cone photoreceptor pigments for these colors are inherited on the X chromosome, of which women, unlike men, possess two copies. Thus, if a woman's red or green pigments do not function adequately in one chromosome, they are likely to work normally in the other chromosome. Persons who are red/green color-deficient find it difficult to distinguish red, orange, and yellow. There are also rare individuals who lack the ability to discern color altogether and, as a result, they see paintings (not to speak of the world at large!) in various shades of gray, much as those with normal vision see the world at night.[17]

But painters too have been known to suffer from deficiencies that affect their art. The best-known example of color blindness among visual artists is the distinguished etcher, Charles Meryon (1821–1868), who, in the course of his art studies, recognized the need to shift to a black-and-white medium. One of his few known works in color, a pastel of a ship in a storm, shows an avoidance of red and green, with a preponderance of yellow and blue.[18] Since color-deficient artists rarely choose to pursue painting, visual defects among painters usually are due to diseases developing in the course of their careers. For example, Marmor has shown the effects that Monet's cataracts and Degas's macular degeneration had on their later paintings: thus, by comparing an earlier and later version of Monet's bridge at Giverny, and earlier and later nude bathers by Degas, he demonstrates how the blurriness of vision and lack of color discrimination emanating from these two diseases made their way into their work ("Ophthalmology"). Monet in his old age was even said to have depended on the labels of his tubes of paint to know what color he was using ("Ophthalmology," 1767).

For a final example of how a defect can hamper one's experience with art, I turn to a condition known as prosopagnosia, or face blindness. Those afflicted by this show impairment in the fusiform gyrus, the brain site, as mentioned earlier, specializing in the recognition of faces. Since I happen to suffer from a mild version of congenital prosopagnosia,[19] I use this opportunity to indicate the limitations that this condition has imposed on my experience with art forms. The present chapter would scarcely be the appropriate venue to discuss the personal difficulties that prosopagnosia has created for me—embarrassments such as facing old friends who, when we meet at professional gatherings, fail to put on their name tags, or my inability to recognize my college-age children when each had been away for several months. I concentrate instead on my problems recognizing faces in works of art. To be sure, I have never found portrait paintings quite the chore that Zeki suggests they are in the chapter of *Inner Vision* on the effects of prosopagnosia (167–182). At least I never had much trouble in college art-history quizzes asking me to identify famous portraits by, say, Bronzino, Rembrandt, or Ingres: even if the details of faces eluded me, I recognized the pictures by the clothing and the color composition, for a defective fusiform gyrus does not affect one's ability to recognize features other than faces.

But films and plays, especially if they have many characters, prove a considerable challenge. To some degree it helps to learn an actor's gait, which, to the extent that one sees whole live bodies moving about the stage, proves easier in plays than in films. Nothing is so difficult for me as films of an earlier time in which male characters are seen wearing similar hats and suits and where only the face is exposed—and where, moreover, there was no color to come to my aid. As a result, I easily confuse heroes and villains as well as major and minor characters. My ability to understand many films remains dependent on the kindness of companions willing to help me identify who is who throughout the picture. Needless to say, I've never been much fun to go to the movies with.

The normative responses to art forms that I have described, together with the various deviations from the norm with which I have concluded, may well arouse suspicions among my colleagues in the humanities about the motives behind this inquiry. Am I trying to divest great art of its inherent mysteries? they may ask. And am I minimizing or even denying the roles that personal taste and cultural difference play in people's understanding and enjoyment of art?

For at least two centuries a certain mystique has ranged around what we dignify with the label art (not crafts, certainly, and not the artifacts of popular culture!). Most of us feel uncomfortable accepting the possibility that our responses to art can be charted by science; even when such charting seems plausible, we prefer to add a *je ne sais quoi* to assert the need to stop before the mystique has totally disappeared. One might remember that the example with which this essay started, Aristotle's theory of catharsis, does not invoke anything mysterious in the experience of drama, but, if we take the term *catharsis* literally, seeks a physical basis in this experience, much in the spirit displayed by neuroscientists today.

The cognitive psychologists and neurobiologists whose work I have used to think out this chapter certainly do not attempt to offer large-scale explanations for art's alleged mysteries. Rather, in accord with traditional scientific procedures, they typically pinpoint a single, narrowly defined phenomenon—the chills that certain musical pieces set off in specific locations within the brain, or the ways that painters create the illusion of movement by juxtaposing equiluminant colors. Unlike many of us in the humanities, they do not reach out for global theories.

Yet despite the modesty of reach that characterizes these individual scientific studies, a larger picture seems to be emerging, a picture that suggests we may come to know more about how we perceive, process, and enjoy art than we suspected before. I do not suggest that science will rob us of our traditional task of conserving the past and placing art works within an appropriate historical context, or, for that matter, of subjecting them to the interpretation and evaluation that their consumers have long depended upon. But what we learn from science may well affect the methods that we choose to exercise our interpretive and evaluative skills, and also, I might add, how we come to look at the various arts in relation to other reward-offering activities.

The studies I have consulted go back little more than two decades, and most date from the past five or six years. Our exploration of the brain (if I may indulge in the sort of analogy endemic within the humanities) may be no more advanced at this point than was the exploration of the New World around, say, the mid-sixteenth century. And if we are still near the beginning of our knowledge of how the arts and the brain engage with one another, we may yet be in for a few surprises.

Notes

1. For mentoring me in scientific matters relevant to this chapter, I am grateful to Dr. Edward D. Huey, NINDS, National Institutes of Health, the Litwin-Zucker Research Center, and the Feinstein Institute for Medical Research, and to Professor Michael F. Marmor, Department of Ophthalmology, Stanford University. I take full responsibility for any errors and misunderstandings.

2. For a review of the amygdala's role in fear and other emotions, see Phelps and LeDoux. See also LeDoux's description of how the amygdala communicates with other sections of the brain (*Emotional Brain* 157–178).

3. It could be debated whether Aristotle meant something like empathy ("I feel your pain," to cite the words of a former U.S. president) or sympathy ("I feel pity for your pain"). Whichever it was, the spectator is meant to experience a close tie to the hero. For the distinction between empathy and sympathy, see Suzanne Keen's chapter in this volume.

4. For popular accounts of this work, see Winerman and Azar. For a detailed discussion of the relation of mirror neurons to the development of empathy in the course of human evolution, see Gallese, who was a member of the group (Rizzolatti et al.) that did the earlier experiments on mirror neurons.

5. The significance of mirror-neuron theory for an understanding of the relationship of text and audience is evident from the fact that four of the contributions to this volume—those by Patrick Colm Hogan, Suzanne Keen, Ellen Spolsky, and Lisa Zunshine—make use of this theory.

6. For a more popular account of this experiment and its relation to Goldstein's and Blood and Zatorre's work on this issue, see Levitin, *This Is Your Brain*, 185–187.

7. The authors used the same musical selections, together with their scrambled versions, in another study that demonstrated that music is processed in areas associated as well with language processing. They conclude that both music and language display similar temporally ordered sequences; the scrambled musical selections, lacking as they do any meaningful temporal order, did not show the same activation as the originals. See Levitin and Menon, "Musical Structure."

8. For a summary of these views, including Levitin's extended argument for music's role in evolution, see Levitin, *This Is Your Brain*, 261.

9. Note that Darwin's statement above also stresses the association of music and speech when "strong emotions" are being expressed.

10. See Zeki's illustrations of different brain responses to particular line orientations (102).

11. For a detailed description of Pointillism "as a psychophysiological process," see Lanthony.

12. To demonstrate the centrality of Schoenberg's "contratonality," Huron compares forty-two tone rows by Schoenberg with randomly generated sets of rows: the latter display a far greater affinity to traditional tonality (341–343).

13. See, for instance, Isabelle Peretz et al., "Dissociations," and Isabelle Peretz et al., "Music." The first of these papers shows that I. R.'s brain damage affected only her perception of music, not of language, and thus argues that music and language are autonomous in processing auditory information. The second examines the possibility that emotional and non-emotional judgments are products of different pathways within the brain. The deficiencies observed in I. R. in these and other studies serve, among other things, as a way of defining the differences between normal and brain-damaged subjects.

14. Huron speaks of "experienced listeners" to modernist music who "adapt to this [the composers'] strategy and learn to expect the unexpected" (348).

15. For a detailed description of *Europeras 1 & 2*, see my chapter, "Regulated Anarchy: John Cage's *Europeras 1 & 2* and the Aesthetics of Opera," in Lindenberger 240–264.

16. Ibid., pp. 182–183, 253–254. For a review of research on Williams, see Meyer-Lindenberg. Although Levitin anecdotally cites a good clarinet player among Williams sufferers he has seen (*This Is Your Brain* 183), Meyer-Lindenberg, citing several studies, concludes that those with Williams syndrome, though "highly interested in music . . . are not musically gifted" (386).

17. For illustrations of what the color-blind "see," see Livingstone 34–35.

18. For a history of Meryon's career and a reproduction of this pastel, see Ravin and Lanthony.

19. For a study of congenital prosopagnosia, see Behrmann and Avidan.

CHAPTER TWO

Narrative as Nourishment

ELLEN SPOLSKY

Tragedy arouses pity and fear, poetry teaches and delights, history keeps us from repeating our mistakes, and bedtime stories soothe drowsy children into sleep. Or do they? The first three claims have clearly expired, and now Elizabeth Kolbert tells us that many recent bedtime books for children seem to be about "figuring out ways to put off going to bed" ("Goodnight Mush," 91). If we are skeptical about the power of texts to work in the ways we are told they were intended, we are still interested, as Kolbert's essay shows, in exploring the possibility that literary texts act in the world.[1] The virtual universality of storytelling would seem to support the even stronger claim that its actions advance the well-being of the individuals and groups who know how to make use of stories.

Narratives, in Jerome Bruner's description (1991), do their work by serving individual understanding, but are also in the service of larger cultural agendas, in all their variety. Narratives seem to colonize human brains with forms that are well suited to negotiate among different brain modules,[2] within sensory and social environments. Bruner says that "we organize our experience and our memory of human happenings mainly in the form of narrative—stories, excuses, myths, reasons, etc. . . . , [and] we use narratives as tool kits to accomplish this" (6). He also calls them "prosthetic devices" that "operate as an instrument of mind in the construction of reality" ("Narrative Construction," 4). But notice the switch in agency between these two claims: at first, "we" use narrative, and then, it uses us, on behalf of "mind." This hint of bi-directionality was picked up by David Herman, a later exponent of the "activity theory" of narrative, whose 2003 article, "Stories as a Tool for Thinking," expands some of the most important suggestions Bruner made about how narrative works at "constructing and representing the rich and messy domain of human interaction" ("Narrative Construction," 4). Herman sensibly continues to fudge the question of which is the user and which the tool.

With the benefit of more than a decade of interdisciplinary cognitive theorizing, the circularity of influence is now widely accepted: narratives exist in dynamic relationships with the minds and imaginations of their creators and audiences.[3] Herman agrees with Bruner that narrative "functions as a powerful and basic tool for thinking" ("Stories as a Tool for Thinking," 163), fitting us to our environment by helping us make satisfying sense of it. But he then also describes how storytellers act upon their environment by "opportunistically exploiting narrative structure" (169) in situations not normally thought of as primarily narrative. They may, for example, write up reports of scientific experiments or medical case histories, intervene in disputes by testifying in court, or otherwise resolve confusions by introducing background stories that change the participants' evaluation of events (163).

Further, by exploring what Bruner calls the unexpected, the non-canonical "breach[es] of legitimacy" ("Narrative Construction," 15) with which stories typically concern themselves (more on this below), storytellers may imagine and describe counterfactual models of social structures, suggesting alternatives. Framing a possibility by means of a revised story is a first step to bringing about change in conventional public narratives, as it is with private narratives. No wonder, then, that narrative is ubiquitous and storytelling universal: it has work to do in the world, in the minds of individuals and in the community. Herman makes clear how both stories and storytellers act and are acted upon, even if the way they act can't any longer be expressed in the aphorisms we associate with Aristotle and Horace.

Once we agree that narratives, broadly understood, have useful work to do in human lives,[4] and if we have established the usefulness of understanding narrative as a tool that serves both individuals and communities, that both acts and is used, the next step theorists may want to take is to enrich the study of narrative acts by investigating their biological function, as suggested by the analogy of narrative as nourishment. It is my goal, in this chapter, to display the benefits of understanding narrative as food for a specific need: food to feed *representational hunger*.[5] I will first suggest some of the directions in which current work in cognitive science and in interpretive theory encourages us to elaborate a parallel between people's need for food and their interest in telling and in hearing stories, and then I will suggest two theoretical concerns that this elaboration serves.

The naturalness—the obviousness even—of understanding narrative as nourishment was suggested by Thomas Cranmer in his preface to the first English Bible printed in England in 1549. Chastising those who objected to the provision of vernacular Bibles for all English Christians, Cranmer remarks: "I would marvel much that any man should be so mad as to refuse in darkness light, in hunger food, in cold fire, for the word of God is light, food, and fire" (Cranmer STC2079). One hundred years later, the argument for free access to texts still had to be made. In the Areopagitica of 1644, Milton writes that "books are as meats and viands." . . . some are healthful, some rotten. Even those that may themselves be unwholesome serve the "discreet and judicious reader" by helping to "confute, forewarn, and illustrate."[6] The comparison had not soured three centuries later when Karl Popper claimed that we are "most active in our acquisition of knowledge—perhaps more active than in our acquisition of food" ("Evolutionary Epistemology," 243). Understanding narrative as metabolism acknowledges the way in which a story and its readers are part of a mutually supportive and self-regulating homeostatic system.[7] Both stories and audiences are enlivened by intentionality; they talk to each other and are mutually arousing. What Herman describes as "opportunistic" storytelling (interventions of narrative strategies in order to organize or clarify a situation in need of management) can be seen from a biological perspective as a self-motivated process that appears parallel in many ways to metabolism.[8] Flagging energy is the need that unbalances the human body, that triggers the release of the chemical that produces hunger pains, that provokes the search for food to replenish energy. It suggests that it may be interesting to consider telling stories, hearing them, and reading them as a process that works in the same homeostatic way. Recent research in cognitive linguistics suggests that eventually neurobiologists will be able to describe narrative activity—both the production and the comprehension of stories—as an evolved, embodied process, like language and like metabolism.[9]

Encouragement from the Cognitive and the Neurosciences

Talking about narrative as an aspect of human biology resonates harmoniously with work already done in several areas of cognitive science and in literary theory, and advances the project of understanding the work of narrative in culture, and the role of representation in cognition. The general theory (of which Bruner's is one version) is a formalization and enlargement of the claim that stories teach because their audiences can learn by analogy the things they need to know in order to survive and the behavioral patterns they need to thrive within their cultural world.[10] But not wanting to claim that all narratives moralize, or that all they do is teach, in the usual sense of the word, I need to import an expanded notion of learning from neurophysiology, where the assumption is that learning is recognized and described by changes in the synaptic connections between neurons driven by sensory experience. This description lets us consider as learning activities such as memorizing a role in a play or riding a bike, through the body's "learned" immune responses, and including changes in behavior that result from fatigue, disease, and age.

While this makes it too broad a definition for many purposes, this is just the breadth I need for the claim that narratives indeed teach us by managing our neuronal/brain/body responses in all kinds of situations. Continuous encounters with narrative, on this view, recursively reorganize an individual brain/mind into a connected set of schemata that represent the self and the situation of that self in its environment, such that the achieved or constructed patterns support both the individual's identity and his or her behavior.[11]

An encounter with narrative, then, as it is a sensory experience, can literally "change your mind." Marc Jeannerod and others have complicated and enriched this view, insisting that these constructed patterns by which we know ourselves are, like other learned patterns, not only in our brains (although the majority of neurons are indeed concentrated there) but are also embodied in our skin, our limbs, and our muscles as well. Important parts of what we know may be considered to reside at the ends of the neurons that connect to muscles and to sense receptors. This knowledge is not abstract, and may be difficult to articulate, but nevertheless it guides not only the actions we perform, but those we recognize others performing.[12] The claim, then, is that when we can analogize aspects of our own bodies and worlds with structures we encounter in stories, the self-constructed narratives that make up our understanding of our position in the world gain strength and coherence. Here we approach an answer to the question of why fictional narratives are satisfying even in situations such as those in which our primitive ancestors found themselves, situations in which one might think truths about poisonous mushrooms, for example, would be more important and thus more satisfying than narrative fiction.

An achieved narrative can be thought of as a representation or schema that satisfies because it enables prediction; when one can represent possible future states, one can decide how to act. The evolutionary basis for the claim that learning how to use narratives has survival value has been argued by Paul Hernadi: "participation in protoliterary transactions" may have made some people "more astute planners, more sensitive mind readers, and more reliable cooperators than their conspecific rivals," thereby increasing their chances of survival and reproduction ("Why Is Literature," 26). Although prediction of danger is surely not the only function of storytelling, that need was important enough to ensure that all human cultures would take advantage of it, with local variations, of course. Once storytelling exists, it can be used for many purposes: to entertain, distract, or console, for example.[13]

This train of thought converges with those of cognitive psychologist Daniel L. Schacter (1996) and clinical neurologist Antonio Damasio (1999), both of whom suggest that there probably isn't very much of a self until a growing child has collected enough narrative memories to allow the construction of that self.[14] From early childhood, on this argument, the ingestion of narratives, linked to the emotions with which they were first encountered, is followed by their digestion—that is, they are connected to schemata already in place or under construction. The emotional values of events allow the individual to answer the question that is the basis for acting or not acting: do I want that to happen again? The accretion of experiences and their connected emotions feeds the dynamic that produces the self, as eating produces the growth of the body. That process continually resupplies the more basic processes that use and deplete energy. Changing balances within the body and between the body and its world keep the processes ongoing and dynamic, and make the need for new organization and for representations of that organization constant.

A recent study from Yale University has shown that the neurohormone ghrelin, best known for its role in appetite and energy metabolism, also influences learning and memory. High levels of ghrelin, which stimulate appetite, also stimulate the growth of dendritic synapses in the hippocampus and (at least in rats) enhance learning.[15] This is one of the first studies to suggest a physical basis for Caesar's recognition of that very co-occurrence in Cassius: "he has a lean and hungry look, he thinks too much; such men are dangerous" (Shakespeare, "Julius Caesar," I.ii.194–195). The claim here is that to be hungry for knowledge is not "just" metaphorical but is properly analogical: in the course of human evolution, organisms analogize to their own profit. As Gould and Lewontin argued (see n. 19), they learn how to reuse already evolved biological processes for other work. The visual system, for example, activated for seeing what is immediately present to the eyes, is also used for producing visual images of things once seen but not present, and for imagining things never seen. The ghrelin experiment hints at a similar kind of reuse. Ghrelin is an appetite stimulant hormone, secreted from the stomach when the stomach is empty. But it also promotes the growth of synapses that have a positive correlation with spatial memory and learning. The hormone may be, the authors of the study suggest, "a molecular link between learning capabilities and energy metabolism" (384). If bodies signal the hunger for knowledge in the same way they signal hunger for food, if they need to know things in the same way as they need food, then the things that satisfy those needs are reasonably considered food. To put it another way, other things besides food can be metabolized for the body's benefit. The parallel bears on how we understand the production and reception of narrative by adding credibility to claims that have been made for the cognitive work of narrative in the world, and undercutting those that understand narratives, particularly fictional narratives, to be cognitively inert—"just for entertainment."

We are thus encouraged to pursue such parallels as suggest themselves, and see which turn out to be productive to the further understanding of narrative acts. Daniel L. Schacter's work on memory, for example, suggests this parallel between the production and intake of food and narrative: though it takes many workers to get food to our tables, no one can eat it for us. Similarly, narratives composed by others are individually, even uniquely, processed. We do not copy them into our brains but reconstruct them by fitting the new narrative into the schemata already present. Those schemata are multisensory personal constructions, made of shreds and patches and modified not only by words and images, but also by a teller's tone of voice, emphasis, and weighting or reliability, as they themselves have ingested previous experiences and values. (See Schacter's *Searching for Memory*.) Even when we think we are reading or hearing someone else's words we are already also analogizing them to similar experiences in our own memories and adapting them to our personal library of schemata.[16] Categorizing, organizing, and sorting the stories we hear, we pay no attention at all to some details, and focus on others. We may redivide the narrative, and recombine smaller patterns into larger ones, responding to the emotional weights of the experiences as well as the plots. In the end, only some parts of the narrative may be used (remembered, transformed, reconnected), while other parts are forgotten. So with the digestive system, the treatment of incoming food differs according to what is already in the system. The digestive system breaks the food down, identifies its parts so as to decide what to use and what to excrete, and sends the needed bits on differing pathways to organs that may use them immediately, or recombine them, or store them for a future time of need. Just as the food as it is eventually used bears little resemblance to what first appeared on the dinner plate, so a story, recalled, is never an exact replica of the original. If all is working well, it will be revised to suit its present context.[17] The identification, categorization, and assignment for further processing of the intake of both food and narrative are accomplished by a set of processes some of which are innate and some of which are learned. Like other biological systems, the digestive system begins life as a minimal set of connections which apparently learns through feedback loops to deal with what it is fed—with the specific food itself, in what quantities and at what intervals. Similarly, the human brain starts out with a set of preferences for semantic interpretation based on its evolved architecture, but which are modified by the child's interactions with the community's standards of interpretation and weighting. A tendency of the brain, for example, to categorize two individual items as analogous, that is, to represent them as alike, even if not exactly alike, is probably innate; a good memory trick, say Schacter and Addis,[18] as is the apparently parallel preference for recognizing pairs of opposites. If these preferences are not innate, neither are they random. Bruner enumerates "ten features of narrative" ("Narrative Construction," 6) as aspects of brain structure, and tries to map them onto aspects of human cognition, presumably evolved universals. Herman makes the search for these matches explicit, proposing to account for "a taxonomy of the core problem-solving abilities supported by stories" ("Stories as a Tool," 165). Although, as Bruner notes, the conventional forms of narrative differ from one culture to another, the brain structures they bond with presumably represent, as Herman notes, "the most basic issues facing human beings—for example, how to divide the manifold of experience into knowable and workable increments, as well as how to reconcile constancy and change, stability and flux" (166).

Keith Stenning, a logician working on the "relation between logic and psychology in cognitive processes," or what he calls "a theory of human information consumption" (note again the terminology of feeding), reminds us that "not any formalism can be internalized, and the process of internalization has to connect it to what is already there" (i, 5). The specifics of what counts as an acceptable match or opposition will be determined in part by the evolved architecture of the mind, but also by the possibilities offered within a specific culture. Both the digestive and the semantic systems distinguish relevant from irrelevant and recognize patterns of connectivity that facilitate further categorization and hierarchalization. Both processes can be described as working to identify causes and effects, and both allow choices of action. A mature interactive narrative system works by supplying satisfying adjustments to the changes of one's own body and to the changes in the fickle world around us. It helps us to maintain a stable understanding of the environment, where understanding means the ability to act satisfactorily within and upon that environment.

Literary theorists usually assume that the causal structures of a narrative have analogical power; its plot, the relationships among the characters, how they act toward each other, and the values attached to these actions, are models against which individuals match patterns in the details of their own bodily as well as mental lives. If so, then reading expands one's treasure chest of patterned experiences and their probable results with minimal risk. The rhetorical strategies of narrative bind these perceived patterns into forms that can be remembered, shared, and consulted at a later time. Narratives become part of our experience, and depending on the cultural weight with which they are tagged, become functionally indistinguishable from our own experience and may even outweigh it. Wisdom literature and prophetic books, for example, are indeed identified and described by their claims for the truth of unseen worlds and powers, unavailable to and yet outweighing normal human experience. Useless or even pernicious narratives may be rejected, either left unread, forgotten, or, in more complicated cases, banned or burned. This again is in parallel to the body's rejection of some food, by excretion, passing gas, vomiting, by food allergies which teach avoidance, or by dietary restrictions authoritatively imposed.

The Contribution of Cognitive Epistemology and Hermeneutic Theory

The argument for considering narrative to be a biological process is based on the claim that all knowledge is embodied—that at least for a start, all we have to use, in order to manage in the world, is our bodies, and thus whatever we do or think must ultimately be subserved by bodily processes. We may fruitfully analogize our thinking to tools or to machines, but the work of thinking is ultimately going to be grounded in proteins, muscles, neurons, and so forth. Andy Clark, a philosopher whose work mediates among robotics, artificial intelligence, and epistemology, argues that internal embodied representations take the form they take because they are produced from and oriented toward action.[19] It is their success or failure in guiding action, rather than their fidelity to structures outside the mind and body, that establishes which representations will be reused and strengthened. This view, of course, converges with the work of neurologists like Antonio Damasio and developmental psychologists who understand children's learning as a result of their physical interactions with their surroundings.[20] Of course, the closer the representation comes to being an accurate account or image of that which it represents, the likelier it is to succeed as a guide to action. But note that the standard of truth or fidelity in the copy has been demoted from a necessary condition of a successful representation to a gradient typicality condition.

The claim, further, is that interactions with the world determine how competing representational patterns (including narratives) are to be weighted and integrated with earlier experiences. The patterns narratives display not only take their form in response to environmental pressures on the writer, but will be interpreted and reinterpreted by an individual according to that individual's local needs for responsive action. A satisfactory representation or pattern of representations is not judged by whether it is true but by whether it is strong enough to support appropriate and habitual (and therefore fast) reactions in familiar situations. It must also be flexible enough to recognize that a new or revised action may be needed in a situation that at first may have seemed familiar, but turns out to be crucially different.

This emphasis on our bodily activity in accumulating experience suggests another way in which reading and eating are parallel, and that is that we are not only ready to consider eating or taking in interesting food or new information when it arrives, but we actively seek both food and information.[21] The narrative parallel to the practiced routines of hunting and gathering, shopping and cooking, may be the rhetorical and hermeneutic systems that ensure the permanent availability of new nourishment to be discovered in whatever texts are available whenever they are needed. And if we hunger for narratives, are there universals about which books satisfy that hunger better than others? Why do we call some books "classics" while others are understood to be ephemeral? Cultural critics have claimed that the hegemonic power of the canonizers is what makes the difference, but I suggest it is, rather, the ability of readers to use the texts, to digest them, that distinguishes those sought after from those neglected. A community of readers generally has a set of conventional ways of interpreting particular kinds of texts.[22]

Bible readers, as I know from my own community in Jerusalem, can reread the same stories endlessly, not only because of the authority they bear, but because they have learned a set of hermeneutic procedures which allow them to produce new readings—virtually without end— from the familiar texts. The respect for the divine status of the biblical texts does not trump current need: the learned hermeneutic chooses from the available supply and responds to current needs to produce new interpretations, whether they be understood as footnotes to older texts, as interpretations of those texts (midrash), or as entirely new works, such as Harold Fisch's suggestion that King Lear is a reinterpretation of the Book of Job.[23] For example, on September 11, 2001, there were Jewish women in New York whose husbands were presumed dead in the collapse of the World Trade Center and who were left not only widowed, but in the halachic (legal) status of being unable to remarry, since there were no witnesses to attest to their husbands' deaths. Halachic scholars, however, were able to read the centuries-old text of the Talmud according to rules that allowed them to resolve the problems of the women's status. The availability of hermeneutic procedures for interpretation allowed the text to be appropriately productive.

A different hypothesis about what makes some texts not only valuable but reusable was made in 1977 by Mary Louise Pratt. According to Pratt, tellable narratives, that is, narratives that are worth the time it takes to tell and perhaps to write down, have to be about events that are not only unusual but also problematic. They report events that don't "go without saying." Something is worth telling because it breaks an expected pattern. Pratt provides a helpful description of what makes an anecdote or narrative tellable:

> A speaker is not only reporting but also verbally displaying a state of affairs, inviting addressee[s] to join in contemplating it, evaluating it, and responding to it. The point is to produce in its hearers not only belief but also an imaginative and affective involvement in the state of affairs represented and an evaluative stance toward it. The teller intends them to share the wonder, amusement, terror, or admiration of the event. Ultimately, it would seem, the goal is an interpretation of the problematic event, an assignment of meaning and value supported by the consensus of teller and hearers. (136)

Pratt's description fits the interactive and the embodied model of narrative as nourishment handily. Narratives, on this view, are most needed when they deal with problematic events, for example, with human failure. When things are going well we don't need analyses. But when situations become difficult, we must have additional resources.

Theory of Mind

One of these resources (see Zunshine in this collection) is our ability to make guesses about other people's understanding and intentions, that is, what developmental psychologists and philosophers have discussed as the Theory of Mind.[24] Somewhere between the ages of three and four on this theory children begin not only to be able to act according to their own needs and desires, but also to be able to take into account what they infer other people to be thinking and wanting. They develop, it is said, a Theory of Mind. By observing faces, gestures, and body language, and by noting tone of voice as well as by listening to words, I attribute minds to others, and on the assumption that others have the same kind of mind I have, I may decide (for example) that if they are showing fear, maybe I should also be afraid.

There is a lot that still is unclearly understood about theories of mind, but it is interesting to note how they dovetail with other neural theories of the brain as active in seeking out information. The neurologist Jean-Pierre Changeux cites work on what is called "autoactivation," that is, the outward-looking character of the nervous system, which not only responds, but spontaneously "explores and tests its physical, social, cultural environment, to capture and register responses from it" (32). Normal neural systems have evolved to receive incoming information about the world, and to integrate it with other bodily knowledge—about internal states, for example—and also to actively seek out such information. Recent work on mirror neuron systems, for example,[25] suggests that learning by analogizing between one's own body and the body of another is fundamental to human development, and begins within hours of birth. Young children, on this theory, recognize that they can analogize from what others are doing (looking at, reaching for) to what they can look at and reach for. Similarly, they may analogize from what they are thinking or intending to what others might be thinking or intending, and can then draw inferences about likely events and about the intentions of others toward themselves.[26]

Not unlike other monitoring systems in our bodies, such as that which produces feelings of hunger, mirror neurons may be part of a system that anticipates the body's needs. Themselves dependent on vision, hearing, touch, smell, and proprioception, and on other as yet unspecified neurological outreach devices, these neurons help us feed on bits of the world beyond our skins with which we need to interact in order to grow into adult humans. We become the people we are by stepping up to the buffet table of life and filling our plates. Simultaneously attending to the narratives we hear, we also ingest the stories and metabolize them.

Theories of mind, then, and the cognitive theories of wired-in connectivity between people on which they are based, are particularly interesting in the study of narrative interactions, not because they promise a reality-based way of understanding characters or narrative that we didn't already have, but because they are situated on the border of the cultural and the personal. They allow us to investigate just that precarious point where an author may place a character in interesting danger by granting her the freedom to break from the conventional or to go beyond it, with the attendant risks and unforeseeable consequences.

Our greatest literary texts explore just these failures of inference and of mind-reading, among characters or between reader and author, or between reader and character. They display mis-attribution of intentionality, failures to recognize the emotional entailments of actions, and the gaps between received communal norms and the values of an individual hero or heroine, often defined precisely by their standing outside the community. The virtual certainty of some failure of mind-reading or inference is not, then, a mark of the human (the human track record being weighted toward success rather than failure), but of the literary— that is, the tellable. Temporary and redeemable failures are comic; the devastating and uncorrectable are tragic. Any display of mind-reading, although it is grounded in the recognition of the internal similarity of different individuals, and based on a set of minimally universal and at least partially involuntary facial expressions and gestures,[27] will rarely be unambiguous, because it is always embedded in complex local events. As with other biological systems roughed in by human genetic structures but then developed for and in specific contexts, the human facility at mind-reading is all the more useful for its flexibility, although it achieves this flexibility at the price of the kind of certainty one might wish for or imagine one might have. The clues to which we sensibly learn to be attentive cannot be relied on absolutely because bodies themselves, the bodies that are evolved to give external expression to internal states, learn to produce these clues within contexts differentiated by cultural categories such as gender, age, social class, and occupation. Not only our interpretations of them but the evolved physical expressions themselves are enriched and/or distorted by social overlays, making both misinterpretation and deliberate deception possible.

Furthermore, knowledge about others that can be had for the looking provides an opening for social arbiters to attempt to control others' looking. Socially dominant groups have been known to take advantage of their power to limit what other groups can know by seeing.[28] Many cultures stigmatize various kinds of looking. Women and children, in some cultures and in some situations, may be taught not to meet the gaze of men or other adults, but to avert their eyes from any such encounter, presumably because there is a lot to be learned in this way.[29] By forbidding the intimate knowing that looking into another's eyes allows, the intimacy itself is deflected. In Western countries children are taught not to "stare" at disabled people, even though staring would be a big help in learning to assess whether a person might need assistance.

The Theory of Mind, then, is one in a series of cognitive hypotheses about the way we feed our cognitive hunger. It holds out the possibility of deduction from form to meaning—the hope that you will be able to decode what you see in a straightforward way. That hope is compromised from the start, however, because cognition is an embodied, biological, and thus context-specific and flexible process, allowing inference but not deduction. What might seem like bad news about the intelligibility of the world, however, is good news in literature. A scene of mind-reading, in a literary text, is particularly tellable because its inbuilt ambiguity, usually at a pivotal point in the plot, demands the kind of display a literary text provides. Among the privileges a literary text affords is permission to stare, as it were. Especially in novels, authors can not only stop for the display, giving it almost any amount of time, but can produce this display by means of all the usual genre affordances not available outside of texts: they can pass through walls, be in several heads at once, leap back and forth in time, all in order to enrich the display.

Representatively Hungry Problems

Before we despair, recognizing our real-world biological limitations, we need to acknowledge again that hunger is what drives us to find the food that can feed our human needs. As noted above, Paul Hernadi has argued that the creation and consumption of fictional narratives provide evolutionary advantages to a group, preparing them to anticipate challenges they may some day face by familiarizing their young with a range of hypothetical scenarios. His claim is that the extended, expanded, even slow-motion re-representation of particularly difficult moments gives people a chance to learn from the mistakes of fictional others. My difficulty with Hernadi's hypothesis is that it naturalizes literature by demonstrating how it is parallel to, say, lion cubs' pretend fighting, but then doesn't answer the question of why art is marked, in all cultures, as "unnatural," as deliberate artifice. Here is where Pratt and Clark converge. What Pratt calls display texts (and, I would add, pictures, films, and statues as well) are reusable, as it were, because they are just those texts and pictures that display representationally hungry situations. They attempt to re-represent situations that don't yield an easy reading, and therefore draw to them a lot of artists' representational energy. Moments of conflict are interesting and tellable because it is moments like these for which we certainly don't want "real life" practice; they may well be moments of conflict for which resolutions are not available.

In life we will meet confusing moments for which prior "off-line" training would be advantageous, and would be lucky if we had been able to train ourselves on narrative. But for some representationally hungry problems—and my claim is that this is the stuff of classic, high art—no amount of re-representation seems to avail; the mystery is not resolved on the last page. Great art marks the boundaries of human understanding of our own basis in the world, the limits of our self-understanding. It has been claimed that narrative themes that are frequently repeated, such as stories that revolve about mating plots, are somehow fundamental to human survival and therefore universally interesting.[30] Hernadi would presumably add that they are also useful for training the young. In fact, I would draw quite a different conclusion.

When a representation, narrative or other, is considered in its historical context, as an embodied theory of knowledge requires it be, it is the threatening changes in local environments that evoke re-representation; biological universals, being constant, are far less disruptive. If eighteenth- and nineteenth-century novels seem statistically to concentrate on mating themes, it will be because cultural changes, including a growing bourgeoisie, made "the old dance," as Chaucer's Wife of Bath calls it, newly problematic. Not only does a theme with a preponderance of re-representations *not* show that the issue is an essential, biological one, but it shows how culturally complicated and thus resistant to solution the problem expressed by that theme may be. Such a constellation testifies, rather, to a clash of universal needs and the local demands of culture, and shows how representationally hungry the issue may be, how in need of re-representation. Many of the works of art that are most highly valued are not those that display solutions, but those that display failures, even if the end of the narrative is "happy."

This perspective, one that views literary experiences as imbricated within both bodily and cultural matrices, allows us to express what is wrong with the claim that literary experiences are not real but escapist, make nothing happen. They let us see that some narratives are useful in allowing the brain and the body to ruminate over particularly difficult social puzzles in good time. But for others—maybe even the best of them—the rumination may end with the acknowledgment that the only solution to the problem is a skeptical or a stoic resignation.[31] This enriched view of tellability, i.e., that a story which is "tellable" is necessarily a narrative embedded in the permanent dynamic between the individual's evolved possibilities for knowing and the information that can be gleaned from the habitus, encourages us in the search for a normative epistemological theory. The theory of mirror neurons and the Theory of Mind look like they might turn out, when further developed, to be part of such a theory. They would be useful, then, in literary interpretation or in narrative theory because they draw readers' attention to the limits of human knowing, and show us how hypothetical narrative texts draw us into attempts to re-represent difficult problems. Yet what great art so often teaches is that even using all we've got, we're going to arrive at the boundaries of our capacity to cope, to understand, maybe even to survive, in our fraught environments.

I'd argue, thus, that when we come upon a text that is constantly re-read to produce an ever expanding cluster of interpretations around the same characters, for example, analyses of Macbeth's marriage; or, when we recognize a cluster of texts on the same theme, like romantic nature poetry, we should assume that the authors and interpreters involved are gnawing away at a hard bone; they are trying to make order in a corner of our minds and also in our society on a subject that needs the forming and organizing powers of dramatic, pictorial, or narrative art, but is also particularly resistant to attempts at such organization or clarification. What does motivate a woman to incite her husband to murder? How are we to understand the relationship of people to nightingales and wind, or to rocks, and stones, and trees? Will we ever get an answer to these questions? The reason for the thematic clusters, then, either of interpretations, or of texts, is that the structuring and organizing job—the representing—is in some cases exceptionally hard to do. For some problems, it may never get done. Even among those of us who love stories and believe that they work constructively in the world, it must be admitted that some of the problems humans face, as individuals or in groups, are apparently beyond our ability to represent in narratives or in the visual arts.[32] They resist the organizing powers of even the most innovative of narrators. One of the most interesting outcomes of thinking of narrative as food, then, is that it allows us to confront *representational hunger.* The words themselves hint at famine. Even starvation is a possibility.

Failure

My hypothesis is this: when we see a cluster of narratives (or other kinds of representation, like plays, or paintings) all hovering around the same subject and at more or less the same time, we normally say that the theme was popular. We would do better to think about the eating disturbance called bingeing. As with overeating, the food that should satisfy, for some reason does not. Literary repetitions suggest this kind of dissatisfaction. Repeated attempts to represent a complicated issue might well be recognized as the repeated failure of artists and storytellers to re-represent the issues in such a way as to make their art or their stories satisfying to the people who need to ingest and understand them. They invest the effort because the problems are real and pressing and because the existing representations don't quite satisfy, or no longer satisfy.

And here again literature has an advantage over life: as we are allowed to stare in literature, we are allowed to binge. We are allowed to create and devour as many representations as we want.[33] I have written elsewhere about the repeated retellings and repaintings of the rape of Lucretia story, for example, in the early modern period in Europe,[34] which attempt again and again to display and make sense of a raped woman's suicide. Not an easy issue, especially since this particular rape and suicide is said to produce a publicly beneficial outcome: the overthrow of tyranny and the founding of the Roman Republic. The later history of narrative in the West exhibits similar clusters. One that comes to mind are the eighteenth- and nineteenth-century stories exploring the issue of "natural piety," questioning whether people need religious institutions, scriptures, and sacred rites to be good, or whether they can learn virtue from their own experiences. In English this theme is taken up in Defoe's *Robinson Crusoe*, then in Fielding's *Joseph Andrews*, and in George Eliot's *Silas Marner*, all of which pit unchurched but good people against more normally socialized foils. The theme is not a motif, if by that term we indicate a fixed abstract idea and its familiarly associated imagery. The point about it is precisely that it is not fixed—not able to display a clear and convincing stance or persuasively resolve the dilemma.

A twentieth-century example is the very large number of postwar German novels about the war, the genocide, and the guilt of the German people. This is clearly an impossible problem for even extraordinarily gifted artists to represent and organize, and it is well described as a cognitively hungry problem. To call these issues popular is merely question-begging. Their repeated re-representation attests rather to disease: dis-ease or frustration—to the fact that so far no one has found a satisfactory way to represent them, to feed the hunger for understanding that they display.

And it is here that we see how changing the terms of discussion can open up a new way of thinking about the relationship of literature to fundamental questions. I would claim that it is the best writers who take up the challenges of cognitively hungry problems, even if they fail to solve them. Further investigation in this direction will help us articulate the distinctions among the texts we call classics, serious literature, high art, and those we recognize as interesting and enjoyable but ephemeral. Once we understand narrative as a source of cognitive nourishment, and understand individuals or groups as finding themselves in situations of greater or lesser cognitive hunger, we can observe how they step up to the challenge.

If it is so, then some of the needs for fiction might be relatively easily satisfied. Others are more complex (or, to put it in evolutionary terms, harder for creatures with our kind of brain to think about or think through), demanding more or more skilled narrative effort. The saving insight here is Clark's hypothesis that brain representation stretches along a gradient. There are situations that are hard to represent and if the issues involved are pressing, evoke repeated efforts at representation.[35] These representationally hungry issues are likely to attract the elaborated representation that especially talented writers and artists can provide. Indeed, the repeated treatment of certain issues by talented artists is itself evidence of the difficulty of representing them in a satisfying way.

The Example of Shakespeare's *Merchant of Venice*

In the cases I have investigated, one of the ways artists attempt to re-frame and re-represent cognitively hungry problems is by retelling or repainting the story in a different genre. It is clear, for example, that Shakespeare's *Merchant of Venice* was written as a comedy, although it is no longer played as one. The main plot, beginning, as comedies do, with a complication, ends with the confounding and conversion of the villainous Jew: the happy ending. The subplot, a romance, also turns out happily for all the young people who marry in love and in good fortune. The comic genre itself endorses the anti-Semitic attitudes expressed in the play: audiences are asked to agree that depriving the Jew of his money, forcing him to convert to Christianity, and rejoicing in the elopement of his daughter with a Christian add up to a happy solution to the problems with which the play opens.[36]

When the play is to be staged now, however, the genre is always changed. The director cuts, adds, and rearranges so that at least some of its heavily weighted conditions of satisfaction are subverted. The 2004 film of *The Merchant of Venice*, directed by Michael Radford and starring Al Pacino as Shylock and Jeremy Irons as Antonio, is a good example. Radford cuts hefty chunks of dialogue and adds scenes at the beginning that make watching the film a predominantly dark and often ominous experience. Opening with rolling titles explaining the disadvantageous situation of Jewish moneylenders in sixteenth-century Venice, Radford immediately inserts, before we begin to hear Shakespeare's words, several almost wordless scenes. We see a close-up of a torch setting alight several Hebrew religious books, and then see a Franciscan standing in a gondola on the canal, using biblical citations against usury to incite a crowd on the Rialto. Local youths take the message and rough up some Jews, and with shouts of "usurer," one is tossed over into the water. Shylock, caught in the agitated crowd, recognizes the merchant, Antonio, and calls out to him—"Antonio!"—as if to ask the merchant, apparently his business acquaintance, to intervene and calm the situation, but Antonio's reaction is to spit at Shylock. Yet another addition, still before we hear Shakespeare's first lines, is a scene in a Venice synagogue, after which Shylock's daughter Jessica is seen to drop a handkerchief as a sign to her secret admirer, the Christian Lorenzo, whom she does not otherwise acknowledge.

When we are finally launched into Shakespeare's play, the threatening scenes of Venice are reinforced by the camera's close attention to Antonio's sorrow, first among his friends, and then with his young friend Bassanio, and all in the dark colors of old Renaissance paintings. We seem to be asked to credit his sadness to the realization that he will have to cede what is clearly a very strong attachment, allowing the feckless youth to compete for an advantageous marriage to the rich Portia. Shylock's mention of his hatred for Antonio, "for he is a Christian" (1.3.32), is cut from Act I, as are his words to his daughter as he leaves for dinner with Antonio, where Shakespeare had him admitting to his daughter that he will "go in hate" (2.5.15). In the movie, though not in the play, Shylock does not express his desire for revenge or his hatred of Christians until he hears that his daughter has run off with Lorenzo and a large amount of his money, in Act 3, scene 1. His motive, thus, for insisting that Antonio repay the defaulted debt with a pound of his flesh, though the bargain seemed earlier to have been proposed as a joke, is not played as "feed[ing] fat the ancient grudge" (1.3.37) that Shylock says he bears toward Antonio but, apparently, as a direct response to the personal and local pain of his daughter's betrayal.

Shakespeare's play concludes with three young newly married couples going off to bed, exchanging delicate witticisms that extend the sexual conceit of the rings that Portia and Nerissa had given to their husbands and are about to give away again. The film, however, concludes with added footage, precluding our understanding of it as a comedy. First, still at Belmont, we are shown, wordlessly, a gloomy view of the marriage of Jessica and Lorenzo, and of the sadness of Antonio, still melancholy at having lost Bassanio. Another wordless scene, back in Venice, shows us the broken Shylock, physically excluded from the synagogue when members of the community bar the doors at his approach. These wordless (i.e., non-Shakespearean) scenes are filmed at night, or in darkened settings.

Although many revisionary productions of the play have been appreciated by critics as both interesting and satisfying, the complex of historical circumstances, plot, characterization, and poetic language in this play continues to resist satisfactory elucidation. The director of a 2007 New York production, Darko Tresnjak, criticized Radford's deletions of Shylock's early expressions of hate for gentiles as sentimental. Tresnjak moves the play's time to "the near future" in the glass and steel surroundings of an investment banking firm. The caskets are laptops—one of many changes intended to obliterate all of the play's original lightness. F. Murray Abraham plays an extremely disagreeable Shylock, corrupted by the climate of hatred in which he lives, "a little more sinister than sympathetic. . . . a soul poisoned by a thirst for vengeance." A reviewer of the production calls it "a soul-searing tragedy" that makes the "play's classification as one of Shakespeare's high comedies seem preposterous."[37]

The very large number of re-visions, of re-representations of the story of Shylock, attests to the indigestible problem of hatred and revenge at the core of the play.[38] No one—no philosopher and no theologian—has been able to help us understand either the anti-Semitism of the Venetians, or Shylock's hatred and rage. When asked by Salerio why he insists on taking the pound of flesh ("What's that good for?"), his anger prompts him to answer: "If it will feed nothing else, it will feed my revenge" (3.1.40). When his desire for revenge remains unsatisfied, he may be said to self-destruct by starvation. Here again we see that the multiplication of versions of the same subject or narrative is itself something more than and different from popularity. It suggests, rather, that subjects that are frequently repainted, re-narrated, and re-dramatized are the thematic evidence of persistent cognitive difficulties around a non-trivial issue, of a remarkable, that is, tellable, challenge to satisfactory representation.

Eating lunch doesn't mean we won't be hungry for supper. And even when an artist has produced a satisfying re-representation, the turning world itself is likely to undo the good work. Satisfying representations of our very hardest problems certainly have been produced, but time passes, and the satisfaction fades or changes its character. If we find satisfaction in George Eliot's *Middlemarch* today, it is unlikely to be because it satisfies the same needs that made it popular when it was first written. Here, then, we may glimpse a second issue that could benefit from rethinking in terms of the model of dynamic interactivity suggested by metabolism, and that is literary periodization.

In addition to considering formal and thematic ways in which the creative works of a local time and place cohere, we may consider the coherence of a group of works around a particular representationally hungry problem. I don't suggest we start from historical descriptions of political issues, because the most crucial of them may not even be articulated, may not even be recognized as problems, until writers and artists begin to attempt representations of them. But once someone produces a truly revolutionary work, the competition is on. The challenge of a new representation, new in style, genre, or theme, will stimulate others to expand the search for the kind of nourishment that could satisfy. Satisfy what? Who decides? Is this the birth of a "movement"? In the loop of need and satisfaction, there are presumably multiple points at which intervention can produce changed forms, and of course there are multiple agents of intervention, critics as well as artists, patrons, or politicians.

New forms, on this view, are not so much responsive to new issues as they are active in acknowledging them, and in bringing their audiences to a similar awareness. There is work to be done here. Let's not toss out the metaphor of narrative as tool use, but let's add another course to the menu of our investigations.

Notes

1. Recent interest in the question can be dated from W. K. Wimsatt and M. C. Beardsley, "The Intentional Fallacy," in *The Verbal Icon* (Lexington: University of Kentucky Press, 1954), first published in 1946.

2. For a discussion of brain modularity and its implications for literary interpretation, see Ellen Spolsky, *Gaps in Nature: Literary Interpretation and the Modular Mind* (Albany: SUNY Press, 1993).

3. For a summary of many of the arguments for the power of narrative see David C. Rubin and Daniel Greenberg, "The Role of Narrative in Recollection: A View from Cognitive Psychology and Neuropsychology," in *Narrative and Consciousness: Literature, Psychology, and the Brain*, ed. Gary D. Fireman, Ted E. McVay, and Owen J. Flanagan (Oxford: Oxford University Press, 2003). Martin Kreiswirth's article in *Routledge Encyclopedia of Narrative Theory*, ed. David Herman, Manfred Jahn, and Marie-Laure Ryan (Abingdon: Routledge, 2005), 377–382, is a helpful summary of how the study of narrative has spread.

4. For some good arguments, see Alan Richardson and Ellen Spolsky, eds., *The Work of Fiction: Cognition, Culture, and Complexity* (Aldershot: Ashgate, 2004).

5. The term *representational hunger* is taken from Andy Clark, *Being There: Putting Brain, Body, and World Together Again* (Cambridge: MIT Press, 1997).

6. *Complete Prose Works*, 8 vols. (New Haven: Yale University Press, 1935–1982).

7. The term *homeostasis* was introduced in 1932 by Walter B. Cannon (*The Wisdom of the Body* [New York: Norton]) to describe the continuous adjustments the body makes to maintain stability. The importance of the concept for literary theory is as an understanding of regulation that doesn't depend on top-down, hierarchical management and control.

8. Bradd Shore is another inadvertent exponent of the biological view of narrative: An "unexpected event is, for normal people, relatively indigestible until it is processed by talk into palatable form." *Culture in Mind: Cognition, Culture, and the Problem of Meaning* (New York: Oxford University Press, 1996), 58.

9. Jerome A. Feldman, *From Molecule to Metaphor: A Neural Theory of Language* (Cambridge, Mass.: MIT Press, 2006), works to link analogical cogni-

tion (metaphor) to its physiological ground, substantiating with neurological evidence the basic cognitive linguistic theories of George Lakoff in a series of studies, beginning with *Metaphors We Live By*, coauthored by Lakoff and Mark Johnson (Chicago: University of Chicago Press, 1980).

10. For discussion of the fundamental place of analogy in cognition, see Dedre Gentner, Keith J. Holyoak, and Boico N. Kokinov, *The Analogical Mind: Perspectives from Cognitive Science* (Cambridge, Mass.: MIT Press, 2001).

11. Daniel Dennett has explored this hypothesis in depth in *Consciousness Explained* (New York: Little, Brown, 1991).

12. *Motor Cognition: What Actions Tell the Self* (Oxford: Oxford University Press, 2006), 169. We recognize the meanings of blushes, of body postures, and of muscle tension, for example. See Ellen Spolsky, "Elaborated Knowledge: Reading Kinesis in Pictures," *Poetics Today* 17, no. 2 (1996): 157–180.

13. Stephen Gould and Richard Lewontin ("The Spandrels of San Marco and the Panglossian Paradigm: A Critique of the Adaptationist Programme," *Proceedings of the Royal Society of London* B 205 [1979]: 581–598) explain how this works and what implications it has for an evolutionary theory.

14. Daniel L. Schacter, *Searching for Memory: The Brain, the Mind, and the Past* (New York: Basic Books, 1996); Antonio Damasio allows for a pre-conscious "proto-self" in an infant, in *The Feeling of What Happens: Body and Emotion in the Making of Consciousness*.

15. Sabrina Diano, et al., "Ghrelin controls Hippocampal Spine Synapse Density and Memory Performance," *Nature Neuroscience* 9, no. 3 (March 2006): 381–388.

16. Charles Schultz has a *Peanuts* strip that demonstrates just this. Charlie Brown is reading to Lucy and Snoopy: "When Sodom and Gomorrah were destroyed, Lot and his two daughters escaped. But his wife was turned into a pillar of salt." The thought balloon over Snoopy's head reads: "What happened to their dog?" (1991, United Features Syndicate).

17. When all is not working well, as in post-traumatic stress syndrome, unwanted stories or parts of them intrude and distress, rather than help a person cope with the present. Some mental disturbances may be described as the inability to use stories helpfully, or the inability to revise them to suit changed circumstances.

18. *Nature* 445, no. 4 (January 2007). For a fuller explanation, see Daniel Schacter, *The Seven Sins of Memory: How the Mind Forgets and Remembers* (Boston: Houghton Mifflin, 2001). A "good trick," according to Daniel Dennett, is a behavior or adaptation that favors survival but is not directly controlled by genes. *Darwin's Dangerous Idea: Evolution and the Meanings of Life* (New York: Simon and Schuster, 1995), 487.

19. *Being There*, 47–51, and *Mindware: An Introduction to the Philosophy of Cognitive Science* (Oxford: Oxford University Press, 2001).

20. See Ulric Neisser, "The Self Perceived," in *Ecological and Interpersonal Sources of Self-Knowledge* (Cambridge: Cambridge University Press, 1993), 3–21. Neisser describes this view as a Gibsonian, or ecological, view of how the understanding self is constructed. "The approach assumes that the perceptual systems of animals evolved to take advantage of objectively existing information"

(6). In the same collection, see Colwyn Trevarthen, "The Self Born in Inter-subjectivity: The Psychology of an Infant Communicating."

21. Jean-Pierre Changeux, *The Physiology of Truth: Neuroscience and Human Knowledge* (Cambridge: Harvard University Press, 2004), 25, 32. Literate people would have great difficulty not reading words that appear in front of them.

22. Of course if the hermeneutic is taught by the hegemonic canonizers I guess the distinction is moot.

23. In *New Stories for Old: Biblical Patterns in the Novel* (Houndmills, Basingstoke: Macmillan, 1998), 86–87.

24. For an early discussion of the Theory of Mind, see D. G. Premack and G. Woodruff, "Does the Chimpanzee Have a Theory of Mind?" *Behavioral and Brain Sciences* 1 (1978): 515–526. For discussions of its applicability to literary study, see Lisa Zunshine, *Why We Read Fiction: Theory of Mind and the Novel* (Columbus: Ohio State University Press, 2006), and Alan Palmer, *Fictional Minds* (Lincoln: University of Nebraska Press, 2004).

25. Gallese and Goldman (1998) and Iacoboni et al. (1999). See also *Perspectives on Imitation: From Neuroscience to Social Science*, ed. Susan Hurley and Nick Chater (Cambridge: MIT Press, 2005), 2 vols., for assessments and critiques of the meaning of the work on mirror neurons.

26. I am optimistically assuming here that further specification of the mirror neuron system will expose its contribution to an enriched Theory of Mind, but it's going to be a while before these multisensory neurons can be implicated in interpersonal cognition.

27. Paul Ekman and Erika L. Rosenberg, eds., *What the Face Reveals: Basic and Applied Studies of Spontaneous Expression Using the Facial Action Coding System [FACS]* (Oxford: Oxford University Press, 2005).

28. I discuss the campaign against religious imagery during the Protestant Reformation in *Word vs. Image: Cognitive Hunger in Shakespeare's England* (Houndmills: Palgrave Macmillan, 2007).

29. Gregory Bateson, *Steps to an Ecology of Mind* (New York: Ballantine Books, 1972), 378.

30. See Brian Boyd, "Jane, Meet Charles: Literature, Evolution, and Human Nature," *Philosophy and Literature* 22, no. 1 (1988): 1–30, and also the essay by John Carroll and another by Daniel J. Kruger et al. in *The Literary Animal: Evolution and the Nature of Narrative*, ed. Jonathan Gottschall and David Sloan Wilson (Evanston, Ill.: Northwestern University Press, 2005). See my review "Frozen in Time?" in *Poetics Today*.

31. For a study of the interplay of art and hermeneutic skepticism in early modern Europe, see Ellen Spolsky, *Satisfying Skepticism: Embodied Knowledge in the Early Modern World* (Aldershot: Ashgate, 2001), chap. 1.

32. Spolsky, "Doubting Thomas and the Senses of Knowing," *Common Knowledge* 3, no. 2 (1994): 111–129.

33. There are, unfortunately, regimes in which this imaginative work cannot be done, or cannot be published if it is done.

34. "Women's Work Is Chastity: Lucretia, *Cymbeline*, and Cognitive Impenetrability," in *The Work of Fiction: Cognition, Culture, and Complexity*, ed. Alan Richardson and Ellen Spolsky (Aldershot: Ashgate, 2004).

35. Clark, *Being There.*

36. For an interesting and persuasive account of how the issues of usury and anti-Semitism might have been understood in the context of the growth of a merchant class and a recognition of the need for lending money in the late-sixteenth-century England in which the play was written and performed, see Murray Roston's *Tradition and Subversion in Renaissance Literature: Studies in Shakespeare, Spenser, Johnson, and Donne* (Pittsburgh: Duquesne University Press, 2007), chap. 1.

37. All citations are from Charles Isherwood in the *New York Times* of 5 February 2007, reviewing the production of *The Merchant of Venice* by the Theater for a New Audience.

38. In Sonia Massai's collection of articles on reworkings of Shakespeare (*World-wide Shakespeares: Local Appropriations in Film and Performance* [New York: Routledge, 2005]), three of the chapters are about *Merchant.* Leah Marcus's edition of the play for the Norton Critical Editions includes a section on "Rewritings and Appropriations," which reprints small selections of the many available, including six poems about Shylock. Kenneth Gross, in his most interesting and original book, *Shylock Is Shakespeare* (Chicago: University of Chicago Press, 2006), reconsiders some of the twentieth-century reworkings of the play, and asserts that Shakespeare must have identified strongly with Shylock, and that "Shakespeare places at the center of his dramatic script a point of stark resistance to performance" (4).

CHAPTER THREE

Narrative Empathy

SUZANNE KEEN

We are living in a time when the activation of mirror neuron areas in the brains of onlookers can be recorded as they witness another's actions and emotional reactions.[1] Contemporary neuroscience has brought us much closer to an understanding of the neural basis for human mind-reading and emotion-sharing abilities—the mechanisms underlying empathy. The activation of onlookers' mirror neurons by a coach's demonstration of technique or an internal visualization of proper form and by representations in television, film, visual art, and pornography has already been recorded.[2] Simply hearing a description of an absent other's actions lights up mirror neuron areas during fMRI imaging of the human brain.[3] The possibility that novel reading stimulates mirror neurons' activation can now, as never before, undergo neuroscientific investigation. Neuroscientists have already declared that people scoring high on empathy tests have especially busy mirror neuron systems in their brains.[4] Fiction writers are likely to be among these high-empathy individuals. For the first time we might investigate whether human differences in mirror neuron activity can be altered by exposure to art, to teaching, to literature.

This newly enabled capacity to study empathy at the cellular level encourages speculation about human empathy's positive consequences. These speculations are not new, as any student of eighteenth-century moral sentimentalism will affirm, but they dovetail with efforts on the part of contemporary virtue ethicists, political philosophers, educators, theologians, librarians, and interested parties such as authors and publishers to connect the experience of empathy, including its literary form, with outcomes of changed attitudes, improved motives, and better care and justice. Thus a very specific, limited version of empathy located in the neural substrate meets in the contemporary moment a more broadly and loosely defined, fuzzier sense of empathy as the feeling precursor to and prerequisite for liberal aspirations to greater humanitarianism. The sense of crisis stirred up by reports of stark declines in reading goes into this mix, catalyzing fears that the evaporation of a reading public leaves behind a population incapable of feeling with others. Yet the apparently threatened set of links between novel reading, experiences of narrative empathy, and altruism has not yet been proven to exist. This chapter undertakes three tasks preliminary to the scrutiny of the empathy-altruism hypothesis[5] as it might apply to experiences of narrative empathy, an argument I develop more fully in *Empathy and the Novel* (2007).[6] These tasks include: a discussion of empathy as psychologists understand and study it; a brief introduction to my theory of narrative empathy, including proposals about how narrative empathy works; and a review of the current research on the effects of specific narrative techniques on real readers.

What Is Empathy?

Empathy, a vicarious, spontaneous sharing of affect, can be provoked by witnessing another's emotional state, by hearing about another's condition, or even by reading. Mirroring what a person might be expected to feel in that condition or context, empathy is thought to be a precursor to its semantic close relative, *sympathy*. *Personal distress*, an aversive emotional response also characterized by apprehension of another's emotion, differs from empathy in that it focuses on the self and leads not to sympathy but to avoidance. The distinction between empathy and personal distress matters because empathy is associated with the moral emotion *sympathy* (also called *empathic concern*) and thus with prosocial or altruistic action.[7] Empathy that leads to sympathy is by definition other-directed, whereas an over-aroused empathic response that creates personal distress (self-oriented and aversive) causes a turning away from the provocative condition of the other. No philosophers who put stock in the morally improving experience of narrative empathy include personal distress in their theories. Because novel reading can be so easily stopped or interrupted by an unpleasant emotional reaction to a book, however, personal distress has no place in a literary theory of empathy, though it certainly contributes to aesthetic emotions, such as those Sianne Ngai describes in her important book *Ugly Feelings*.

In empathy, sometimes described as an emotion in its own right,[8] we feel what we believe to be the emotions of others.[9] Empathy is thus agreed to be both affective and cognitive by most psychologists. Empathy is distinguished in both psychology and philosophy (though not in popular usage) from *sympathy*, in which feelings *for* another occur. So, for instance, one may distinguish empathy from sympathy in this fashion:

Empathy:
I feel what you feel.
I feel your pain.

Sympathy:
I feel a supportive emotion about your feelings.
I feel pity for your pain.

These examples emphasize negative emotions—pain and pity—but it should be noted from the outset that although psychological and philosophical studies of empathy have tended to gravitate toward the negative, empathy also occurs for positive feelings of happiness, satisfaction, elation, triumph, and sexual arousal.[10] All of these positive kinds of empathy play into readers' pleasure, or *jouissance.*[11]

Experts on *emotional contagion*, the communication of one's mood to others, have done a better job of studying the full range of emotional states that can be shared through our automatic mimicry of one another.[12] Indeed, primitive emotional contagion, or "the tendency to automatically mimic and synchronize facial expressions, vocalizations, postures, and movements with those of another person and consequently, to converge emotionally" (Hatfield, Cacioppo, and Rapson, 81) offers a compelling explanation of a component of our empathy as arising from our physical and social awareness of one another, from birth. Inherited traits play an important role in our disposition to experience emotional contagion,[13] but our personal histories and cultural contexts affect the way we understand automatically shared feelings.[14]

So, for instance, emotional contagion comes into play in our reactions to narrative, for we are also story-sharing creatures. The oral storyteller not only takes advantage of our tendency to share feelings socially by doing the voices and facial expressions of characters, but also tacitly trains young children and members of the wider social group to recognize and give priority to culturally valued emotional states.[15] This education does not create our feelings, but renders emotional states legible through their labels, and activates our expectations about what emotions mean. Narratives in prose and film infamously manipulate our feelings and call upon our built-in capacity to feel with others. Indeed, the early history of empathy as a subject of study emphasized both emotional contagion and aesthetic responses.

The word *empathy* appeared as a translation of *Einfühlung* in the early twentieth century. In 1909, the experimental psychologist E. B. Titchener translated as "empathy" aesthetician Theodor Lipps's term *Einfühlung* (which meant the process of "feeling one's way into" an art object or another person).[16] Notably, Titchener's 1915 elaboration of the concept in *Beginner's Psychology* exemplifies empathy through a description of a reading experience: "We have a natural tendency to feel ourselves into what we perceive or imagine. As we read about the forest, we may, as it were, become the explorer; we feel for ourselves the gloom, the silence, the humidity, the oppression, the sense of lurking danger; everything is strange, but it is to us that strange experience has come" (198). In the beginning of the twentieth century, the English novelist Vernon Lee brought *Einfühlung* and empathy to a broader literary audience. In a public lecture followed by a magazine piece in a popular journal,[17] Lee advanced a theory of aesthetic perception of form involving empathy, though not (at first) so named. Originally Lee's aesthetics focused on bodily sensations and muscular adjustments made by beholders of works of art and architecture and downplayed emotional responsiveness. By the time she revised and expanded her ideas for presentation in book form, however, Lee had adapted Lipps's understanding of empathy, a parallel development from common sources in German aesthetics.

Defining the purpose of art as, in part, "the awakening, intensifying, or maintaining of definite emotional states" (Lee 99–100), Lee makes empathy a central feature of our collaborative responsiveness (128). In an account that combines motor mimicry, memory, and psychological responsiveness to inanimate objects, Lee argues that empathy enters into "imagination, sympathy, and also into that inference from our own inner experience which has shaped all our conceptions of an outer world, and given to the intermittent and heterogeneous sensations received from without the framework of our constant and highly unified inner experience, that is to say, of our own activities and aims" (68). No sooner had the term been announced and situated so centrally in aesthetic theory for an English-language audience, however, than it received brisk challenge from high modernist quarters. The disdain of

Bertolt Brecht for empathy (and his advocacy of so-called alienation effects), the embrace of difficulty by modernist poets, and the dominance of New Criticism, which taught students to avoid the affective fallacy, all interfered with the integration of empathy into literary theory until recently. Novelists and novel readers who prized experiences of emotional fusion cultivated narrative empathy throughout periods when the term was in eclipse.

How Is Empathy Studied?

The focus on our embodied experience in feminist criticism, disability studies, cognitive approaches to narrative, and some ecocriticism, draws literary studies closer to disciplines that accept the use of making measurements, doing tests and experiments, and interpreting empirical evidence. This section explains some of the methods being used by neuroscientists and developmental and social psychologists to study empathy. Developing the conversation between literature and psychology ought to benefit both disciplines, however, and the subsequent comments on what is known and especially what has not yet been tested about the effects of narrative techniques contribute to a more nuanced application of psychonarratology to questions of interest to social and developmental psychologists.

Psychologists test and record empathy in a variety of ways. Physiological measures, sometimes combined with self-reports, can show the strength or weakness (or presence and absence) of empathic responses.[18] Psychologists measure changes in heart rate and skin conductance (palm sweat). They collect data on perceptible and imperceptible facial reactions, the latter captured by EMG (electromyographic) procedures.[19] They ask subjects how they feel or how they would act in certain situations, gathering responses through self-reports during or immediately after experiments and through surveys. Specialized surveys known as "empathy scales" are used to assess subjects' strength of empathic feeling.[20] Recently, Functional Magnetic Resonance Imaging (fMRI) has had a profound impact on brain science, including the study of empathy.[21]

Tania Singer and her colleagues have recently published a study in *Science* documenting empathetic responses to witnessing another's pain, supported by fMRI data. This study broke new ground in demonstrating why a person perceives that she feels another's pain, while not literally experiencing the identical sensations. Singer compared what happened in a subject's brain when she was actually shocked, when pain regions in the limbic system (the anterior cingulate cortex, the insula, the thalamus, and the somatosensory cortices) lit up on the fMRI, with what the brain looked like during observation of another's pain. When watching a loved one in the same room receiving a sharp shock, subjects showed active responses in the *affective* parts of the brain's pain matrix (in the anterior insula and anterior cingulate cortex, the lateral cerebellum, and the brain stem), but not in the somatosensory cortices of the brain. The affective brain areas responded to both real and imagined pain. A person not actually experiencing pain but observing a loved one being shocked showed brain activation of matching emotional areas, though not the sensory areas. Empathy alone did not light up the sensory areas for pain. Singer and her colleagues conclude that empathy is mediated by the part of the pain network associated with pain's affective qualities, but not its sensory qualities (Singer et al., 1157). They observed that subjects with higher scores on general empathy scales[22] "showed stronger activations in areas significantly activated when the subjects perceived their partner as being in pain" (1159). They also discovered that the same empathetic effects could be elicited without an emotional cue—in other words, subjects did not need to see their partners grimacing in pain in order to show empathic responses. An "arbitrary cue" signaling the feeling state of another was sufficient to elicit empathy (1158). This set of results affirms what neuroscientists working on mirror neurons on monkeys have theorized and what philosophers since David Hume have been saying about empathy for centuries. For the first time, brain images supporting the long-standing introspective account of empathy have been recorded.

The questions of how and why empathy works in the bodies and brains of human beings can still only be answered with theoretical speculations about the physiological substrate,[23] though the fMRI-based research described above and recent neuroscience on the shared manifold for intersubjectivity get researchers closer than they have been before. Stephanie Preston and Frans de Waal propose that witnessing or imagining another in an emotional state activates automatic representations of that same state in the onlooker, including responses in the nervous system and the body. They write that "empathy processes likely contain fast reflexive sub-cortical processes (directly from sensory cortices to thalamus to amygdala to response) and slower cortical processes (from thalamus to cortex to amygdala to response). These roughly map onto contagious and cognitive forms of empathy, respectively" (Preston and de Waal, 12). The advantages of automatic responses lie in their speediness. Joseph LeDoux has written about how fear responses in the amygdala provide a quick-and-dirty, possibly life-saving response to environmental threats, which can then be evaluated as the slightly slower cognitive evaluation of a threat kicks in (Le Doux, 168–178). What is sometimes called "primitive empathy" may work in the same way, providing a first, fast, feeling response to seeing or learning about another's emotional state, before cognitive evaluation through deliberate role taking occurs.[24]

The human capacity for primitive empathy, or the phenomenon of spontaneously matching feelings, suggests that human beings are basically similar to one another, with a limited range of variations. Psychologist Martin Hoffman, for instance, believes that the structural similarities in people's physiological and cognitive response systems cause similar feelings to be evoked by similar events (Hoffman, *Moral Development*, 62). However, Hoffman would be the first to concede that similarity itself is not enough to guarantee an empathic response. Singer and her colleagues believe that our survival depends on effective functioning in social contexts, and that feeling what others feel, empathizing, contributes to that success. They suggest that "our ability to empathize has evolved from a system for representing our internal bodily states and subjective feeling states" to ourselves (Singer et al., 1161). In other words, empathy as Singer's group understands it participates in a theory of mind that links second-order re-representations of others to the system that allows us to predict the results of emotional stimuli for ourselves. Recent research suggests a mechanism at the neural level that would enable such representations of others' actions, including facial expressions and bodily postures that may convey emotional states.[25] Contemporary neuroscience theorizes a system for representation of others' feelings that participates in the task of enabling us to understand the motives, beliefs, and thoughts of others. This work on empathy thus supports the theories of evolutionary psychology that emphasize the adaptive function of our social relations.[26] Given this basis in human shared intersubjectivity, empathy thus appears to many to be a key element in our responsiveness to others.

My work seeks to clarify why the link between narrative empathy and altruism is nonetheless so tenuous. For a novel reader who experiences either empathy or personal distress, there can be no expectancy of reciprocation involved in the aesthetic response. The very nature of fictionality renders social contracts between people and person-like characters null and void. Unlike the hostage children in Beslan who wished that Harry Potter would come to their rescue, adult readers know that fictional characters cannot offer us aid. Similarly, we accept that we cannot help them out, much as we may wish to intervene: Don't marry him, Dorothea! We may feel intense interest in characters, but incurring obligations toward them violates the terms of fictionality. That is, an empathetic response can be diverted from a prosocial outcome through interfering cognition.

The treatment of emotions and rationality as separate and dichotomous features of our experience has been challenged in recent decades. Thinking and feeling, for Antonio R. Damasio, are part of the same package.[27] In a series of academic articles and popular books, he has shown that clinical patients suffering from emotional disorders have cognitive difficulties. Ronald DeSousa has also advocated recognition of the rationality of emotions, and Joseph LeDoux's cognitive neuroscience focuses on *The Emotional Brain*. Evolutionary psychologists Leda Cosmides and John Tooby[28] speak for a growing group of scientists who believe that "one cannot sensibly talk about emotion affecting cognition because 'cognition' refers to a language for describing all of the brain's operations, including emotions and reasoning (whether deliberative or nonconscious), and not to any particular subset of operations" (Cosmides and Tooby, "Evolutionary Psychology," 98).[29] In the relatively recent field known as Cognitive Approaches to Literary Studies, where the work of LeDoux and Damasio has virtually canonical status, matters of affect are generally considered to fall under the umbrella of the term *cognitive*. Few literary cognitivists acknowledge that a psychologist might not readily accede to the centrality of emotions to cognition. The sub-disciplinary boundaries within the extremely diverse field of psychology result in different emphases and perspectives on the place of the emotions. Empathy studies have from the start challenged the division of emotion and cognition, but they have also been altered by the convictions and disciplinary affiliations of those who study empathy.

I set aside for the moment the view that emotions and cognition describe different processes of the central nervous system, for empathy itself clearly involves both feeling and thinking. Memory, experience, and the capacity to take another's perspective (all matters traditionally considered cognitive) have roles in empathy. Yet the experience of empathy in the feeling subject involves the emotions, including sensations in the body. In any case, narrative empathy invoked by reading must involve cognition, for reading itself relies upon complex cognitive operations. Yet overall, emotional response to reading is the more neglected aspect of what literary cognitivists refer to under the umbrella term *cognition*. This does not need to be so. The discipline of aesthetics, which has historical ties both to philosophy and to psychology, as well as to literary studies, has been interested for over a century in empathy as a facet of creativity and an explanation of human response to artworks.[30] In its strongest form, aesthetics' empathy describes a projective fusing with an object—which may be another person or an animal, but may also be a fictional character made of words, or even, in some accounts, inanimate things such as landscapes, artworks, or geological features.[31] The acts of imagination and projection involved in such empathy certainly deserve the label "cognitive," but the sensations, however strange, deserve to be registered as feelings. Thus I do not quarantine narrative empathy in the zone of either affect or cognition: as a process, it involves both. When texts invite readers to feel, they also stimulate readers' thinking.[32] Whether novel reading comprises a significant enough feature of the environment of literate people to play a critical role in their prosocial development remains to be seen. Even the leap between reading and empathizing can fall short, impeded by inattention, indifference, or personal distress. Readers' cognitive and affective responses do not inevitably lead to empathizing, but fiction does disarm readers of some of the protective layers of cautious reasoning that may inhibit empathy in the real world.

Narrative theorists, novel critics, and reading specialists have already singled out a small set of narrative techniques—such as the use of first-person narration and the interior representation of characters' consciousness and emotional states—as devices supporting character identification, contributing to empathetic experiences, opening readers' minds to others, changing attitudes, and even predisposing readers to altruism. In the course of reviewing the available research on this subject, I point out the gaps in our knowledge of potentially empathetic narrative techniques. No specific set of narrative techniques has yet been verified to override the resistance to empathizing often displayed by members of an in-group regarding the emotional states of others marked out as different by their age, race, gender, weight, disabilities, and so forth.[33] Human beings, like other primates, tend to experience empathy most readily and accurately for those who seem like us, as David Hume and Adam Smith predicted.[34] We may find ourselves regarding the feelings of those who seem outside the tribe with a range of emotions, but without empathy.[35] If empathetic reading experiences start a chain reaction leading to mature sympathy and altruistic behavior, as advocates of the empathy-altruism hypothesis believe, then discovering the narrative techniques involved matters. It is one thing to discover, however, that high empathizers report empathetic reading experiences, and quite another to show that empathetic reading experiences can contribute to changing a reader's disposition, motivations, and attitudes. If novels could extend readers' sense of shared humanity beyond the predictable limitations, then the narrative techniques involved in such an accomplishment should be especially prized.

A Theory of Narrative Empathy

Character identification often invites empathy, even when the fictional character and reader differ from one another in all sorts of practical and obvious ways, but empathy for fictional characters appears to require only minimal elements of identity, situation, and feeling, not necessarily complex or realistic characterization. Whether a reader's empathy or her identification with a character comes first is an open question: spontaneous empathy for a fictional character's feelings sometimes opens the way for character identification. Not all feeling states of characters evoke empathy; indeed, empathetic responses to fictional characters and situations occur more readily for negative emotions, whether or not a match in details of experience exists. Finally, readers' experiences differ from one another, and empathy with characters doesn't always occur as a result of reading an emotionally evocative fiction.

Several observations help to explain the differences in readers' responses. Most important, readers' empathic dispositions are not identical to one another. Some humans are more empathetic to real others, and some feel little empathy at all. (Some research suggests that empathizers are better readers, because their role-taking abilities allow them to more readily comprehend causal relations in stories.)[36] The timing and the context of the reading experience matters: the capacity of novels to evoke readers' empathy changes over time, and some novels may only activate the empathy of their first, immediate audience, while others must survive to reach a later generation of readers in order to garner an emotionally resonant reading. Readers' empathy for situations depicted in fiction may be enhanced by chance relevance to particular historical, economic, cultural, or social circumstances, either in the moment of first publication or in later times, fortuitously anticipated or prophetically foreseen by the novelist.

Novelists do not exert complete control over the responses to their fiction. Empathy for a fictional character does not invariably correspond with what the author appears to set up or invite. Situational empathy, which responds primarily to aspects of plot and circumstance, involves less self-extension in imaginative role taking and more recognition of prior (or current) experience. A novelist invoking situational empathy can only hope to reach readers with appropriately correlating experiences. The generic and formal choices made by authors in crafting fictional worlds play a role in inviting (or retarding) readers' empathic responses. This means that for some readers, the author's use of the formulaic conventions of a thriller or a romance novel would increase empathic resonance, while for other readers (perhaps better educated and attuned to literary effects), unusual or striking representations promote foregrounding and open the way to empathic reading.[37]

Novelists themselves often vouch for the centrality of empathy to novel reading and writing, and express belief in narrative empathy's power to change the minds and lives of readers. This belief mirrors their experiences as ready empathizers. Yet even the most fervent employers of their empathetic imaginations realize that this key ingredient of fictional worldmaking does not always transmit to readers without interference. Authors' empathy can be devoted to socially undesirable ends that may be rejected by a disapproving reader. Indeed, empathic distress at feeling with a character whose actions are at odds with a reader's moral code may be a result of successfully exercised authorial empathy. Both authors' empathy and readers' empathy have rhetorical uses, which come more readily to notice when they conflict in instances of empathic inaccuracy (discordance arising from gaps between an author's intention and a reader's experience of narrative empathy). Experiences of empathic inaccuracy may contribute to a reader's outraged sense that the author's perspective is simply wrong, while strong concord in authors' empathy and readers' empathy can be a motivating force to move beyond literary response to prosocial action. The position of the reader with respect to the author's strategic empathizing in fictional worldmaking limits these potential results. I theorize that *bounded strategic empathy* operates with an in-group, stemming from experiences of mutuality and leading to feeling with familiar others. *Ambassadorial strategic empathy* addresses chosen others with the aim of cultivating their empathy for the in-group, often to a specific end. *Broadcast strategic empathy* calls upon every reader to feel with members of a group, by emphasizing common vulnerabilities and hopes through universalizing representations.

Empathetic Narrative Techniques

Consider the commonplace that first-person fiction more readily evokes feeling responsiveness than the whole variety of third-person narrative situations. Even a college sophomore with a few weeks' training in theoretical terms can report that within the category of first-person narratives, empathy may be enhanced or impeded by narrative consonance or dissonance, unreliability, discordance, an excess of narrative levels with multiple narrators, extremes of disorder, or an especially convoluted plot. Genre, setting, and time period may help or hinder readers' empathy. Feeling out of sorts with the implied readership, or fitting it exactly, may make the difference between a dutiful reading and an experience of emotional fusion.[38] Contrasting first-person with third-person puts the question too broadly, with too many other variables, to reach a valid conclusion. Narrative theorists can contribute specificity and subtlety to the research into narrative empathy.

A variety of narrative techniques have been associated with empathy by narrative theorists and discourse processing experts carrying out empirical research into literary reading. The formal devices themselves are regarded as empathic in nature by some theorists and researchers, while for others the disposition of the reader toward the text can be measured by inquiring about particular consequences of literary reading. The observations made by this latter group often lead to speculations about narrative technique. Mapping these ostensibly empathic narrative techniques draws attention to the many aspects of narrative form that have *not yet* been associated with readers' empathy, but which ought not to be ruled out without careful consideration.

The most commonly nominated feature of narrative fiction to be associated with empathy is *character identification*. Specific aspects of characterization, such as naming, description, indirect implication of traits, reliance on types, relative flatness or roundness, depicted actions, roles in plot trajectories, quality of attributed speech, and mode of representation of consciousness, may be assumed to contribute to the potential for character identification and thus for empathy.[39] The link between readers' reports of character identification and their experiences of narrative empathy has not yet been explained.

A close second for formal quality most often associated with empathy would be *narrative situation* (including point of view and perspective): the nature of the mediation between author and reader, including the person of the narration, the implicit location of the narrator, the relation of the narrator to the characters, and the internal or external perspective on characters, including in some cases the style of representation of characters' consciousness.[40] Many other elements of fiction have been supposed to contribute to readers' empathy, including the repetitions of works in series,[41] the length of novels,[42] genre expectations,[43] vivid use of settings,[44] metanarrative commentary,[45] and aspects of the discourse that slow readers' pace (foregrounding, uses of disorder, etc.).[46] The confirmation of many of the hypotheses about specific narrative techniques and empathy has yet to be undertaken in most cases, but the work that has been done *as often* fails fully to support the commonplaces of narratology as it authenticates them.[47] Whether this has to do with faulty experimental design, insufficient grasp of the nuances of narrative theory, or verifiable confutations of theory has yet to be discovered.

Character Identification

To begin with the necessary clarification, character identification is not a narrative technique (it occurs in the reader, not in the text), but a consequence of reading that may be precipitated by the use of particular techniques of characterization.[48] These qualities have not yet been investigated in a comprehensive fashion. Peter Dixon and Marisa Bortolussi emphasize aesthetic qualities of narrative that open the way to personal involvement.[49] In contrast, Jèmeljan Hakemulder suggests that readers experiencing strong admiration of an author's writing style may engage less readily with the fictional world and its inhabitants (Hakemulder, *Laboratory*, 73–74). Readers' personal involvement with a fictional character may (or may not) be contingent upon the use of a particular technique or the presence of certain representational elements that meet with their approval.[50] Keith Oatley believes that readers' personal experiences of patterns of emotional response provoke sympathy for characters, especially as readers identify with characters' goals and plans.[51] David S. Miall and Don Kuiken argue that emotional experiences of literature depend upon the engagement of the literary text with the reader's experiences,[52] but they emphasize foregrounding effects at the level of literary style that shake up conventions, slow the pace, and invite more active reading that opens the way for empathy.[53] Don Kuiken's research shows that readers who linked themselves to story characters through personal experiences were more likely to report changes in self-perception, if not actual empathy.[54] Max Louwerese and Kuiken suggest that empathy may work as a gap-filling mechanism, by which a reader supplements given character traits with a fuller psychologically resonant portrait.[55] Readers' judgments about the realism of the characters are supposed to have an impact on identification,[56] and the similarity of the reader to the character is widely believed to promote identification.[57] None of these phenomena, however, inhere in particular narrative techniques contributing to character identification.

A few techniques of characterization have actually been tested for their relation to readers' emotional responsiveness or empathy. Characters' involvement in a suspenseful situation provokes physiological responses of arousal in readers even when they disdain the quality of the narrative.[58] Plot-laden action-stories have been shown to promote faster reading than narratives focusing on characters' inner lives,[59] which may suggest by contrast greater reflectiveness on the part of character-focused readers, as Hakemulder supposes (Hakemulder, *Laboratory*, 74). However, this does not account for the quick, apparently involuntary responses to particular plot situations inspired by trashy novels. Speedy reading may be a token of involvement in a character's fate, identification, and even empathy. With the exception of appraisal of causality, virtually nothing about the role of plot structure has been associated with readers' empathetic responses, or tested in controlled settings.[60] Aspects of plot structure and narration that might have a role in invoking readers' empathy include the control of timing (pace), order (anachronies), the use of nested levels of narrative (stories within stories)[61], serial narrative, strong or weak closure, the use of subsidiary (supplementary, satellite) plot events, repetition, and gaps. Since each one of these structural categories contains an array of possibilities for characterization, their neglect leaves us with an incomplete picture of the devices whose use makes character identification possible.

Many aspects of characterization familiar to narrative theorists have not yet been tested in controlled experiments, despite their nomination by theorists. The naming of characters (including the withholding of a name, the use of an abbreviation or a role-title in place of a full name, or allegorical or symbolic naming, etc.) may play a role in the potential for character identification. The descriptive language through which readers encounter characters is assumed to make a difference (content matters!), but what about grammar and syntax? Does the use of present tense (over the usual past tense) really create effects of immediacy and direct connection, as many contemporary authors believe? The old "show, don't tell" shibboleth of creative writing classes remains to be verified: direct description of a character's emotional state or circumstances by a third-person narrator may produce empathy in readers just as effectively as indirect implication of emotional states through actions and context.[62] David S. Miall has suggested in "Affect and Narrative" that characters' motives, rather than their traits, account for the affective engagement and self-projection of readers into characters, though it remains unclear when, and at which cues, readers' emotional self-involvement jump-starts the process of interpretation. Bortolussi and Dixon believe that "transparency," or the judgment of characters' behavior as sensible and practical, contributes to identification (Bortolussi and Dixon 240). This may be too simple: even traditional novels are complex, polyvocal, and various, and Wayne Booth offers this sensible caution: "What we call 'involvement' or 'sympathy' or 'identification,' is usually made up of *many reactions* to author, narrators, observers, and other characters" (Booth 158, my emphasis). Some way of accounting for the multiplicity of reactions making up a normal novel-reading experience needs to be devised in order to study the transition from distributed characterization in narrative fiction and readers' everyday synthesis of their reactions into an experience of character identification.[63]

This may mean setting aside some common value judgments about techniques. For instance, the critical preference for psychological depth expressed by the "roundness" of characters "capable of surprising in a convincing way" (Forster 78) does not preclude empathetic response to flat characters, minor characters, or stereotyped villains and antagonists. Drawing on the literature of cognitive social psychology, Richard J. Gerrig has suggested that readers are likely to make category-based judgments about fictional characters, and to emphasize attributed dispositions of characters over their actual behavior in situations.[64] This theory suggests, as Forster intuited, that flat characters—easily comprehended and recalled—may play a greater role in readers' engagement in novels than is usually understood. Fast and easy character identification suffers in theorists' accounts of the reading process, which often privilege more arduous self-extension and analogical reasoning. Patrick Colm Hogan, for instance, regards categorical empathy (with characters matching a reader's group identity) as the more prevalent form, while situational empathy, the more ethically desirable role taking, depends upon a reader's having a memory of a comparable experience, which is never guaranteed.[65] If Hogan's situational empathy alone leads to the ethics of compassion, as he has it, then quick-match categorical empathy looks weaker and more vulnerable to bias through ethnocentrism or exclusionary thinking. We do not know, however, that categorical empathy does *not* lead to compassion, no more than we know the ethical results of situational empathy for fictional characters. Neither hypothesis has yet been tested. While literary critics and professionals value novels that unsettle convictions and contest norms, readers' reactions to familiar situations and formulaic plot trajectories may underlie their genuinely empathetic reactions to predictable plot events and to the stereotyped figures that enact them.[66] The fullness and fashion by which speech, thoughts, and feelings of characters reach the reader are very often supposed by narrative theorists to enhance character identification, as I discuss below, but relatively externalized and brief statements about a character's experiences and mental state may be sufficient to invoke empathy in a reader. Novelists do not need to be reminded of the rhetorical power of understatement, or indeed of the peril of revealing too much. Indeed, sometimes the potential for character identification and readers' empathy *decreases* with sustained exposure to a particular figure's thoughts or voice.[67]

Narrative Situation

It has been a commonplace of narrative theory that an internal perspective, achieved either through first-person self-narration, through figural narration (in which the third-person narrator stays covert and reports only on a single, focal center of consciousness located in a main character), or through authorial (omniscient) narration that moves inside characters' minds, best promotes character identification and readers' empathy. Wayne Booth, for instance, writes, "*If* an author wants intense sympathy for characters who do not have strong virtues to recommend them, *then* the psychic vividness of prolonged inside views will help him" (Booth 377–378, emphasis in original). The technique also works for characters in which readers have a natural rooting interest, such as Jane Austen's heroines. Booth's detailed account of how Austen uses the inside view to promote sympathy for the flawed Emma is a classic of narrative theory (245–256). Booth asserts, "By showing most of the story through Emma's eyes, the author insures that we will travel with Emma rather than stand against her" (245). Austen, one of the early masters of narrated monologue to represent characters' consciousness, crafts smooth transitions between her narrator's generalizations about characters' mental states (psychonarration) and transcriptions of their inner thoughts, in language that preserves the tense and person of the narration.[68] Also called free indirect discourse, narrated monologue presents the character's mental discourse in the grammatical tense and person of the narrator's discourse.

Subsequent theorists have agreed that narrated monologue has a strong effect on readers' responses to characters. David Miall specifically mentions the means of providing "privileged information about a character's mind," free indirect discourse, as especially likely to cue literariness and invite empathic decentering (Miall, "Necessity," 54). Sylvia Adamson arrives independently at a similar point, arguing that narrated monologue should be understood as "empathetic narrative." In Adamson's language the representational technique and its ostensible effects fuse. *Quoted monologue* (also called interior monologue, the direct presentation of characters' thoughts in the person and tense of their speech) also has its champions, who regard the move into first-person as invariably more authentic and direct than the more mediated or double-voiced narrated monologue. *Psychonarration,* or the narrator's generalizations about the mental states or thoughts of a character, has fewer advocates, perhaps because it is associated with traditional narratives such as epics. However, both Wayne Booth and Dorrit Cohn suggest that psychonarration can powerfully invoke character identification, and Cohn points out that both poetic analogies and metaphors for feeling states (as Virginia Woolf often employs) require the use of psychonarration.[69] Despite the frequent mention of narrated monologue as the most likely to produce empathy,[70] quoted monologue and psychonarration also give a reader access to the inner life of characters. Most theorists agree that purely externalized narration tends not to invite readers' empathy.[71]

In addition to these speculations about modes of representing inner life, the person of the narration often seems likely to effect readers' responses to narrative fiction and its inhabitants. In particular, first-person fiction, in which the narrator self-narrates about his or her own experiences and perceptions, is thought to invite an especially close relationship between reader and narrative voice. For instance, Franz Stanzel believes that the choice of internal representation of the thoughts and feelings of a character in third-person fiction and the use of first-person self-narration have a particularly strong effect on readers. Novelist and literary theorist David Lodge speculates that historical and philosophical contexts may explain the preference for first-person or figural third-person narrative voice: "In a world where nothing is certain, in which transcendental belief has been undermined by scientific materialism, and even the objectivity of science is qualified by relativity and uncertainty, the single human voice, telling its own story, can seem the only authentic way of rendering consciousness" (Lodge 87). However, the existing experimental results for such an association of technique and reaction are not robust. In several studies of Dutch teenagers, W. van Peer and H. Pander Maat tested the notion that first-person narration creates a "greater illusion of closeness . . . allowing the reader a greater and better fusion with the world of the character."[72] They conclude, "It remains unclear why point of view has no more powerful and no more overall effect on readers, given the effort devoted by authors in order to create these devices that produce a point of view" (van Peer and Pander Maat 152). While noting that readers certainly express preferences about point of view and prefer consistency over inconsistency, they found that enhancement of sympathy for protagonists through positive internal focalization actually weakened as teenagers matured (152–154).

Lodge concedes that the first-person voice "is just as artful, or artificial, a method as writing about a character in the third-person," but he insists that it "creates an illusion of reality, it commands the willing suspension of the reader's disbelief, by modeling itself on the discourses of personal witness: the confession, the diary, autobiography, the memoir, the deposition" (Lodge 87–88). In my book I argue the opposite, that paratexts cuing readers to understand a work as fictional unleash their emotional responsiveness, in spite of fiction's historical mimicry of non-fictional, testimonial forms. My research suggests that readers' perception of a text's fictionality plays a role in subsequent empathic response, by releasing readers from the obligations of self-protection through skepticism and suspicion. Thus they may respond with greater empathy to an unreal situation and characters because of the protective fictionality, but still internalize the experience of empathy with possible later real-world responsiveness to others' needs. While a full-fledged political movement, an appropriately inspiring social context, or an emergent structure of feeling promoting change may be necessary for efficacious action to arise out of internalized experiences of narrative empathy, readers may respond in those circumstances as a result of earlier reading.

How Narrative Empathy Works: Authors and Audiences

The dispositions and beliefs of novelists themselves also belong in a thorough study of narrative empathy. Fiction writers report looking at and eavesdropping on their characters, engaging in conversations with them, struggling with them over their actions, bargaining with them, and feeling for them: characters seem to possess independent agency. In a remarkable study of fifty fiction writers, Taylor and her collaborators discovered that 92 percent of the authors reported some experience of the illusion of independent agency (IIA) and that the more successful fiction writers (those who had published) had more frequent and more intense experiences of it. Taylor hypothesizes that IIA could be related to authors' expertise in fantasy production (Taylor 361, 376–377), suggesting that it occurs more easily and spontaneously with practice, or that writers naturally endowed with creative gifts may experience it more readily. Though clearly novelists still do exercise their authority by choosing the words that end up on the page, they may experience the creative process as akin to involuntarily empathizing with a person out there, separate from themselves. Several tests administered by Taylor to her subjects support this connection. Taylor found that the fiction writers as a group scored higher than the general population on empathy (361). Using Davis's Interpersonal Reactivity Index (IRI), a frequently used empathy scale, Taylor measured her subjects' tendency to fantasize, to feel empathic concern for others, to experience personal distress in the face of others' suffering, and to engage in perspective taking (369–370). Both men and women in her sample of fiction writers scored significantly higher than Davis's reported norms for the general population, with females scoring higher in all four areas than males. Fiction writers of both genders stood out on all four subscales of Davis's IRI, but they were particularly off the charts for fantasy and perspective taking. Taylor speculates that "these two subscales tap the components of empathy that seem most conceptually related to IIA and might be seen as 'grown-up' versions of variables associated with children who have imaginary companions (pretend play and theory of mind skills)" (377).

Taylor's discoveries lead to speculation about the function of narrative empathy from the authors' perspective: *fiction writers as a group may be more empathetic than the general population.* However, we must also consider the difficulty of pinning down the difference between innate dispositions and results of practice and habitual use in groups of people; thus, *the activity of fiction writing may cultivate novelists' role-taking skills and make them more habitually empathetic.* These proposals do not imply that the actual behavior of fiction writers is any better than that of the population at large. Even the most ardent advocates of narrative ethics hesitate to argue that being a novelist correlates with being a better person, and novelists known to be nice people sometimes also exercise their empathy on behalf of nasty characters.

Most theories of narrative empathy assume that empathy can be transacted accurately from author to reader by way of a literary text (critiques of literary empathy disparage this goal as an unwholesome fantasy or projection). The comments of writers about their craft suggest the formation of a triangulated empathic bond. In this model authors' empathy contributes to the creation of textual beings designed to elicit empathic responses from readers. In fact there is no guarantee that an individual reader will respond empathetically to a particular representation. Because real people—for instance, objects of pity presented by charitable organizations—might find their position in that empathic triangle discomforting, critics of empathy claim it results in misunderstanding or worse.

From the failures of empathic individuals to question whether their assessments of the other's feelings could be off base comes a great deal of the negative reputation of empathy as a particularly invasive form of selfishness. (I impose my feelings on you and call them your feelings. Your feelings, whatever they were, undergo erasure.) Contrary to this fearful scenario, the research on empathic accuracy records the remarkable degree of correctness in human mind-reading abilities, though to be sure more cross-cultural verification of these findings would be welcome.[73] Whether from expert reading of facial cues, body language, tone of voice, context, or effective role taking on the part of the empathizer, ordinary subjects tend to do pretty well in laboratory tests of empathic accuracy. Verification can be achieved readily enough through interviews cross-checked with physical measurements and observations.

When we respond empathetically to a novel, we do not have the luxury of questioning the character: we cannot ask, Is that how you really felt? The text, however, may verify our reactions even as it elicits them. No one narrative technique assures readers that our empathic reaction precisely catches the feelings embedded in the fictional characters. For this reason, extratextual sources, such as interviews with authors, become important tools in assessing literary empathic accuracy. My term *empathic inaccuracy* describes a potential effect of narrative empathy: *a strong conviction of empathy that incorrectly identifies the feeling of a liter-*

ary persona. Empathic inaccuracy occurs when a reader responds empathetically to a fictional character at cross-purposes with an author's intentions. Authors also sometimes evoke empathy unintentionally. This accident contributes to empathic inaccuracy. Unlike in real-world, face-to-face circumstances, the novel-reading situation allows empathic inaccuracy to persist because neither author nor fictional character directly confutes it. Indeed, literary studies privilege against-the-grain interpretations of fiction that may be founded on deliberate acts of role taking that subvert the authors' apparent intentions and increase empathic inaccuracy. A reader persuaded that she has felt with a fictional character may defy the stated or implicit intentions of an author. When the author's intention matches the reader's feelings and the agreement resonates with empathic accord, then the introduction of alternative perspectives on the matter at hand may meet with disbelief or outrage. *Empathic inaccuracy,* to craft a proposal out of this circumstance, *may then contribute to a strong sense that the author's perspective is simply wrong.* This is by no means an unproductive critical stance.

For those writers who hope to reach readers with emotionally resonant representations, the struggle against empathic inaccuracy thus has two component liabilities, failure and falsity. On *failure* of narrative empathy, I propose that *while authors' empathy may be an intrinsic element of successful fictional worldmaking, its exercise does not always transmit to readers without interference.* A second form of empathic inaccuracy occurs when authors represent a practice or experience that unintentionally evokes empathy in readers, against authors' apparent or proclaimed representational goals.

Focus on the *falsity* of narrative empathy expresses the concern that experiencing narrative empathy short-circuits the impulse to act compassionately or to respond with political engagement. In this view, narrative empathy is amoral (Posner 19), a weak form of appeal to humanity in the face of organized hatred (Gourevitch 95), an obstacle to agitation for racial justice (Delgado 4–36), a waste of sentiment and encouragement of withdrawal (Williams 109), and even a pornographic indulgence of sensation acquired at the expense of suffering others (Wood 36). To some feminist and postcolonial critics, empathy loses credence the moment it appears to depend on a notion of universal human emotions, a cost too great to bear even if basic human rights depend upon it.[74] The fearful view of authors' empathy as corrupting readers by offering them others' feelings for callous consumption leads in some quarters to the depiction of empathy itself as a quality that weakens humans and makes them vulnerable to others' cruelest manipulations. Narrative empathy becomes yet another example of the Western imagination's imposition of its own values on cultures and peoples that it scarcely knows, but presumes to feel with, in a cultural imperialism of the emotions. Empathic inaccuracy, in this quarrel with moral sentimentalism, then becomes evidence of the falsity of the whole enterprise of sympathetic representation.

Rather than attempting to eliminate empathic inaccuracy by arguing with or correcting readers' feeling responses, recognizing the conflict between authors' empathy and readers' empathy opens the way to an understanding of narrative empathy as rhetorical. *Both authors' empathy and readers' empathy have rhetorical uses, which may be more noticeable when they conflict in instances of empathic inaccuracy.* By using their powers of empathetic projection, authors may attempt to persuade readers to feel with them on politically charged subjects. Readers, in turn, may experience narrative empathy in ways not anticipated or intended by authors. When those readers articulate their differences with a text's or an author's apparent claims, they may call upon their own empathetic responses as a sort of witness to an alternative perspective. Arguments over empathic differences between authors and readers, or among readers with different emotional reactions to a shared text, give feeling responsiveness to fiction a status it has not often been granted in academic analyses of literature. Narrative empathy can impede or assist arguments staged in the public sphere. Indeed, the existence of empathic novel reading experiences, whether accurate or not, often enters into debates covertly. More self-consciousness about our own experiences of narrative empathy depends in part upon identifying where we stand as members of the diverse audiences reached by authors' empathic representations.

Narrative empathy intersects with identities in problematic ways. Do we respond because we belong to an in-group, or can narrative empathy call to us across boundaries of difference? Even this formulation could be read as participating in a hierarchical model of empathy. The habit of making the reactions of white, Western, educated readers home base for consideration of reader response has not yet been corrected by transnational studies of readers, though narrative theorists such as Peter J. Rabinowitz offer subtle ways of understanding the various audiences narrative fiction may simultaneously address. When the subject positions of empathizer and object of empathetic identification are removed from the suspect arrangement that privileges white Western responses to subaltern suffering, the apparent condescension of empathy can be transformed by its strategic use.

In advancing a theory of *authorial strategic empathizing* I hope to add to the theoretical understanding of the relationship among authorial audiences (comprised of those ideal readers imagined and hoped for by authors) and actual, historic audiences made up of a variety of real readers. The resources of rhetoric, with its traditional emphases on authorial tactics, textual strategies, and audience responses, appropriately match the critical needs of a theory of narrative empathy, which must account for the empathy of both authors and readers, as well as the potentially empathetic tropes and techniques of narrative texts, as discussed above. My term *strategic empathizing*, coined in slant rhyme with Gayatri Chakravorty Spivak's strategic essentializing,[75] indicates the intentional (not always efficacious) work of narrative artists to evoke emotions of audiences closer and further from the authors and subjects of representation. (That many narratives have more than one built-in audience, as Peter J. Rabinowitz and Brian Richardson have argued, I assume.[76]) Like strategic essentializing, strategic empathizing occurs when an author employs empathy in the crafting of fictional texts, in service of "a scrupulously visible political interest."[77] These interests make different demands on the various audiences that may read and respond to the invitation to emotional resonance. Thus I enquire, How does strategic narrative empathy reach readers in immediate, more distant, and totally remote audiences, where the metaphor of nearness/distance also correlates with familiarity/strangeness and sameness/otherness?

Strategic empathizing, a variety of authors' empathy, describes how authors attempt to direct an emotional transaction through a fictional work aimed at a particular audience, not necessarily including every reader who happens upon the text. *Bounded strategic empathy occurs within an in-group, stemming from experiences of mutuality, and leading to feeling with familiar others.* This kind of empathy can be called upon by the bards of the in-group, and it may indeed prevent outsiders from joining the empathic circle. Though we are prone to notice the delineations of class, gender, sexuality, and ethnicity that mark a text as for one audience and not so much for another, an effort of strategic empathizing may also be bounded by the toggle switch of experience. Either you have lost a partner, or you have not; either you have battled a life-threatening illness, or you have not; either you have been assaulted or imprisoned, or you have not. The gate swings open to invite you in, or it stays closed. Nothing prevents a reader who has not shared an experience from exercising the role-taking imagination while reading a remote or forbidding work, but some works do not extend the invitation to all potential readers. The persistent reader *may* or *may not* join the authorial audience, in

Rabinowitz's term. The disinvited reader may not choose or be able to live up to the terms of the ideal audience projected by a narrative.

As Brian Richardson observes, "divergent audiences are regularly addressed in a literary work, and in many cases the disparate groups fail to mesh."[78] Sometimes a novelist writes to more than one audience simultaneously, with a hope of bridging the gap between them. Novelists with a purpose may set out to reach—and change—the attitudes and beliefs of a target audience. This ambition encourages the use of *ambassadorial strategic empathy*. It also directs an emotional transaction through a fictional work aimed at a particular audience, not necessarily including every reader who happens upon the text. Successfully exercised ambassadorial strategic empathy on the part of authors can be a powerful rhetorical tool in reaching and swaying the feelings of audiences. It may teach readers to feel with otherwise alien fictional others. *Ambassadorial strategic empathy addresses chosen others with the aim of cultivating their empathy for the in-group, often to a specific end.* Appeals for justice, recognition, and assistance often employ it. If identity and experience mark the audience of texts employing *bounded* strategic empathy, the relationship between the time of reading and the historical moment of publication delimits *ambassadorial* strategic empathy. Ambassadorial strategic empathy is time sensitive and context and issue dependent, marked by the period when a text's dissemination performs its ambassadorial duty by recruiting particular readers to a present cause through emotional fusion. Accompanying paratextual statements of intention on the part of writers and organizations make the attempt to use ambassadorial strategic empathy legible. A text's reception history, including enthusiastic accounts of rapport and conversion to the cause as well as hostility, also preserves the traces of ambassadorial strategic empathy at work. Polemical, didactic, biased, purposeful: narratives employing ambassadorial strategic empathy in order to reach and even change target audiences also meet resistance.

The judgments of literary merit that critics make about narrative fiction do not reliably correspond with evidence of a text's transaction of strategic empathy from author to target audience(s). Indeed, a narrative's popularity may simultaneously interfere with its reputation among tastemakers and vouch for its empathetic intensity for its readership. Indeed, narratives that deploy *broadcast strategic empathizing* in an attempt to reach the widest possible audience most often invite derisive judgments from cultural watchdogs. *Broadcast strategic empathy calls upon every reader to feel with members of a group, by emphasizing our common human experiences, feelings, hopes, and vulnerabilities.* Narrative empathy in the form of an author's broadcast strategic empathizing employs a universal tool (language) to reach distant others and transmit the particularities that connect a faraway subject to a feeling reader. While not all novelists seek to influence as many and as various a set of readers as their books can reach, some do, and we reliably find in their fiction opportunities for character identification emphasizing the commonalities of our embodied experiences, our psychological dispositions, and our social circumstances. The fact that many postcolonial novelists aspire to extend readers' sense of our shared humanity suggests that broadcast strategic empathy deserves attention more nuanced than refusal of empathy as an impossible goal of representation.

Unanswered Questions

In the book from which this chapter is derived, I subject to critical scrutiny the literary version of the empathy-altruism hypothesis, which holds that novel reading, by eliciting empathy, encourages prosocial action and good world citizenship. If indeed such a link could be substantiated (it has not yet been verified), then investigation of the effects of narrative techniques on real readers would have to extend beyond generalizations about character identification and a small subset of narrative situations. To the questions currently under investigation, many more may be added.

What effect (if any) does consonance (relative closeness to the related events) and dissonance (greater distance between the happening and the telling) have on readers of first-person, self-narrated fictions? Does a plural, communal narrative voice, a "we" narration, bring the reader into a perceptive circle where empathetic reactions are more readily available? Does the use of second-person "you" narration enhance the intimacy of the reading experience by drawing the reader and narrator close, or does it emphasize dissonance as it becomes clear that "you" cannot include the reader? In third-person fiction, does the use of a figural reflector, rather than an authorial (omniscient) narrator, make any difference in readers' emotional responsiveness to situations and characters?[79] Does the location of the narrator inside (or outside) the storyworld affect readers' reactions to the content of the narration? Does a covert narrator, who scarcely does more than provide cues about characters' movements and speech, disinvite empathy for those characters, or invite readers to see the action with a greater sense of immediacy, as if it were a play, as Bortolussi and Dixon suggest (Bortolussi and Dixon 202)? In the most fully polyphonic novels, in which a single narrative perspective is simply not available to the reader, does readers' empathy increase, dwindle, or vary according to the page they are on?

Finally, to bring the questions back to what happens in actual readers, if a narrative situation devised to evoke empathy fails to do so, does the fault lie in the reader, or in the overestimation of the efficacy of the technique? While I am inclined to agree with Wayne Booth that no one ethical effect inheres in a single narrative device, the commentary on narrative form often asserts (or assumes) that a specific technique inevitably results in particular effects—political, ethical, emotional—in readers. These views, in my opinion, should be subjected to careful empirical testing before any aspect of narrative technique earns the label of "empathic." To persist in the nomination of favored techniques as empathic without attention to the full range of techniques that may be contributing to empathic effects renders the study of narrative empathy an impressionistic endeavor at best.[80]

Notes

1. On mirror neurons, see Keysers et al., "Demystifying Social Cognition" (501), and Gallese et al., "A Unifying View of the Basis of Social Cognition" (396).

2. For an overview of this research in neuroscience, see Gallese, "'Being Like Me': Self-Other Identity, Mirror Neurons, and Empathy."

3. On the neural effects of hearing narrative, see Tettamanti et al., "Listening to Action-Related Sentences" (273). Though most neuroscientists working on mirror neurons agree that the effects are strongest in real life, face-to-face interactions, what Gallese calls the "shared manifold for intersubjectivity" still operates when subjects see videos, experience virtual reality through computer interfaces, and simply hear narration about others. See Blakeslee (F1, F4).

4. Dr. Christian Keysers, cited in Blakeslee (F1, F4).

5. Social and developmental psychologists, philosophers of virtue ethics, feminist advocates of an ethic of caring, and many defenders of the humanities believe that empathic emotion motivates altruistic action, resulting in less aggression, less fickle helping, less blaming of victims for their misfortunes, increased cooperation in conflict situations, and improved actions on behalf of needy individuals and members of stigmatized groups. See Batson et al., "Benefits and Liabilities of Empathy-Induced Altruism" (360–370), for a discussion of the recent research on each of these results of empathy. For a warm, agent-based virtue ethics view of empathy and sympathy, see Slote, *Morals from Motives* (109–110).

6. Feeling with fictional others does not guarantee changes in the capacity

to feel with real others and does not predict altruistic behavior on real others' behalf. For my argument that the transposition of the "empathy-altruism" hypothesis to experiences of narrative empathy lacks experimental support, see Keen, *Empathy and the Novel* (90–92, 99, 116, 145).

7. On personal distress as aversive and empathy as a precursor to sympathy or empathic concern, see Batson, *The Altruism Question* (56–57), and "Altruism and Prosocial Behavior" (282–316); see also Eisenberg, "Emotion, Regulation, and Moral Development" (671–672), and "The Development of Empathy-Related Responding."

8. Charles Darwin's treatment of sympathy in *The Expression of the Emotions in Man and Animals* clearly includes empathy, though he does not use the term. Paul Ekman, the leading authority on facial expressions as indicators of universal human emotions, does not treat empathy as a core emotion, but as one of the nine starting points for emotional reactions (when we feel what others feel). See Ekman, *Emotions Revealed* (34, 37). Neuroscientist Jaak Panksepp argues that emotional systems in the brain involve central affective programs comprised of neural anatomy, physiology, and chemicals. Panksepp considers empathy one of the higher sentiments (mixing lower, reflexive affects and higher cognitive processes) emerging out of the recent evolutionary expansion of the forebrain. See "Emotions as Natural Kinds within the Mammalian Brain" (142–143). For philosopher Martha C. Nussbaum, empathy comes into play as a part of compassion, which she treats as a human emotion. See *Upheavals of Thought* (327–335). For John Deigh and those working at the intersection of ethics and cognitive science, empathy is one of the moral emotions.

9. For working definitions of different vicariously induced emotional states, see Eisenberg and Fabes, "Children's Disclosure of Vicariously Induced Emotions" (111). I follow Eisenberg in differentiating empathy, aversive personal distress, and sympathy. Empathic response includes the possibility of personal distress, but personal distress (unlike empathy) is less likely to lead to sympathy, if it proceeds beyond evanescent shared feeling.

10. Positive forms of empathy are drastically underemphasized in the literature. See Ainslie and Monterosso, "Hyperbolic Discounting Lets Empathy Be a Motivated Process."

11. See Barthes for the distinction between the relatively easy pleasure of the readerly text and the bliss that comes when the demanding writerly text helps readers break out of their subject positions.

12. See for instance the treatment of happiness, joy, and love in Hatfield, Cacioppo, and Rapson, *Emotional Contagion*. Theodor Lipps, an important early theorist of empathy, proposed motor mimicry as an automatic response to another's expression of emotion. See Lipps, "Das Wissen von Fremden Ichen."

13. For genetic influences on prosocial acts and empathic concern, see Zahn-Waxler et al., "Empathy and Prosocial Patterns in Young MZ and DZ Twins."

14. Cultural differences implicate differences in the nature of emotional experience. Our understanding of what it means to be a person in our cultural context affects the way we experience daily emotions of pleasantness and unpleasantness, or whether we feel entitled as individuals to express a particular emotion. See Shields.

15. Oral storytelling is not isolated to preliterate cultures. Children in lit-

erate cultures also absorb cultural values and narrative styles through collaborative storytelling. See the comments on rapport and empathy in Minami and McCabe.

16. See Titchener, 181–185. See also Lipps, *Zur Einfühlung.*

17. See Lee and Anstruther-Thomson for the original journal articles.

18. Evaluation of patients who show changes in behavior as a result of brain injuries, ailments, or surgery contributes to the understanding of empathy. See Gratton and Elsinger.

19. Physiological measures have the advantage of being unaffected by the subjects' desire to present themselves favorably, as may occur in surveys, interviews, or self-reports. See Eisenberg et al., "Physiological Indices of Empathy." On deceleration of heart rate in response to negative experiences of others, see Craig. On the measurement of palmar skin conductance and heart rate in response to images of people in pain, see R. S. Lazarus et al. For a skin conductance study suggesting that empathetic arousal occurs when subjects believe a person is receiving a painful shock, see Geer and Jarmecky. On facial or gestural responses as indications of empathy, see Marcus. See also Hoffman, "The Measurement of Empathy." On EMG and other physiological measurements of emotional responses, see Cacioppo and Petty, "Just Because You're Imaging the Brain."

20. A number of empathy scales developed since the 1950s are still in use by psychologists. The Sherman-Stotland scale includes a factor (VI) measuring "fantasy empathy" for fictional characters in stories, plays, and films. See Stotland (135–156). More recent tests of emotional intelligence include the Balanced Emotional Empathy Scale (BEES) and Davis's Interpersonal Reactivity Index (IRI). On BEES, see Mehrabian. For the IRI, which has subscales in Empathic Concern, Perspective Taking, Fantasy (including narrative empathy), and Personal Distress, see Davis, "A Multidimensional Approach" and "Measuring Individual Differences." Survey methodology has its limitations, as psychologists acknowledge. Eisenberg has repeatedly observed that cultural influences such as sex-role differentiation show up more in the kinds of tests that rely on surveys and interviews and much less (or not at all) in tests using physiological methods. See Lennon and Eisenberg.

21. For a salutary caution on the interpretation of these fMRI studies, which feature such dazzling pictures and often receive quite credulous promotion in the press, see Cacioppo and Petty, "Just Because You're Imaging the Brain."

22. Singer and her colleagues employed two empathy scales, Mehrabian's Balanced Emotional Empathy Scale and Davis's Empathic Concern Scale ("Empathy for Pain," 1159).

23. The amygdala, anterior temporal cortex, and orbital frontal cortex (as well as physiological synchrony of the autonomic nervous system) are probably involved in empathy, as the evidence of emotional impairment in brain-damaged or diseased patients suggests. See Brothers; Levenson and Ruef; and Rosen et al.

24. This account is consistent with the emotion theory of neuroscientist Edmund T. Rolls, who hypothesizes that human brain mechanisms provide two routes to action, one a quick, unconscious prompt for a behavioral response

(which we share with other mammals) and the other a slower, language mediated, rational planning faculty. The two routes can produce conflicting results.

25. Mirror neurons fire not only when carrying out an action but also when observing another carrying out the same action. They provide a basis for understanding primates' mind-reading, and by extension, human empathy. See Gallese et al., "The Mirror Matching System." See also Iacoboni et al.

26. See for instance Cosmides and Tooby, "Cognitive Adaptations for Social Exchange."

27. See for instance Antonio Damasio et al., "Somatic Markers and the Guidance of Behavior: Theory and Preliminary Testing." Damasio's works for general readers (*Descartes' Error, The Feeling of What Happens*, and *Looking for Spinoza*) have resulted in wide dissemination of his theories.

28. See Cosmides and Tooby, "Evolutionary Psychology and the Emotions."

29. Some neuroscientists informally refer to "cogmotions" to emphasize the fusion of the two concepts in their research. (My informant is neuroscientist Dr. Tyler S. Lorig.) Nonetheless, many experts in cognition carry out their work without regard to the emotions, and basic textbooks on cognition rarely refer to emotions. See for instance Reed's introductory college text, *Cognition: Theory and Applications*, 6th ed. Emotional states receive fleeting mention on just three pages of this text. The younger hybrid discipline of Social Cognition is more likely to reflect the understandings of affect and cognition as intertwined. See for instance Forgas.

30. The core elements of the modern concept of empathy in aesthetics can legitimately be traced to Lee, who was also a novelist. As with several key dates in psychology, rival claimants to earliest usage appear. Lipps's 1897 work on *Einfühlung* gets translated in 1909 by experimental psychologist Titchener as *empathy*. Lee drew on Lipps's work for *The Beautiful*. Freud also had Lipps's books in his library and adopted the term *Einfühlung*. See Wispé.

31. For speculations on the role of aesthetics in human evolution, see Cosmides and Tooby, "Does Beauty Build Adapted Minds?"

32. Philosopher Lawrence Blum believes that insofar as emotions of sympathy and empathy promote perspective taking, they may result in better prosocial responses than rationality alone. See Blum (122–139).

33. Some studies suggest that people with very empathetic dispositions respond more positively to members of out-groups than less empathetic people do, but, for most people, perceived similarity encourages empathy. For a classic study affirming similarity's relationship to higher empathy scores, see Krebs. On out-groups, see Sheehan et al. On similarity, see the literature review in Davis, *Empathy* (15, 96–99, 105–106, 109, 116–118).

34. On evolutionary bases for empathy for those who are like us, see Kruger; see also Hoffman, *Empathy and Moral Development* (4, 13, 206).

35. See the account of empathy's potential to replace egocentrism with ethnocentrism in Sherman.

36. On empathy as a precursor to reading comprehension, see Bourg.

37. See Miall and Kuiken, "What Is Literariness?"; see also Miall, "Beyond the Schema Given."

38. For this catalog of helps and impediments to empathetic reading of first-

person fiction, I draw upon the in-class essays of the students in English 232, The Novel, composed on 20 February 2006, answering this question: "How does your recent reading experience in this course square with the notion that first-person narration is especially productive of empathetic reading? What differences in technique in the variety of first-person narrative situations might alter readers' responses?"

39. Very little empirical research has been attempted to verify the theoretical speculations about aspects of characterization that operate in readers' character identification. Bortolussi and Dixon's pioneering study *Psychonarratology* reports their findings that character actions contribute to readers' assessments of character traits, while self-evaluations provided by the narrator (description) do not. However, the test stories employed first-person narrators, so narrators' evaluations of characters in third-person fiction cannot be included in this preliminary conclusion (160–165).

40. Schneider represents narrative situation as a factor in eliciting readers' empathy, and lack of representation of inner life as a likely inhibitor of it.

41. On affective responses to serial fiction, see Warhol (71–72). See also Hakemulder, *Moral Laboratory* (93, 143), drawing on Feshbach's observations of the effects of repetitive role-taking.

42. Nussbaum's empathy-inducing novels are invariably long. Writing about the character David Copperfield's reading habits in Dickens's novel of that name (1849–1850), Nussbaum comments, "He remains with [books] for hours in an intense, intimate, and loving relationship. As he imagines, dreams, and desires in their company, he becomes a certain sort of person." For Nussbaum the length of the immersion is a vital component of the process, permitting intensity, dreaming, and desiring that develops the reader's loving heart. See Nussbaum, *Love's Knowledge* (230–231).

43. Canonically, see Jameson. See also Zwaan, who compares readers' behavior when processing texts labeled as "news stories" or "narratives." Bortolussi and Dixon aptly caution that research in discourse processing has focused on broad generic distinctions rather than on narrative fiction's subgenres (253–254). For evidence of emotional responses to fictional subgenres in television, see Bryant and Zillmann, eds. Literary genre critics have been reluctant to adopt findings from mass communications research (to the extent that they are aware of them), perhaps because audiovisual (iconic) representations are assumed to be more emotionally stimulating than the verbal representations of prose narrative fiction. This assumption, however, has not been investigated systematically.

44. Feminist criticism often celebrates the power of women's writing's vividly represented spaces and places, in tandem with identity themes, to work out boundary-crossing potentials for connection, communication, and change. See for instance Friedman.

45. See the account of Nünning's remarks on empathy-inducing functions of metanarration, in Fludernik, "Metanarrative and Metafictional Commentary" (39).

46. Relevant to slower pace as potentially fostering empathy is Zillmann, who hypothesizes that the fast pace of television news stories and dramas may

impede empathetic response ("Empathy," 160–161). Miall's work on fore-grounding and empathy in literary texts correlates a slower reading pace with enhanced empathy.

47. See for instance van Peer's judgment in "Justice in Perspective."

48. Character identification thus exemplifies what Bortolussi and Dixon identify as readers' mental constructions, as opposed to textual features (*Psychonarratology* 28). Bortolussi and Dixon systematically measure how particular readers process specific textual features in narratives, but the experimental results bridging disciplines of discourse processing and narrative theory are still quite scanty. On their narratology, see Diengott.

49. See Dixon et al. ("Literary Processing," 5–33).

50. Writing about identification with dramatic characters, Zillman argues that the audience member's disposition precipitates empathic and counterempathic reactions and suggests that audiences must be made to care about characters one way or another. He believes that enactment of good or evil deeds by protagonists and antagonists, with opportunities for the moral appraisal of their actions, promotes strong emotional reactions. See Zillman, "Mechanisms of Emotional Involvement."

51. See Oatley, "A Taxonomy of the Emotions of Literary Response."

52. In this respect Miall and Kuiken are in accord with earlier work that demonstrates a relationship between a subject's prior similar experiences and empathy felt for another in the same situation. See Stotland.

53. See Miall and Kuiken, "What Is Literariness?" (121–138), and Miall, "Beyond the Schema Given."

54. See Kuiken et al., "Locating Self-Modifying Feelings."

55. See Louwerse and Kuiken ("Effects of Personal Involvement," 170). Their research confirms some of what Iser proposes about active reading as gap-filling (168–169).

56. For critiques of this assumption, see Konijn and Hoorn, from a discourse processing angle; see also Walsh.

57. See Klemenz-Belgardt (368); see also Jose and Brewer. Hakemulder reports on recent studies confirming the importance of personal relevance for intensity of reader response. See *Moral Laboratory* (71).

58. Wünsch cited in Hakemulder, *Moral Laboratory* (73).

59. For this reading-speed research, see Chupchik and Lázló.

60. Research into the empathy evoked by various genres of television advertisements suggests that discontinuous, nonlinear "vignette" ads discourage empathy, whereas classical, character-centered dramatic form in ads evokes viewers' empathy. See Stern.

61. For a good application of cognitive theory on levels of embedding to readers' capacity to comprehend embedded accounts of characters' mental states, see Zunshine, "Theory of Mind."

62. For a subtle treatment of the variety of techniques by which sympathy for characters may be cultivated, see Booth, *Rhetoric of Fiction* (129–133, 243–266, 274–282, 379–391). Ultimately Booth prefers the use of an "inside view" for invoking sympathy, but he describes the full range of strategies that authors from the classical period to the modernists actually employ.

63. For an excellent description of readers' imaginative construction of characters, see Cohan.

64. See Gerrig, "The Construction of Literary Character." See also his account of participatory responses to fiction in *Experiencing Narrative Worlds*.

65. See Hogan, "The Epilogue of Suffering." Though this article suggests a preference for the cognitive role-taking Hogan associates with situational empathy, his later very brief treatment of readers' empathy in *Cognitive Science, Literature, and the Arts* improves on his theory by describing how emotion triggers invoke quick-and-dirty responses, as well as imaginative role-taking, neither of which need be denigrated as egocentric (186–187). For work confirming the role of lived experience in spontaneous situational empathy with characters on film documentaries, see Sapolsky and Zillmann, "Experience and Empathy." Women who had given birth responded to a medical film of actual childbirth with more intense physiological reactions; otherwise, gender and related experiences had a negligible effect on empathy.

66. See Jauss (152–188), especially his summary figure, "Interaction Patterns of Identification with the Hero" (159). See also Hogan on emotions and prototypes in narrative, in *The Mind and Its Stories*.

67. For preliminary confirmation from film studies, see Andringa et al. (154–155).

68. For narrated monologue, psychonarration, and quoted monologue, see Cohn, *Transparent Minds*.

69. See Booth's discussion of traditional literature's use of "telling" in *Rhetoric of Fiction* (3–16); see also Cohn on psychonarration in *Transparent Minds*.

70. See for instance Palmer, *Fictional Minds*.

71. Three good starting points for recent work on the representation of consciousness are Fludernik's magisterial *The Fictions of Language and the Languages of Fiction*; Zunshine's *Why We Read Fiction: Theory of Mind and the Novel*; and George Butte's *I Know That You Know That I Know*.

72. Van Peer and Pander Maat designed experiments using five versions of stories, rewritten to test the relationship between positive internal focalization and readers' allocation of sympathy. See "Perspectivation" (145).

73. For a variety of essays verifying human beings' tendency accurately to identify others' feelings and states of mind, see Ickes, ed.

74. Indeed, human rights are not exempted from criticism. Some regard "the whole idea of 'universal' human rights" as a "gigantic fraud, where Western imperialist or excolonial powers try to pass off their own, very specific and localized idea of what 'rights' should be as universal, trampling roughly over everyone else's beliefs and traditions." See Howe (3).

75. See Spivak (214).

76. See Rabinowitz, *Before Reading*, and two essays by B. Richardson, "The Other Reader's Response" and "Singular Text, Multiple Implied Readers." Rabinowitz and Richardson offer subtle ways of understanding the various audiences narrative fiction may simultaneously address, and the conditions that pertain when we choose to join them.

77. See Spivak (214).

78. See Richardson, "The Other Reader's Response" (38).

79. Bertolussi and Dixon have done the best work on this subject, though they phrase the question differently: To what degree do readers fuse narrators and characters as a result of perceptual access to a particular character's perspective, and thus develop a rooting interest in that character and making assumptions about the narrator's and author's gender? See *Psychonarratology* (166–199).

80. This chapter is a revision of an essay that originally appeared in *Narrative* 14, no. 3 (2006): 207–236. The chapter is reprinted with permission from *Narrative*.

CHAPTER FOUR

The Biolinguistic Turn:
Toward a New Semiotics of Film

JAVIER GUTIÉRREZ-REXACH

The current state of affairs in film analysis can be characterized as one of certain theoretical dispersion or even confusion. There does not seem to be a dominant paradigm for the interpretation and understanding of film and, furthermore, the pervasive view is that this state of affairs is probably desirable. On the one hand, the complexity of the cinematic media precludes the "one-bullet" approach to hit the explanatory target. Film, as a cultural phenomenon, has obvious social, historical, and psychological dimensions. Previous attempts for a "general theory" have failed to properly explain cinema without downplaying or reducing the importance of one or more of its intrinsic or intended manifestations. On the other hand, the strong theoretical programs of the seventies and eighties have been subject to increased scrutiny and criticism.[1] So much so that some authors argue for a new period in film studies in which the dominance of Theory (or "grand theory") will be effectively challenged and replaced by historiography, cognitive theories of perception and interpretation, theories of representation, and empirical research.

In their edited volume *Post Theory*, Carroll and Bordwell conclude that Film Theory—the variant of Film Theory predominant in the North American academic milieu—depends uncritically and dogmatically on psychoanalytic theory as an analytical tool, failing to interrogate the problematic areas of this field, choosing instead to disregard/dismiss writers who refrain from employing their somewhat cryptic and self-referential vocabulary. Film Theory in general is based on the assumption that a theory about film should correspond to or entail a political or social program. Film theorists tend to assume that film analysis is equivalent to active political resistance. The systematic confusion between film analysis and political activism leads to a misunderstanding of "theory," where description, interpretation, and analysis are substituted for the tasks of formulating hypothesis and theorizing. Carroll and Bordwell conclude that psychoanalytic or feminist theories are not required to make many of the points advocated in the name of these fields. For Bordwell, culturalism and subject-position theories "are framed within schemes which seek to describe or explain very broad features of society, history, language and psyche" (3). These Grand Theories are unfavorably contrasted with a "third, more modest trend which tackles more localized film-based problems without making overarching theoretical commitments." The alternative "middle-level theory" or, in Carroll's terms, "piecemeal theory" that is proposed as the exemplary model for Film Studies would presumably be less theoretically ambitious but more flexible. Both Carroll and Bordwell have endorsed alternative cognitive theories, but they have refrained from formulating a full-fledged approach of their own.

Although I believe that those scholars who envision a post-theory stage of the discipline as a desirable and probably overdue phase of Film Studies are correct in their criticism of psychoanalytic, Marxist, or culturalist approaches—the received view in Film Theory—I think that the following step has to be constructive in nature. It is not enough to dismantle or question the theoretical constructs of the seventies and eighties. What is needed is to build and test theoretical frameworks that do not suffer from the shortcomings of culturally based models of analysis. This constructive task is undeniably more difficult but also more rewarding than the preceding destructive step.

I will argue that a contemporary linguistic—or, more generally, cognitive—approach does not represent a traumatic break with some of the explicit and implicit goals of the semiotic program and can pay unexpected additional dividends. It does not represent a sharp detour from certain theoretical analyses from previous decades and capitalizes on current debates within the cognitive sciences. If this approach is successful, it would allow for empirically testable analyses of films.

One Film, One Author?

The crisis of the models for the interpretation of the cinematic text and spectatorial experience originated during the 70s and 80s is not restricted to theoretical frameworks of "continental" origin—Marxism and psychoanalysis in their "retooled" French versions. Classic frameworks such as those based on the role of the author (director) or the spectator/viewer also have been questioned in the last two decades. For example, *auteur* theory—an analytical movement originated from the writings of *Cahiers du Cinéma* critics (Godard, Truffaut, etc.) and advocated in America by Andrew Sarris[2]—views film as the manifestation of an individual's viewpoint, most generally the director's. This view—which is prevalent and practically goes unquestioned in the milieu of film criticism in general—analyzes a film as the result of the creative effort of one author (Caughie 1981). This leading assumption introduces a normative component in the analysis of cinema, where part—if not a majority—of the critical assessment of a film is related to how it conforms to or deviates from a certain author's "canon": the set of his most characteristic thematic, narrative, and stylistic devices. Certainly, a thin line has to be walked between what films actually are and what they should be. Such normative assessment leads to the under-appreciation of authors whose work does not fall within certain aesthetic or narrative parameters, for example as represented in the high culture/low culture dichotomy: Why is it that filmmakers such as Tod Browning, Terence Fisher, and John Carpenter do not receive the same critical recognition as Josef von Sternberg, Billy Wilder, and Ingmar Bergman, to name a few? Is this related to the former directors' ascription to certain "low" narrative modes (fantastic/horror cinema)? On a different note, it is clear that one can talk about a Hitchcock film or a Bergman film, but can that assertion be extended to a vast majority of contemporary commercial cinema? Can we talk about a "Michael Bay film" (director as author), or, more properly, about a "Jerry Bruckheimer film" (producer as author)? Additionally, attributions of authorship are sometimes rather complicated. For example, no one can deny the importance of a film such as *Gone with the Wind* in the history of cinema. Nevertheless, determining who the author of this film was is not a straightforward matter: Victor Fleming and George Cukor (directors) or David O'Selznick (producer). One could ignore or label as "non-artistic" those films which are not the manifestation of a clear *auteuristic* voice or whose creative purpose is not believed to be the manifestation of an individual creative will, leading to the significant problem of who decides what to include or not as artistic or *auteuristic* cinema—a debate related to the pertinence of "the canon."

Theories of spectatorship, viewership, or audience behavior also face several paradoxes, such as the tension between the individual spectator who is engaged in the film or viewing experience and the notion of an audience, which is a collective entity and probably a social construct.

Can psychological notions such as "viewer's reaction" or accounts in terms of emotions and the cognitive mechanisms derived from and associated with them (see Plantinga and Smith 1999) be reconciled with the audience-mediated decisions that are decisive in the creative process of most contemporary Hollywood films—based on group profiles? The reception of a film involves a complex array of cognitive and social mechanisms—which are partially understood and targeted by those responsible for creating a film—but should also incorporate certain cognitive strategies on the part of the viewer.

The Structuralist View

The "strong" semiotic program advocated by Metz and his followers during the late sixties and early seventies envisioned the possibility of analyzing film as a text with grammatical/syntagmatic structure. The first wave of semiology produced an important achievement, the *grand syntagmatique*, a typology or classification of shots, which could be used to map the syntagmatic structure of a film. Nevertheless, this first wave was only short-lived and was ultimately perceived as a worthy but not completely successful attempt, since it "was difficult to apply with uniform results; and it was not clear that those results were significantly informative" (Carroll 1988). The second wave—which developed during the second half of the 70s and early 80s—received its theoretical inspiration from Althusserian Marxism and the psychoanalytic theory of Jacques Lacan. Whereas semiologists viewed the second wave as superseding and correcting the inadequacies of the theoretical ideas of the first wave,[3] other theorists uncovered the shortcomings of this approach (Carroll 1988) and advocated a cognitive approach (Bordwell 1989a, 1989b, 1990). This is why it has been very difficult to identify one theoretical framework as dominant in the "post-theory" environment of the nineties and during the present decade.[4]

Nevertheless, looking retrospectively, some of the goals of the initial semiotic program can be argued to be still valid, when implemented with what we know about grammar and language, and their interconnection and dependence with other cognitive systems. The failure of the semiotic program was probably more the result of the shortcomings of its theoretical and analytical tools than a product of the basic assumptions of film semiologists, especially the claim that cinema should be studied as a semiotic/linguistic entity with internal structure, by determining the structure of cinema or visual communication in general and the proper balance among psychological, linguistic, and social factors.

Saussure did not deny the cognitive internal aspects of language but was also aware of its social nature. He made a distinction between language (*langue*) and speech (*parole*) to capture this dual identity (Saussure 1905). Language as a system or code, in the terms of the mathematical/semiotic theory of communication, is inherently social. The use of the code by an individual in a concrete situation belongs to the *parole* and contains aspects that are not reducible to social analysis. Structuralism was ultimately concerned with codes as systems of signs. Signs are ubiquitous, pervasive, and come in different fashions. Semiotic coding extends to other forms of communication and socially created regulations. Concepts, ideas, images, and anything that can hypothetically be coded and decoded belong to sign systems, each with a specific structure. The role of the structuralist is to decide whether the relevant constellation of semiotic entities is a system of its own and decipher the components of such a system. All these possibilities inherent to the fundamental assumptions and hypotheses of structuralism were not implemented simultaneously in the area of film analysis, and they did not generate a comprehensive framework for cinema either, despite certain assertions stating this as an ultimate goal. The first stage of structuralism took a more formal or linguistic stand, looking at the internal structure of the cinematic text and its relationship with other code-dependent textualizations. The second stage of structuralism focused on the psychological and sociological dimensions of cinema. In what follows I will briefly review the contributions of linguistic structuralism.

Language and Cinema

The most influential film structuralist is the French theorist Christian Metz, who developed a series of frameworks for the analysis of cinema. In his first writings (Metz 1974a, 1974b) he explored the issue of whether cinema was language (*langage*) or a language system (*langue*). He concluded that cinema was indeed a language but could not be characterized as a language system, unlike natural languages. Language (semiotic) systems are systems devised for two-way communication: The speaker can become the hearer and the other way around. Cinema allows only for deferred communication. Such deferral or delay is double in nature. There is a delay between the production of a film and its reception, and a second independent delay in the ensuing response, if there is any. Metz's caveat with respect to cinema could be transferred to literature.

The second reason for rejecting cinema as a language system is that it lacks arbitrary signs. What is arbitrary in the linguistic sign is the connection between signifier and signified. There is no motivated connection between the sequence of phonemes in the word *table* and the intended referent: a table. The relationship between signifier and signified in the filmic sign is completely different. The signifier is a mechanic or photographic reproduction of the signified. The relationship between signifier and signified is not arbitrary but motivated or even analogical.

This conception of the linguistic sign led Metz to an endorsement of André Bazin's defense of realism in cinema (see Wollen 1972).[5] Metz relativized this position in *Language and Cinema* (Metz 1974b), situating the pertinent analogical link not between signifier and signified but between the parallel perceptual situation of real-life experience and the cinematic experience. The filmic sign is the product of coding, and as such it must belong to organized meaningful arrangements (semiotic structures). The filmic sign lacks the degree of arbitrariness of the natural-language sign but—unless one endorses a version of Bazinian realism—the relationship between signifier and signified is clearly mediated.

What Metz seems to be hinting at derives from the standard contrast between the syntactic structure of a text—its coding—and the semantic conventions that are being followed. The filmic text undoubtedly has syntactic structure: a set of discrete units that are arranged together in non-arbitrary fashion, following certain concatenation rules. The semantic content of such an arrangement is not always completely dependent on structure. The minimal units of cinematic discourse are not discrete items whose inventory can be described; that is, they do not form a lexicon or vocabulary. Rather, they are created as iconic representations. According to Metz, "a close up of a revolver does not signify 'revolver' (a purely potential lexical unit) but signifies *as a minimum*, leaving aside its connotations 'Here is a revolver.' It carries with it its own actualization, a kind of 'here is'" (67). A cinematic minimal unit—a shot—directly refers to what is represented in such a shot. It behaves like a demonstrative or a pronoun in natural language. A shot—a minimal film fragment—lacks any content by itself. Only when it is actualized does it represent something and acquire such a meaning.

A problem for the indexical or iconic theory of the filmic sign is that there is a potential difference in the actualization of natural language deictic expressions and cinematic indexicals. Natural-language indexicals acquire their content at the moment of utterance, when they are inserted in a sentence and associated with a pointing gesture.[6] The filmic sign has the potential for indexical actualization at two different points in time: (1) during film production, when the movie is actually shot and the fragment of film represents a direct "reflection" of what has been captured by the camera, or (2) when the actual shot is part of the actual film. At this point the meaning or content of the shot might be altered or modified. For example, a shot of the director's hand wearing a glove might be inserted in a narrative sequence and become a shot of a killer wearing gloves.[7] There are reasons to actually maintain that the moment of "utterance" for the cinematic sign is the point of insertion in the cinematic narration—that is, when the shot is edited in the film. A given shot might be altered or modified, by blue screen techniques or using computer-generated imagery, so its indexical meaning is created at several stages, including extensive post-production work. Additionally, the final content of the shot may well represent something that does not exist in our actual world, as with the groundbreaking dinosaurs of *Jurassic Park* (Steven Spielberg); or not even in the past or a conceivable future, as with the terminator that morphs and reconstructs itself in *Terminator II* (James Cameron). The facts that such shots can be created to represent non-realistic meaning with such a degree of verisimilitude and that filmmakers are able to create alternative fictional worlds that effectively suspend the spectator's disbelief constitute powerful counterarguments to a Bazinian realistic theory of cinema. Furthermore, post-production tools such as editing and music scoring create additional conventionalized layers of meaning.

The *Grand Syntagmatique*

Metz's *grand syntagmatique* attempts to apply to the cinematic text the discovery procedures that structuralists applied to natural language with the goal of isolating its minimal combinatorial units. The procedure of substitution or commutation allows linguists to identify the minimal units in the dimensions of speech sound and meaning. Metz claimed that the filmic sign is not doubly articulated, so cinematic texts only have units that are arranged in one dimension. Metz's goal is to identify the autonomous segments or syntagmas: units of narrative autonomy within a text. He distinguished eight syntagmas. The first type of syntagma is the autonomous shot: a syntagma consisting of just one shot. The remaining seven syntagmas can be either achronological or chronological. There are two achronological syntagmas: the bracket syntagma and the parallel syntagma. The parallel syntagma's organization is based on alternation.[8] The shots in the bracket syntagma are organized according to a concept or an external way of arranging reality. There are five types of chronological syntagmas. In the descriptive syntagma, shots are displayed in an order suggesting spatial coexistence; this syntagma describes a "fragment" of reality. The four remaining types are narrative syntagmas. The alternate or alternating syntagma alternates shots from two (or more) events or spatial locations that are chronologically related. The chronological relation between events is what separates the alternate syntagma from the parallel syntagma. Examples of alternating syntagmas are the cross-cutting of shots reflecting a car chase—one shot of cops alternates with one of the fleeing robbers, etc.; a shootout, where we get alternating shots of the participants; etc. The remaining three syntagmas reflect linear arrangements of an event. A scene is an arrangement that is chronological and linear, offering spatio-temporal continuity. The episodic sequence represents a succession of events that are chronologically related and are glued together by other narrative mechanisms such as music, theme, dissolves, narrative voice-over, photographic effects (faded images), etc.[9] Finally, the ordinary sequence differs from the episodic one in representing a continuous action scenario, in which unimportant fragments have been elided. For example, a trial can be condensed into a few shots capturing the most important events in it.

The *grand syntagmatique* was subject to several criticisms, especially when viewed as a universal typology without exceptions. Metz restricted its scope to the cinematic code of classical narrative cinema (from the early 30s to the late 50s). It can be considered an inductive typology of the syntagmas used in classical cinema, but not a theoretical inventory of the hypothetically possible arrangements in a cinematic text—whether actually produced or not. The new avant-garde movements of the 60s and 70s—beginning with the French *Nouvelle-Vague*—introduced arrangements that did not clearly instantiate any of the modes of composition of the *grand syntagmatique*. Metz's typology follows the structuralist inductive methodology in that structural arrangements are postulated only if they are empirically found in an actual text. The Cartesian methodology that served as the basis for the Chomskyan revolution in linguistics and in cognitive psychology questioned the need for strict empirical induction in theory and hypothesis formation. This change paved the way for alternative approaches to the structural analysis of film. Even during the 80s and 90s, those scholars who directly embraced the Chomskyan linguistic methodology or related, but broader, cognitive approaches still took Metz's typology as a starting point or reinterpreted the *grand syntagmatique* within the new frameworks. The work of Michel Colin (Colin 1995) is significant in this respect, since he frames Metz's typology within the main tenets and assumptions of generative grammar.

The Generative Enterprise

Following Chomsky's pioneering work, the central object of research in contemporary linguistics is "the language faculty," a postulated mental organ which is dedicated to acquiring linguistic knowledge and is involved in various aspects of language use, including the production and understanding of utterances. The aim of linguistic theory is to describe the initial state of this faculty and how it changes with exposure to linguistic data. Chomsky (1981) characterizes the initial state of the language faculty as a set of principles and parameters. Language acquisition consists in setting these open parameter values on the basis of linguistic data available to a child. The initial state of the system is universal grammar: a property of the mind/brain shared by all human beings as part of their biological endowment. Grammars constitute the knowledge of particular languages that result when the parametric values are fixed (Fodor 1983). It is not sufficient to inductively generalize a descriptive typology from the relevant data. What is needed is to propose universal principles, which are abstracted from individual grammars by a mixture of inductive and deductive procedures. This higher level of abstraction is to some extent at odds with Metz's structuralist inductive procedures. The *grand syntagmatique* represents a classificatory typology of certain attested arrangements in the period of classical cinema. It does not strive to be an abstract inventory of the principles that regulate the concatenation of shots in film.

Chomsky (1957) distinguished three levels of adequacy of a theory with respect to the data that are being characterized: observational adequacy, descriptive adequacy, and explanatory adequacy. Theories are observationally adequate if they are able to generate the observable data.

Theories are descriptively adequate if they "adequately" characterize the grammars (and hence the mental states) attained by native speakers. Theories are explanatorily adequate if they show how descriptively adequate grammars can arise on the basis of exposure—the data children use in attaining their native grammars. In ulterior developments of generative grammar—mainly the theory of principles and parameters (Chomsky 1981)—explanatory adequacy rests on an articulated theory of universal grammar: a detailed theory of the general principles and parameters that define the initial state of the language faculty.

Colin (1995) correctly argues that the main purpose of Metz's *grand syntagmatique* is descriptive adequacy, which makes the theory interesting from a "pedagogic" point of view, but still raises the question of what an explanatory theory of filmic discourse is. Since the *grand syntagmatique* attempts to explain rules and procedures, the problem arises of their explanatory adequacy from the point of view of "spectatorial" competence. Colin develops a restatement of the *grand syntagmatique* as an explicit set of rules. For example, natural languages share the following sentential rule: "S = NP + VP"—a sentence S has ("rewrites as") two constituents: a Noun Phrase and a Verb Phrase. Colin proposes to formalize Metz's *grand syntagmatique* as a set of "deep structure" rules. Each one of the syntagmatic types would be the terminals in the phrase structure tree. The main difference with natural-language structural descriptions or trees would be that the constituents or branches of the tree would not be linked by the concatenation connector (+)—or the conjunction connector (&)—but by the disjunction connector (◇). The branches in the tree would indicate a relation of inclusion. Thus, one rule would be "Syntagma = Chronological ◇ Achronological."

As Colin points out, it is not obvious what is gained from this formalization from the viewpoint of explanatory adequacy. The role of the generating rules would be clearly different in natural language texts and in filmic texts. The rewriting rules of the standard theory of generative grammar represent a model of the linguistic competence of a speaker. The application of those rules would generate the grammatical sequences of a given natural language. The non-generated sequences would be considered ungrammatical. The generative formulation of the *grand syntagmatique* by Colin has a clearly different role. First, the rules do not generate sequences of terminals, since we are substituting the disjunction (inclusion) relation for the concatenation relation of natural language rules. Given a series of shots, such rules do not provide a univocal recursive procedure for generating the relevant constituents (syntagmas) and, more importantly, a structural analysis of this sequence: How minimal constituents combine to derive intermediate and maximal filmic constituents. We are clearly not modeling the spectatorial competence of a viewer, his ability to process and understand the filmic text as a film and not as a random succession of moving images. This leads to a reconsideration of the framework of choice to address these issues: Is the rule format the most appropriate to model the filmic competence of a spectator? How are images processed into a narratively coherent whole that can be genuinely called "a film"?

The Biolinguistic Approach to Visual Computation

The relationship between concepts and language, as defended in the Chomskyan perspective, contrasts in several aspects with the structuralist view. The centrality of language resurfaces in a much more technical perspective, as the centrality of a language's syntax (Chomsky 1957, 1965), which in part corresponds to the structuralists' notion of a code. Language is considered to be an organ, one specialized for the language faculty. The overall perspective on language is psychological and biological.

More recently, a new enterprise has emerged with a clear biological focus. Biolinguistics attempts to frame linguistic research as a direct answer to biological questions (Jenkins 2000). Language is a property of the human brain and its most important aspects are genetically encoded. What is distinctive about our species is the ability to develop the language faculty on the basis of limited exposure to linguistic data. Other species, even if they are exposed to similar data, are not able to reach the same level of linguistic maturity. The faculty of language can be regarded as a language organ in the sense in which scientists speak of the visual system or immune system, as organs of the body (Chomsky 2000). The language organ is assumed to be like others in that it is a manifestation of our genetic configuration. Each language is the result of the interplay of two factors: the initial state and experience. Our innate language acquisition device takes the initial state and experience as input and gives adult language as an output; an output that is internally represented in the mind/brain.

Chomsky explicitly establishes a parallelism between the "language organ" and the visual system or "visual organ." Such an organ does not comprise just the physical organ responsible for perception. It is a system responsible for the integrated perception, categorization, and processing of images. The task of this organ is not the static registration of images—that would make it similar to a camera. Rather, the processing of visual stimuli and the integration of the information extracted from them is its main task. Marr (1982) analogizes visual perception taking place in the brain to a complex information-processing system in a computer. He distinguishes between computers and computations, and points out that studying either one is insufficient for understanding how a machine (brain, computer) carries out a computation (visual processing). He proposes that, in order to understand this process, we have to understand it at three independent levels: computational theory, representation and algorithm, and hardware implementation levels. The top level, the computational-theory level, is where one tries to explain *what* is computed (e.g., properties of the world from an image) and *why*. The second "representation and algorithm" level specifies the representation of input and output, and the algorithm, which outlines how the computations that map the input to the output will be carried out. Algorithm is similar to a computer pseudo-code that abstracts away from specific programming languages. The third "hardware implementation" level specifies how algorithm and representation are realized physically. Once the appropriate representation is identified, we have to choose possible algorithms and constrain how they could be implemented.

The levels-of-description framework simplifies complex cognitive phenomena by abstracting away from details at each level. However, it is not a catchall for understanding different models of behavior. Theories of behavior are concerned with how cognitive phenomena emerge from underlying biological mechanisms, and not with optimizing behavior or with static representations of behavior. Parallel distributed processing in the brain does not derive from or create a higher-level description. These levels derive and generate several interactions. Given that human behavior unfolds over time, any explanation that does not take this dynamic aspect of behavior into account faces serious challenges.

Integrating Marr's computational model of vision with a theory of cinematic cognition seems to have several advantages. First, it bridges the gap between structural/generative models based on natural language and the characteristic features of cinematic text, which is predominantly visual in nature. Marr's approach helps make the explanatory connection between grammatical models and visual data. Visual processing shares certain core elements with natural language processing at the computational and representational levels, because they both instantiate human cognitive abilities. It also raises the issue of whether an even more abstract framework that not only attempts to find linguistic elements in visual data is possible. The research question is the following one: What abstract patterns or operational mechanisms are intrinsic to both visual and linguistic processing? One shortcoming of the Chomskyan model is its insistence on the "autonomy of syntax." Grammatical representations are syntactic in nature. Semantic or pragmatic elements (content features) are viewed as emergent properties. They are only significant inasmuch as they are grammatically represented. This point has been criticized by numerous researchers, from cognitive or model-theoretic standpoints.[10]

Universal Grammar and Beyond

The relationship between language and thought is also tackled in the approaches belonging to the cognitive tradition in a radical way. Fodor (1975) formulated the Language of Thought Hypothesis: Language and thought cannot be separated because thoughts are encoded and expressed through language (Fodor 1981, 1983). Our thoughts not only have linguistic form but also linguistic functions. The analysis of the most relevant aspects of concepts and ideas can adopt the methods and techniques posited by the formal theory of the language faculty, as defended by Schatz and Gutiérrez-Rexach (2002) for ideological concepts and Hauser (2006) for moral and ethical judgments. Generative grammar and biolinguistics—the overarching terms that capture the essence of the Chomskyan enterprise to linguistic description and explanation—characterize the structure of language by a set of formal procedures that attempt to capture its intrinsic computational and biological nature. The Chomsky-Fodor cognitive approach to the theory of language, thought, and their interconnection surpasses the structuralist theory in cognitive impact and scientific ambitions but downplays the role of social factors in cognition.

The obliteration of the social is motivated by Chomsky and his followers as a methodological strategy and also as a scientific hypothesis. The science of language should attempt to explain how an infant with limited exposure to linguistic data is able to acquire linguistic structure in a relatively short period of time. This can only be explained if humans are genetically endowed with the language faculty. This faculty is a property of the mind/brain and is triggered by exposure to linguistic data. The fundamental structural principles that can be observed uniformly across languages arise from abstract properties of the faculty of language. These properties are innate and are genetically encoded, as the results of biological evolution (Nowak 2006; Nowak et al. 2001, 2002).

The study of language has to be a study of universal grammar, as the set of principles that govern linguistic structure. Similarly, it can be hypothesized that some of the principles, structures, and patterns of organization that are subjacent to other products of the human mind (concepts, narratives, etc.) are also innate. The ability to process and represent abstract thoughts, stories, or complex audiovisual arrangements in a symbolic fashion has to be part of the language organ (faculty) or, more likely, a natural extension of it, if we accept the Language of Thought Hypothesis. We can then posit the existence of several "organs" or cognitive faculties working in an interrelated fashion to produce the computation and representation of ideologies, works of art, etc. According to Nowak et al. (2002), understanding how Darwinian evolution gives rise to human language requires the integration of formal language theory, learning theory, and evolutionary dynamics. Formal language theory would characterize the abstract properties of language in a systematic fashion; learning theory would explain how individual languages are acquired from the options initially set by universal grammar; finally, evolutionary dynamics describes the cultural evolution of language and the biological evolution of universal grammar. Similarly, understanding how complex products of human elaboration such as cinema are created and processed would require understanding the basic properties of their audiovisual grammar, as derived from universal principles. It would also require explaining how humans are biologically apt to understand complex narratives integrating mixed audiovisual representations.

Parallel to the concept of universal grammar, there must be a universal system of concepts, most of which are instantiated in our mental lexicon, that is, in the mental dictionary stored as part of our cognitive faculty (Schatz and Gutiérrez-Rexach 2002). The term *concepts* should be understood here in a cognitive fashion, as basic or atomic conceptual elements that may be prioritized according to several hierarchies. These prioritizations generate rankings of ethical values, political ideologies, or artistic preferences. The associated hierarchies are subject to learning and evolution, so their study would also be under the umbrella of evolutionary dynamics. Thus, studying the grammar of film would require not only explaining its formal syntax[11] but also its conceptual grammar and associated value hierarchies.

Competence and Performance

Our linguistic competence is the knowledge that we have about linguistic systems. We know, for instance, that some sequences of expressions do not belong to our language. Those sequences that do not belong to the set of well-formed expressions of a language are called ill-formed and can be detected by any speaker of a given language. Our internal grammar determines in an algorithmic fashion whether a sentence cannot be parsed using its rules and principles. Knowledge of a language by its speakers is the internal knowledge of the grammatical principles and procedures that shape its internalized grammar. Knowledge and understanding of ideologies also indicate the existence of a conceptual or ideological competence (Schatz and Gutiérrez-Rexach 2002). Similarly, the ability to produce and act following moral judgments shows that there is a moral competence (Hauser 2006). The idea of a spectatorial competence (Colin 1995) would entail addressing several issues, which are idiosyncratic to visual cognition and cannot be replicated in moral or ideological cognition. First, the spectator of a film not only understands and parses the "grammatical" sequences of images and organizes them in constituents. He also has to be able to understand plot/narrative arrangements, and to decode the association of musical cues to images and the resulting emotional effect. The spectatorial competence is the product of a mixed or hybrid organ, one that combines linguistic capacities with visual and auditive cognition. Positing a hybrid-competence model for cinematic cognition does justice to the complexity of the cinematic text: a succession of images representing a certain narrative "plot" structure with verbal and auditive enhancements (music, dialogue, sound effects).

We may then ask ourselves where individual variation lies within this framework for cinematic analysis. The concept of performance attempts to capture this essential aspect of linguistic behavior. Performance corresponds to the individual use of competence by a speaker, conditioned by external factors. The distinction between competence and performance also applies to the domain of thought and conceptualization. The conceptual competence of an individual is the ability to form and combine thoughts. The extension of the notion of competence to the conceptual realm has not been explored in depth but has many intriguing possibilities. Speakers of different languages have partly similar competences, those that pertain to the knowledge of universal grammar, and partly specific competences, pertaining to the idiosyncratic properties of a different language. The same property can be attributed to our conceptual or moral competence. As I stated before, there is a common core of conceptual tools, procedures, and concept-formation mechanisms. Those would form part of a universal conceptography, a notion that plays in the domain of concepts the same role as the notion of universal grammar plays in the area of linguistic competence. On the other hand, individual agents in different social communities would acquire different conceptual systems, which would be a mixture of the universal conceptography and local conceptual systems (Schatz and Gutiérrez-Rexach 2002). Hauser (2006) emphasizes the universality of certain moral notions as a proof of a common moral competence.

The competence/performance distinction can also be applied to visual cognition, as suggested by Marr. Visual competence would comprise the perceptual and cognitive tools responsible for perceiving images, categorizing them, and integrating them into figures. Much work has been done recently on this aspect of visual cognition (Solso 1994). Stephen Palmer (1999) raises a number of questions with regard to mental representation in visual perception. For Palmer, a representation is a model that refers to a particular state of the visual system standing for an environmental property (an object or event). The relevant model is part of a larger representational system and shares common properties with other representational subsystems of the human mind. The critical property is the referential property of these representational systems: internal representations refer to the external reality.

The core of the representational system would be what we can call the universal grammar of visual cognition. Performance issues arise in the processing and generation of actual token-representations of external reality. For example, a darkened environment might induce some representational deficits, etc. Artists can actually manipulate performance errors or "optical illusions" to create certain effects. Mallen (2003) argues that Picasso's cubism represents the manifestation and/or violation of certain syntactic operations of syntactic composition of the pictorial image.

We have just hinted at another potentially critical issue: Are representational systems in general (conceptual, visual) acquired in the same fashion as linguistic systems? At age six an individual has acquired most of the principles and structural rules that make up an adult grammar. It is not clear whether there is a maturation age, or a period where an agent can be considered to have a mature state of the conceptual or moral system, although empirical studies based on damage to certain brain areas strongly suggest that the principles underlying these systems are acquired.[12] For the visual representational system, the maturation age seems to take place at an earlier age than the language system, most likely because the representations associated with perception are not symbol-based and do not require the acquisition of a lexicon.[13] We probably cannot talk about a maturation process or an adult visual/cinematic system, but the conceptualization faculty certainly undergoes a development.

The Spectatorial System: Parameters, Interfaces, Modules

Let us continue pursuing the analogy between linguistic and visual representations as products of a cognitive system. In recent years, Chomsky has advocated a stronger computational metaphor to explain the behavior of the linguistic system. In the so-called Minimalist Program (Chomsky 1995), the computational conception based on derivation takes center stage. In previous models, linguistic competence was viewed as a system that could be described by a phrase-structure grammar based on rules or as a system with very powerful rules restricted by a series of constraints. This is the model of grammatical competence that several scholars have proposed for the spectatorial or cinematic competence (Colin 1995, Buckland 2000).

In the Chomskyan generative model of the 80s and early 90s—the Principles and Parameters/Government and Binding framework (Chomsky 1981)—the emphasis is placed on characterizing universal grammar and its properties as a modular system, where different subsystems or modules interact. In the Minimalist Program, modules are viewed derivationally. The different principles and constraints that have to be satisfied are satisfied in a serial fashion. A view of cognitive systems as modular components that are intrinsically parameterized solves several paradoxes of the *grand syntagmatique* model and related proposals. The issue of which modes of production instantiate or not certain arrangement orders can be viewed as an issue of variation along parameters. For example, cinematic narration can vary along the chronological/achronological dimension in temporal order. When events are not represented chronologically, several narrative arrangements are possible: flashback, flash-forward, parallel stories, etc. There are semantic parameters, such as the parameter of actuality or connection to the actual world. A "realistic" film only depicts events that are anchored to the actual world. A non-realistic film depicts events that the spectator recognizes as not susceptible of taking place in his real world. Mixed options would allow for narratives that combine real and non-real events.

Applying the computational metaphor to the conceptual system also opens a variety of possibilities. On the one hand, there is the issue of how the spectatorial system would be organized. By spectatorial system we understand the set of cognitive abilities needed to process and understand cinema. The concepts of generation or derivation play an important role in linguistic description and explanation. A structural description of a sequence is a labeled tree that captures the derivational history of that sequence at a certain level of representation. If film narratives—especially complex instances such as theories represented by actual films—are generated derivationally, then an adequate description of a film would be an analysis or structural description that captures the derivational history of that film and the interface with its semantic properties. By derivational history we should understand here a way of arranging or composing shots into bigger units in a discrete, non-arbitrary fashion. Such a derivation is the product of the application of a small number of operations: most relevantly, concatenation, or the attachment of one shot after another. In current grammatical theory such concatenation operation is labeled Merge.[14]

The German philosopher Gottlob Frege developed a similar idea as applied in the semantic domain. He posited that the semantic analysis of a complex linguistic expression, that is, the complex logical expression translating the linguistic expression, should be subject to the principle of compositionality: The meaning of a complex expression is a function of the meaning of its parts and the way in which they combine. If we know the structure of an expression, we can assign a meaning to it by looking at how its constituent parts combine. Similarly, complex strings of shots are articulated compositionally. Editing a film is not an arbitrary process of arranging shots. Cinematic constituents should be compositionally articulated units. If the compositionality principle holds, it must be the case that the meaning or visual articulation of a complex unit—a syntagma, in Metz's terms—is the composition of the meaning of its parts and how they are combined. A related enterprise arises to analyze how we can determine what the primitives of cinematic composition are and to establish the modes of combination or the relevant procedures that give rise to complex constituents.

The view of cognition as a modular system also casts light on the structure of cinematic segments. The cinematic organ is comprised of connected subsystems or cinematic modules. There are very general principles that operate on each one of these cinematic modules or, alternatively, principles that function locally in only one of them. These principles may be parameterized, giving rise to different syntagmatic arrangements. Thus, the *grand syntagmatique* or alternative typologies would be the by-product of contrasting parameter settings. One principle that seems to be active is that a sequence representing an event has to have spatio-temporal continuity. Otherwise, the spectator would not be able to identify the merged shots as belonging to the same sequence. If we merge two shots of an individual walking where the backgrounds are different, the spectator would perceive the resulting sequence as depicting two different events. Another principle is visual consistency. Suppose we merge one general shot of an individual walking and an insert of his legs walking. If the insert was shot from the opposite side, the result of the merger will be that such individual is shown walking in one direction in the general shot and walking in the opposite direction in the insert. Thus, even if the filmed event was a single one, there will be a merger violation, since the concatenation of the two shots will be perceived as an individual walking first in one direction and then reversing his course. This would be a violation of the syntax of film or, more properly, an incorrect representation of an event; the concatenation of the two shots described would not be associated with the intended meaning (an individual walking in a single direction). Syntax and semantics cannot and should not be dissociated. The syntax of a film matters when it is paired with its intended denotation. Apparent "grammaticality" violations should be viewed as mismatches between a merged cinematic constituent and its intended denotation.

Finally, the idea that structure is ultimately computational, as defended in Chomsky's recent Minimalist Program, also has important consequences in its application to the analysis of cinema. The internal structure of a complex cinematic segment is given by the derivational steps that produce the computation of that segment. In this respect, several elements related to computational efficiency arise: Which process of constituent formation is more economical? Which one is safest or more efficient in representing the intended meaning? And so forth.

The Social Challenge

In past remarks, Chomsky (1977) relativized the interest of linguistic studies that take a social approach with regard to the main goals of cognitive science. A sociological approach to linguistic problems based on tests or statistical studies of sample population groups cannot reach the level of generality that is needed to establish the universal and abstract properties of the human mind/brain. This brings us to the issue of what the relevant data for cognitive research are. Chomsky claims that the exploration of the universal properties of grammar, that is, what constitutes the core of the language faculty, has to have recourse to introspection and grammaticality judgments, which are not subject to social variation.

The strength of Chomsky's argument comes from the restrictiveness of his approach. Only properties ascribable to general patterns of data that emerge from grammaticality distinctions are considered. We agree that whether one sentence is grammatical or not is not going to depend on the age or gender group of the speaker. Nevertheless, there are certain linguistic properties that are socially conditioned. We can attribute lexical, intonational, and sometimes even syntactic differences to social groups, as shown by William Labov. Still, these deviations do not constitute a different language, and socially differentiated groups do not have completely different grammars. They may instantiate different dialects of a language. The grammar of a given dialect of a language can differ from the grammar of other variants only to the point of comprehensibility. Speakers of English recognize British, southern American, or midwestern varieties as such—that is, as different dialects of English. The distinction between language and individual idiolects is also a cognitive distinction that applies universally. A speaker's idiolect might instantiate different registers (formal, colloquial) as a function of several factors. Different registers and mechanisms for register change should also be considered part of the cognitive system, given that speakers uniformly recognize such strategies as linguistic subcodes appropriate for certain situations.

These distinctions are also reflected at the cinematic level. A film director contributes stylistic and thematic variations that are dialectal and idiolectal in nature. Approaches to film based on generations, country of origin, or social groups can be said to focus on dialectal differences, as represented by "movements" such as feminist cinema, black cinema,

the *Nouvelle-Vague*, etc. Neorealism as a dialect had several properties that made immediately identifiable the works of filmmakers such
as Rossellini, De Sica, the first Visconti, etc. Register variation in film
is instantiated as different narrative modes: realistic film, avant-garde
film, videoclips, TV shows, etc. These different narrative registers have
different parametric values for certain principles: spatio-temporal continuity, use of extra-diegetic music, etc. They are registers and not dialects because a single filmmaker can produce works in more than one
register. Similarly, spectators can process and understand different registers, as long as they have internalized the set of properties or parameters that characterize them. On the other hand, dialectal parameters
are not stackable: A director or a spectator contributes his own dialectal
properties to the process of creating or decoding a film. Nevertheless,
it is not possible to internalize and be ascribed to several dialects at the
same time in a random fashion.[15]

For a cognitivist, the notion of competence is essential, since it reflects internalized knowledge of cognitive properties. Chomsky's followers focused on syntactic competence. There are other areas of grammar
of which speakers have knowledge: phonology (sound structure), semantics (meaning), and pragmatics (use). The semantic/pragmatic arena reflects very clearly the interface of cognition with the social properties
that we are exploring. Language is used to transmit and process information and to express attitudes about facts and situations (Barwise and
Perry 1983).

Social cognition goes well beyond the linguistic realm, precisely
in what pertains to rationality, decision-making processes, and mass
and individual-level beliefs and ideas. A proper analysis of the social-
cognitive elements in cinema has been recently undertaken, for example,
by studying the cognitive elements inherent to the spectator's emotional
reactions to a movie or to certain elements of the cinematic text (music,
etc.), as shown by Plantinga and Smith (1999).

Conclusion

Recent developments in the field of linguistics, and in the cognitive
sciences in general, seem to be of potential interest for the analysis of
certain structural properties of film. Although the idea that film has
language-like ingredients was already present in the structuralist tra-

dition, reconsidering it from a generative and biolinguistic perspective allows for a more fruitful understanding of these aspects as related to general cognitive devices.

Notes

1. With the label "strong programs" we refer to explanatory enterprises attempting to offer a univocal characterization of all aspects of film. Most strong programs have well-defined philosophical underpinnings (Marxist, Freudian, Saussurean, etc.). Weak programs would be those that deliberately refrain from reducing facts to a strong theoretical explanation. Although deconstructionist and postmodern approaches could be characterized as weak programs, this is only partly accurate, since they also seem to have a clear philosophical agenda—based on certain ideas from Nietzsche, Heidegger, and Derrida, among others. The strong/weak distinction should not be reduced to the rationalist versus empiricist debate either. Deconstructionist approaches are not more empirically grounded than their alternatives. Sometimes the opposite is the case, given that deconstructionists tend to focus on a text or fragment and disregard comprehensive empirical coverage as misguided or fruitless.

2. Andrew Sarris's views, as represented in Sarris (1968), reflect the *auteur* theory in all its greatness and misery. Some of his analytical judgments are obviously misguided or have been superseded, but they represent a clear attempt to separate real "authors" from those who are not worthy of this label.

3. Stam et al. (1992) dissect the main trends and conceptual elements of the semiotic movement.

4. The second-semiology—in particular its variety with the strongest or more orthodox Lacanian roots—has experienced a revival in the works of the philosopher Slavoj Žižek; see his study of David Lynch's *Lost Highway* (Žižek 2000), where he situates himself as strongly opposed to the post-theory assessments of Bordwell and Carroll.

5. For Bazin, only realistic filmmakers are true to the core spirit of film. Films should be made from "fragments of raw reality, multiple and equivocal in themselves, whose meaning can only merge a posteriori thanks to other facts, between which the mind is able to see relations" (Wollen 1972, 132). The films of Rossellini or Renoir should then be defended as true cinematic masterpieces, whereas the works of formalists such as Eisenstein should be criticized. Camera movement (*recadrage* or re-framing) should be preferred to Eisenstein's principle of montage, because lateral camera movements are able to capture a continuous fragment of reality. Montage, on the other hand, distorts or fragments reality and creates effects that are non-realistic. Bazin made clear his disdain for other non-realistic movements, such as Expressionism (*The Cabinet of Doctor Caligari*).

6. Such pointing may be physically explicit, as in a pointing gesture, or an inferential association that the hearer has to make in order to gain access to the entity referred to by the speaker (Bühler 1934).

7. In the classic *giallos* directed by Dario Argento, the inserts of the killer's hands wearing gloves are normally shots of Argento's hands.

8. For example, in *The Birth of a Nation* Griffith alternates images of war and peace, without suggesting any temporal or chronological connection between them.

9. A typical example of an episodic sequence is one representing the rise (or fall) of a character in a succession of shots. For example, the episodic sequence at the beginning of *Citizen Kane* (dir. Orson Welles) narrates the rise to power of Charles Foster Kane (based on William Randolph Hearst).

10. See also Gutiérrez-Rexach (2003) for a compilation of different views on the analysis of meaning, and the interrelation of syntactic, semantic, and pragmatic properties from linguistic and philosophical perspectives.

11. Arijon (1991) presents a very detailed characterization of the grammar of film understood as pure syntax.

12. Anderson et al. (1999) and Driscoll et al. (2004) show that damage to the orbitofrontal cortex in infancy blocks the acquisition of a moral sense but no other higher cognitive functions, including language, planning, and reasoning.

13. See Kanizsa (1979), Rock (1983), and Pinker (1985), among others, for the logical ingredients of perception and visual cognition, from different points of view.

14. A critical issue is to determine at what point the cognitive merger of shots into syntagmas takes place. Preliminary shot composition takes place during the storyboarding process which, for some directors, already gives a very precise idea of the actual look of a film. In the editing process, during the post-production stage of film creation, the editor and the director determine the actual concatenation of shots, their tempo, and potential rearrangements. Finally, when the spectator watches a film, he processes the images into meaningful units. Theoretically, both the arrangement by film creators (director/editor) and the decoding by the film spectator should be instances of the same cognitive system.

15. The parallel with linguistic dialects is not absolute. A director may be ascribed to feminist and black cinema movements simultaneously. Nevertheless, multidimensional ascription is not possible along a single axis of variation. It would not be possible to be characterized as a feminist and non-feminist director at the same time.

CHAPTER FIVE

Voice and Perception:
An Evolutionary Approach to
the Basic Functions of Narrative

KATJA MELLMANN

The distinction of *voice* (who speaks?) and *perception* (who sees/hears/ smells?) (Genette 1986:186; 1988:64) can be said to be the egg of Columbus in Gérard Genette's analysis of narrative discourse. Whereas in traditional models of literary narrative we had to deal with typologies mainly (for instance, of "narrative situations"; see Stanzel 1971, 1984; Fludernik and Margolin 2004; Genette 1980), we now possess a systematic description of the imagination evoked by a text, which takes into account the quasi-ontological (see Bortolussi and Dixon 2003) status of its constituents. In this chapter I search for the *cognitive functions* that correlate with the text features of "voice" and "perception" and for how they bring about such a "layered" imagination in the reader. The aim is to explain how and why literary narratives can run properly in the human mind—which is another way of asking how humans could develop narrative discourse as a way of communication at all.

To make more graspable what I am trying to do, let me start with a consideration by Genette that I regard as crucial for a profound understanding of the interplay between narrative texts and human cognition. Genette says: "Unlike the director of a movie, the novelist is not compelled to put his camera somewhere; he has no camera" (Genette 1988:73). Vice versa, it can be said: Unlike the novelist, the director of a movie is not compelled to talk to the audience; he has no voice. If these propositions are true (I shall discuss them later), it can be stated that there are *two distinct narrative functions*, as I will call them, which can occur independent from one another in principle, although they used to occur in combination very often.

Literary narratives—especially since the age of realism, but also in the Homeric epics and all through history—*can* (but need not) possess longer or shorter "focalized" passages, and motion pictures *can* (but need not) employ title cards, subtitles, or voiceovers, and thus establish an enunciating instance for a moment. But the typical literary narrative is understood to be a (wholly or dominantly) *telling* medium ("voice"), and the typical cinematic narrative is understood to be a (wholly or dominantly) *showing* medium ("perception"). How can it be that one can work without the other? My proposition is that they rely on different strategies of information gathering which evolved separately in the history, and prehistory, of human evolution. Different modes of presentation, different media, make stronger use of the one information system or the other.

In the first section below, I analyze selected passages from Virginia Woolf's *Flush* (1933) in order to illustrate the distinction of voice and perception and explain the concept of focalization. The second section attempts a first linkage between the textual features of "voice" and "perception" and distinct cognitive programs, as well as a rough assessment of their historical realizations. Taking up again Genette's comparison of literature and film in the third section, I shall then discuss the question of whether one can plausibly speak of filmic "narratives," although there is no obligatory voice instance in cinematographic representations.

Voice and Perception as Two Distinct Functions in Literary Narrative

Unfortunately—or fortunately—Genette did not care too much about giving a holistic system of narrative theory. Rather, he formed occasional concepts for what he needed to analyze Proust's *Recherche*, and there are slight shifts between his concepts in *Discours du récit* (1972) and *Nouveau discours du récit* (1983).[1] This is why I do not attempt to give a scholastic exegesis of the "real" Genette here and by this establish a narratological "catechism" which Genette himself refused (1988:74). I just pick up what I conceive as one of his basic ideas: the distinction of voice and perception and the concept of focalization. The structuralist approach was the best one we ever had to extract persistent patterns from a diversity of works of art, and persistent patterns we require when correlating text structures with cognitive functions. One such pattern in literary narrative is that "someone talks to us" ("voice"), another one that fictional events cannot be shown but from a particular angle or position ("perception").[2] In order to avoid "the too specifically visual connotations" of terms like *point of view* and *perspective*, Genette has suggested "the slightly more abstract term *focalization*" (Genette 1986:189; 1988:64). I doubt whether the visual metaphor in terms like *point of view* or *perspective* has ever led to serious problems. I appreciate the term *focalization* rather for omitting connotations of "attitude" which the former terms were charged with and which, in the first instance, refer to the concept of "narrator."

Preliminary Note on the Concept of "Narrator"

The mental attitude (set of values and convictions, intentions, cultural knowledge, etc.) we conceive by reading a story is due to what I am calling a "psycho-poetic effect" (Mellmann 2006:99–103; forthcoming) on the part of the reader, that is, the effect of "making psyche" (where there is none) by imagining any kind of "subject" in response to certain text structures. Because Homo sapiens is an eminently social animal, our social cognition systems are eager to extract relevant information wherever possible. There are many kinds of text features which are well suited to serve as input to particular social algorithms (for examples see Mellmann forthcoming), and once triggered, those social algorithms entail, as a kind of by-product, mental images of quasi-personal unities (like "character," "narrator," "author," or Stanzel's "figural" narrator). Such implied subjective entities can *combine* with the speaker instance or the perceptual instance[3] in the imagination of the reader, but they are not identical with them. However, in literary narratives the sustained "voice" provokes a permanent psycho-poetic effect of "narrator." What has often been criticized in narrative theory under the title of "anthropomorphic fallacy" (see Bortolussi and Dixon 2003:174) is but the symptom of a universal human bias, "a metaphor we live by" in Lakoff and Johnson's term (see Smith 2000:159). The fact that "someone talks to us" simply entails the image of *somebody* talking to us (see Carroll 1995:157). This is why, in literary narratives, the default position to ascribe certain attitudes to is located on the side of the "voice" rather than that of "perception." I give an example from Woolf's *Flush* as an illustration:

Flush darted to the sofa.
"Oh, Flush!" said Miss Barrett. For the first time she looked him in the face. For the first time Flush looked at the lady lying on the sofa.
Each was surprised. Heavy curls hung down on either side of Miss Barrett's face; large bright eyes shone out; a large mouth smiled. Heavy ears hung down on either side of Flush's face; his eyes, too, were large and bright; his mouth was wide. There was a likeness between them. As they gazed at each other each felt: Here I am—and then each felt: But

how different! Hers was the pale worn face of an invalid, cut off from air, light, freedom. His was the warm ruddy face of a young animal; instinct with health and energy. Broken asunder, yet made in the same mould, could it be that each completed what was dormant in the other? She might have been—all that; and he—but no. Between them lay the widest gulf that can separate one being from another. She spoke. He was dumb. She was woman; he was dog. Thus closely united, thus immensely divided, they gazed at each other. Then with one bound Flush sprang on to the sofa and laid himself where he was to lie for ever after—on the rug at Miss Barrett's feet. (Woolf 1983:20)

The charm of this scene is due to its ironic parallelism with romance. The narrator, by his diction, cites romantic discourse and thereby reveals a humorous attitude towards this quite ordinary occurrence between a human and a dog. That is, we draw conclusions about what the narrator's attitude may be from looking at his way of telling the scene. Word choice, imagery, manner of speaking, and their aptness or inaptness in respect of the events, are the clues by which we learn about the narrator's attitude. Independent from this mental "standpoint"—the amusement about the scene as a whole—we are guided through multiple perspectivations of this scene which the narrator employs to show the event itself. For instance, when we are told about the curls/ears, the bright eyes, and the large mouth twice, these seem to be seen through the eyes of Flush/Miss Barrett—in other words, from in front, as in a face-to-face communication. Only from this perspective can we see the curls/ears "*on either side*," *both* eyes, and that the mouth is "*large.*" These perspectives "through the eyes of a character" (Genette's "internal focalization") are not those of the narrator, if we imagine the narrator as a person external to the story (Genette's "heterodiegetic narrator").

I will return to the above quoted passage in more detail below. For the moment it is important to notice that "attitude" in literary narrative mostly is a matter of "voice" (word choice, imagery, manner of speaking), and that it should not be interfused with "perspective," which is a matter of "perception" (of what is perceived before it is told).[4] So what does "perspective," or "focalization," as a matter of "perception," mean exactly?

Focalization

Genette explicated his concept of focalization as "a restriction of 'field'—actually, that is, a selection of narrative information [. . .]. The instrument of this possible selection is a *situated focus*, a sort of information-conveying pipe that allows passage only of information that is authorized by the situation" (Genette 1988:74). Whenever a text passage suggests such a "situated focus" as viewpoint, it is, in Genette's terms, "focalized," which means that there is some restriction of information due to situational conditions within the fictional world. The opposite would be an "unfocalized" passage, showing (theoretically) unlimited information about the narrated world. For an example let us have a look at the very first sentences of the novel:

> It is universally admitted that the family from which the subject of this memoir claims descent is one of the greatest antiquity. Therefore it is not strange that the origin of the name itself is lost in obscurity. Many million years ago the country which is now called Spain seethed uneasily in the ferment of creation. Ages passed; vegetation appeared; where there is vegetation the law of Nature has decreed that there shall be rabbits; where there are rabbits, Providence has ordained there shall be dogs. (Woolf 1983:7)

The narrator begins with telling something out of his general *knowledge* ("It is universally admitted . . . Therefore . . ."), not something which is actually *perceived*. One may argue that there is some restriction of information too, for the narrator's knowledge may be not unlimited either. But this would be no "restriction of 'field'" in the sense suggested above. The limits of the narrator's knowledge are due to reasons external to the storyworld, not to a particular "focus" situated in it.

A slight change takes place with the sentence "Many million years ago . . ." At this point, a kind of story begins (as with "Once upon a time . . ."), and we start imagining a specific fictional world (wherein a future Spain seethes uneasily in the ferment of creation, and so on). We have the impression of a camera recording the spectacle of creation, which is due to the spatio-temporal data in the text. If we ask how the situation of perception is defined, we can answer: somewhere in the "ferment of creation," where Spain is about to emerge *(where?)*, and: many million years ago *(when?)*. As soon as we are given spatio-temporal coordinates of a fictional situation we face the germ of focalization.

However, in this example the spatio-temporal scope is very wide; a larger spatial frame than that of the tohubohu and a larger temporal frame than that of several ages, in comparison to which these could pass as a "restriction of field," are hard to imagine. Thus, the grade of focalization is very low here. It is much higher in the above-quoted sentences: "Flush darted to the sofa. 'Oh, Flush!' said Miss Barrett." Here the events are told as they might be heard and seen by someone who is present in the room. This is a focalization strategy so frequent in modern literature that Franz K. Stanzel (1971; 1984) made it one of his "prototypes" of "narrative situations." It evokes the impression of an invisible witness, always walking about with the characters (Stanzel's "figural" narrator, or Henry James's "reflector person"), who functions as the "information-conveying pipe" between the fictional world and the storyteller. I call it *anthropomorphic focalization* for its humanlike dimension of spatio-temporal access.[5] However, the spectrum of possible focalizations is still much broader: Between the scope of wideness of a camera set at such a great distance that it can record the universe, and of such a permanence that it can record millions of years, and the scope of wideness of a camera set among characters, changing places with them, *any gradation is possible.* Think for instance of the initial zooming (e.g., from country/age across town/season to house/day) in the beginnings of many a novel. And also *below* the level of anthropomorphic focalization one can go down endlessly into microscopic time and space. Moreover, an author can combine narrow time scopes with wide space scopes, and vice versa. There are *infinite forms of focalization.*[6]

The best way to describe a particular focalization is to determine the "situated focus," i.e., the "point from which the narrative is perceived as being presented at any given moment" (O'Neill 1992:333). Its spatio-temporal coordinates define the perceptual situation. Normally, the situation of the perceiving instance (camera) is not identical with the situation of the voice instance. There are some particular cases (like interior monologue or stream of consciousness) where both coincide and the narrator appears to be the perceiver at the same time. Another particular case is the autobiographical schema, where there are several years between the perceptual situation and the situation of utterance, but the (anthropomorphic) focalization can be attributed to the "I" of the narrator and his/her presence in the fictional world. Yet the default, or classical, case of modern literary narrative is that of many realist novels, where the situation of enunciation remains more or less unspecified, whereas perception varies between several particular situations within the fictional world.

So far, I have illustrated the difference between focalized and unfocalized narration, and the continuum between them, which can be described as different grades of focalization. But the grade of focalization (scope wideness) is not all which needs to be taken into account; we have to consider also the grade of autonomy of voice and perception, that is, their specific interplay at work in a particular passage.

Different Grades of Autonomy of Voice and Perception

My conception of voice and perception as two more or less autonomous narrative functions attempts to integrate the classical distinction of showing and telling (or mimesis and diegesis) as two different modes of narration into Genette's model.[7] Let us assume, as default position, an integrated narrative system in which the functions of voice and perception are well balanced and tightly intertwined, so that the reader is not forced to reflect on one of them in particular. Then the showing mode can be reformulated as a deviation which is marked by the dominance of the perceptual function, the telling mode as a deviation with dominant voice function. I initially provide two examples that shall illustrate the extreme poles of this opposition.

The first extreme case is akin to what Roland Barthes described as "reality effects." Those effects are brought about by detailed descriptions, as for instance in a text by Flaubert, which comprise even apparently useless details like a barometer hanging on the wall. If the perceptual data given in a text thus exceed the demands of telling a consistent story and the voice appears to will-lessly follow the input from the perceptual system by just reflecting what is perceived, then the perception system differentiates from the integrated default system and develops into a distinct system of its own, which works in accordance to its own logic (indiscriminately recording everything), independent from external needs and intentions; to show what is there seems to be an end in itself. The voice function, in contrast, is subordinated to the perceptual function in that it does not comment, add, select, or stylize anything, but confines itself to a mere reflector function. This is why in passages with dominant perception and subsidiary voice function we easily forget that the perceived events are mediated through a voice instance. The fictional events seem to "tell themselves" (Genette 1986:164) and "finally say nothing but this: *we are the real*" (Barthes 1989:148).

The other extreme pole can be illustrated by a passage from Diderot's *Jacques le Fataliste:*

Vous voyez, lecteur, que je suis en beau chemin, et qu'il ne tiendrait qu'à moi de vous faire attendre un an, deux ans, trois ans, le récit des amours de Jacques, en le séparant de son maître et en leur faisant courir à chacun tous les hasards qu'il me plairait. Qu'est-ce qui m'empêcherait de marier le maître et de le faire cocu? d'embarquer Jacques pour les îles? d'y conduire son maître? de les ramener tous les deux en France sur le même vaisseau? Qu'il est facile de faire des contes!

The narrator here leaves no doubt about his power to tell what he wants to. That is, by saying that something is that way, he would make the story go that way; the perceiving instance would have to follow every turn of the voice instance. Passages like those fancied by Diderot's narrator, with dominant voice and subsidiary perception function, make the reader particularly aware of the artifact character of a narrative, the fictiveness of its content, and thus entail the very opposite of reality effects.

Now let us have a look at more subtle combinations between voice and perception and turn back to the above-quoted passage about the first encounter of Flush and Miss Barrett. In the first two sentences, we face the default version of narrative texts:

Flush darted to the sofa.
"Oh, Flush!" said Miss Barrett.

Our impression of the passage is that of a homogenous narrative act. The two narrative functions of voice and perception together constitute one integrative system. It would be hard to tell if the words "Flush darted to the sofa" determine the perceptual image of the event or if, vice versa, the event as it is perceived determines the wording. The two narrative functions of voice and perception seem quite equivalent here.

Not so in the next two sentences:

For the first time she looked him in the face. For the first time Flush looked at the lady lying on the sofa.

The perceived event ("looked") is not the only information given in these sentences (like "darted" and "said" were); in addition we are told that this happens "for the first time." This is something which cannot be seen, something which the narrator simply *knows*. One could say that in these two sentences different sources of information are blended on the discourse level. Imagine the text would go like this: "She gazed at his face for a while. It was the first time that she had a look at him." Then the presentation of the perceived event ("gazed") and the additional comment that "it was the first time" would be more separated than in the original sentences. By interdigitating one with the other through a particular way of wording, the voice somewhat dominates the perception, that is, the latter cannot fully unfold as an autonomous dimension of the text (with stronger, activity-indicating words such as *gaze* instead of *look* perhaps, or with proper spatio-temporal data such as "for a while"); the perceived event is reduced to an only shortly introduced fact in order to fit together with the cardinal information that it was the first time. Similarly, the beginning of the novel (the creation scene): I said the spatio-temporal data would provide only the *germ* of a focalized passage, because beside the all too wide scope of focalization we have here a voice instance so dominant again that we can hardly assume a strong perceptual function. If the text went like this: "Where there is vegetation the law of Nature *decreed* that there shall be rabbits; where there are rabbits, Providence *ordained* there shall be dogs"—i.e., if the epic preterit of "seethed uneasily" was kept on, the illusion of ongoing events (Nature and Providence entering the stage, so to speak) would be maintained. But by switching to the present perfect ("has decreed"/ "has ordained"), Woolf evokes the impression of a narrator talking out of his general knowledge (about accomplished laws of nature) again. If perceptual elements are throttled down like this to a mere mention by the voice, which seems to have control over how much of the perceived world gets on through to the discourse level, the perception system has not differentiated into a distinct system of its own but in a way pertains to, or subsidiarily subserves, the voice system. In terms of system theory such non-differentiated systems can be called "trivial machines" (see von Foerster 1984:9f.), which means that they have not yet established themselves as operatively closed systems and developed their own logic, but function after the rules of another and produce a rather predictable output.

The above-quoted passage then goes on as follows:

Each was surprised. Heavy curls hung down on either side of Miss Barrett's face; large bright eyes shone out; a large mouth smiled. Heavy ears hung down on either side of Flush's face; his eyes, too, were large and bright; his mouth was wide. There was a likeness between them.

The first sentence affiliates properly with the antecedent in that it retains the subsidiary perceptual function: Although we feel that the fact that "each was surprised" must have been perceived somehow (by looking into Flush and Miss Barrett) before it is told, we would settle for the abridged version that it simply is that way and would do without a detailed description of how this surprise unfolds in Flush's and Miss Barrett's minds. That is, we would follow the claim of dominance of the voice instance. But then we are guided in this internal perspective exactly, when we gain insight of how Flush and Miss Barrett see one another, so that the perceptual function now gains autonomy and the voice function is throttled down to a subsidiary trivial machine (reflector function). The following sentence, saying that there was a likeness between them, now does not seem to be a comment on the part of the narrator, evaluating these perceptions, but a thought Flush and Miss Barrett (at least she) might have had and which is just replicated (by means of free indirect discourse) by the voice instance.

The dominance of perception, on the one hand, seems to be maintained until the end of the passage. On the other hand, we more and more feel that, even if it was nothing but Miss Barrett's thoughts that is reflected here, it is not done in accordance with her attitude:

> As they gazed at each other each felt: Here I am—and then each felt: But how different! . . . Broken asunder, yet made in the same mould, could it be that each completed what was dormant in the other? She might have been—all that; and he—but no. Between them lay the widest gulf that can separate one being from another. She spoke. He was dumb. She was woman; he was dog. Thus closely united, thus immensely divided, they gazed at each other.

We clearly feel shining through the spirit of the narrator, his attitude, his own thoughts and feelings toward the event, expressed by a particular way of phrasing. Thus, also the voice function establishes a differentiated system here. This autonomy of the voice instance goes without stopping the above-declared autonomy of the perceiving instance. You can catch its persistence by the voice's real-time reaction to ever new events in the perceived scene, indicated by the dashes: "She might have been—all that" indicates that the voice ties itself down to the real-life time scale of Miss Barrett's thoughts by performing the same pause of reflection she makes; the same with "and he—but no," which follows her self-interruption of thought. Thus it is granted enough room to the perceptual function to unfold completely, though we are quite aware that the definite way of utterance (employing the vocabulary of romantic love encounter) is due to the narrator rather than to Miss Barrett (whose thoughts may be of a rather nonverbal nature).

I have now typified four ways of combining voice and perception in literary narratives:

1. integrated narrative default system with balanced voice and perceptual function;
2. autonomous voice instance, with competence to discretionarily abridge the output of the perceptual device;
3. autonomous perceptual system, coming to the fore as the inner film of imagination in the reader, and thus pushing back the voice to a mere reflector function;
4. autonomous voice *and* perception, producing a truth of its own each.

In options (2) and (3) one of the two functions shows clear dominance and differentiates out of the narrative default system as an autonomous system itself, while the other function is reduced to a subserving trivial machine. In option (4) we face a kind of double-coding brought about by two fully developed narrative systems, which might irritate one another because of their "structural coupling"[8] but for the rest work autonomously.

I used these rather abstract terms from system theory in order to make more transparent the very complex design of narrative discourse by reconstructing its underlying principle. Moreover, the conception of voice and perception as of potentially autonomous systems—together with Genette's comparison of novels and films which I introduced above—can help to firm up the supposition that these two different dimensions of narrative texts correspond with two equally different faculties of the human mind, which are able to be performed independently from one another in principle.

Voice and Perception as Inputs for Two Different Systems of Information Gathering

John Tooby and Leda Cosmides, two exponents of evolutionary psychology, have given a notable explanation of narrative by suggesting that socially communicated information is often formatted in a way that mimics firsthand experience, in order to cooperate with our evolutionarily elder, more basic mechanisms of information gathering:

Indeed, we evolved not so long ago from organisms whose sole source of (non-innate) information was the individual's own experience. Therefore, even now our richest systems for information extraction and learning are designed to operate on our own experience. It seems therefore inevitable, now that we can receive information through communication from others, that we should still process it more deeply when we receive it in a form that resembles individual experience, even though there is no extrinsic reason why communicated information needs to be formatted in such a way. That is, we extract more information from inputs structured in such a form. What form is this? People prefer to receive information in the form of stories. Textbooks, which are full of true information, but which typically lack a narrative structure, are almost never read for pleasure. We prefer accounts to have one or more persons from whose perspective we can vicariously experience the unfolding receipt of information, expressed in terms of temporally sequenced events (as experience actually comes to us), with an agent's actions causing and caused by events (as we experience ourselves), in pursuit of intelligible purposes. (Cosmides and Tooby 2001, 24)

What Tooby and Cosmides refer to as quasi-experiential structure or form is, in terms of narratology, the focalization of a text, especially what I called anthropomorphic focalization above. It is this situated focus, adhering to human dimensions of time frame and spatial movement more or less, which evokes the impression of vicarious experience. Hence, we can state that the narrative function of perception correlates with our ability to extract information from environment, that is, it makes use of basically the same information systems that are employed in normal environmental experience of all organisms capable of autonomous movement. Whereas the narrative function of voice—the fact that someone talks to us—correlates with information systems that evolved together with human language, that is, in a much later stage of evolution. The voice aspect of narratives thus should involve mainly those cognitive mechanisms which are associated with verbal communication (semiotic faculties, memory, syntax logic, and the like), whereas the perception aspect should run a simulation on the perceptual system of our brains and focalized passages should be processed by the same second-order circuits as are involved with processing sensory inputs.

So far, the one has nothing to do with the other, and, as Tooby and Cosmides put it, "there is no extrinsic reason" to combine these two systems of information gathering by telling a story. But it is true that, in ancient times, most of our knowledge was bound and passed on in the form of stories. Think of the Gilgamesh epic, the Homeric epics, Greek myths, the Old Testament, medieval novels, or simple folktales. They contain a good deal of knowledge about our ancestors, foreign countries, fabulous animals, the beginning of the world, and, moreover, about vice and virtue, love and hate, bliss and sorrow, danger and salvation, and so on. There are several studies on narrative that focus on its capacity for storing adaptive information (Sugiyama 1996; 2001; 2006; Eibl 2004:257–272), but, as far as I can see, the particular role that focalization plays in this way of storing information has not been analyzed yet. This may be due to the fact that in most ancient myths and simple folktales the voice function preponderates and focalized passages are the exceptional case.[9] A persistent autonomous perception system seems to be more typical for modern literature. But even a subsidiary perceptual function can be a good device for preserving information in the form of stories. Imagine I was a Pleistocene hunter-gatherer saying to you:

1. There is a waterhole a half day's hike away from here in this direction.

You would be informed well enough for an immediate departure to the waterhole. But imagine I tell you about it as follows:

2. You take the path we went yesterday to the stone pit. When you see the big oak tree, you climb down the scarp at the right-hand side and cross the plain toward its leafy end. There you will find a waterhole.

This instruction maps out an imaginary way, that is, a little (proto-) *narrative* of you (as situated focus) going through a particular landscape. This imaginative map of the story makes you independent of a particular direction that I show you from where we are standing in the moment of our conversation. So if you decide not to go for the waterhole until the next day, you will find it from wherever you might be then.

Now imagine you want all your kin to know about this very good waterhole in case you will not be with them when coming to this place again the next year. Then you might tell a story like:

3. Once upon a time there was a little boy looking for a waterhole. He walked on a rocky hill and looked out as far as the eye could see. Beside the hill there was a great plain extended to the horizon. At its end, the boy noticed a small stripe of greenness, and he said to himself: where there is grass there must be some water. He climbed down the scarp and walked ahead toward the green stripe. He marched for half a day, the sun was shining hot, and he was getting tired and more and more thirsty. But coming nearer he could clearly see now

the green grass. He had not deceived himself. When he reached the grassy ground he had not to search long ere he found a wonderful waterhole.

Now we have a classical third person narrative that suits for information storage over longer periods of time and can be communicated from one individual to another several times. (You can adorn it further with several adventures the little boy has to get through, obstacles he must overcome, and so on, to make it more interesting[10] and easier to keep in mind.)

What is particular for human language is that the reference to facts (Karl Bühler's "representation function" of language) can be isolated from the reference to persons (Bühler's functions of "expression" and "appeal"), that is, from its situational anchorage; whereas animal languages, and presumably the proto-language of the early hominid species, are undifferentiated "tri-functional" languages (Eibl 2004:209–275). To flesh out a message (1) by a conversational first- or second-person proto-narration (2) or to transform such proto-narratives into a purely representational third-person story (3) might have been important steps in this process of linguistic sophistication. The narrative format of a message deliberates its descriptive content from the original context of utterance and thus makes it a quasi-objectified, transportable thing. In this sense, the isolation of the descriptive function of human language was described as a "Vergegenständlichung" (reification) of information by Karl Eibl (2004:209–275). What Cosmides and Tooby refer to as the "text book" type of communication is nothing but a radicalization of the same principle: an absolute detachment from any situational meaning, even from vicarious situations. I give a possible textbook version of our story:

4. Waterholes are often found amid leafy places.

This proposition contains the most generic information essence of the story, omitting any additional data about a particular time or area or person. However, successful application of information detached from any situational context is difficult, because the individual has to know by herself under what conditions such an abstract proposition is true. But obviously we are adapted to this problem: We are able to handle abstract information because of a cognitive "scope syntax, that tag[s] and track[s] the boundaries within which a given set of representations can safely be used for inference and action" (Cosmides and Tooby 2001:20). This is how humans came to "live with and within large new libraries of representations" (Cosmides and Tooby 2001:20), which allowed them to store and standardize a greater amount of information than one single individual would be able to gain in a life span, and to pass it on to the next generation. That is, every new generation can build on the knowledge achieved by the last one, which leads to the specifically human, "cumulative" or "cascading" type of cultural evolution (cf. Eibl 2004:236).

However, these great "libraries of representation" have to be literal libraries most of the time. Except for some proverbs and similarly small packages of abstract information, oral cultures are mainly *mythical* cultures, which rely on *storied* lore on a much larger scale than on abstractly coded knowledge. Mere textbook types of cultural patrimony appear rather late in human history and depend on the existence of social elites possessing the required education and media.

The fact that an *autonomous* perception system in literary narrative occurs only seldom before modern times may also be due to cultural evolution and the evolution of media. I would not preclude completely the possibility that high differentiation grades of the narrative perceptual function have always been there in *oral* storytelling. Imagine a gifted storyteller at the campfire of a hunter-gatherer tribe, doing the best he can to entertain his audience, to thrill them, to move them. He would be well advised to flesh out the perceptual situation of his narrative with concrete detail to the utmost. But when oral traditions are recorded in media, this should be done as economically as possible, for most of the time new media are very expensive in the beginning. On this basis, it is not surprising that most of the narrative literature written down until one or two hundred years after the invention of letterpress does not make great use of a differentiated perceptional function. Furthermore, written stories in former times often were not designed to be *read*, least of all alone, silent, and in an immersive attitude of reading, but they served as an aid to memory for an ultimately oral performance again. Conditions were completely different in the realist age, when a most sophisticated art of narrative book-fiction emerged, tapping the full potential of the narrative perceptual function.

With regard to that rough historical sketch, it should have become palpable that the novelist is not compelled to put his camera somewhere in the strong sense of a high grade of focalization or a fully differentiated perceptual system. Low grades of focalization as in a bird's-eye view or through frequent switches (Genette's "multiple focalization") and a merely subsidiary perceptual function (i.e., merely mentioned perceptions) are sufficient to evoke the feeling of narration. In novels, we even tolerate longer voice-only passages (e.g., reflexive commentaries), although they, strictly speaking, interrupt the narration. Thus, storytelling per se is not particularly different from standard communication, where little proto-narrations like that of text (2), or conversational first-person narrations, with low differentiation grade of the perceptual function, occur now and then. To have the impression not only that somebody talks to us but that, moreover, we really see the told events, as if we were there ourselves, is an additional effect brought about by artful storytelling, which is not limited to a particular age or media but can emerge everywhere along the histo-cultural continuum under various conditions.

This seems quite plausible if we assume that the cognitive program which supports the perceptual function of narrative texts is much older than our linguistic faculties and was already there when humans first invented verbal storytelling. But then the question arises whether there is—if perception is what distinguishes narration from standard communication—something like a non- or preverbal storytelling making use only of the characteristic one of the two narrative functions. H. Porter Abbott (2000:248–252), for instance, suggests that the gift of imitation and a thereupon based "mimetic culture" (Donald 1990) gave way to an at least rudimentary form of narrative. Michelle Scalise Sugiyama (2005:181–183), on the other hand, prefers to confine the notion of narrative to verbal accounts, because "nonverbal expressive media (for example, visual art, dance, music) are actually quite inefficient narrative devices," which fail, especially, "in representing the thoughts, beliefs, and motives" of literary characters and in communicating necessary background information (182). Scalise Sugiyama does not, among her examples, discuss dramatic art, which Abbott seems to consider in the first place and which, in the form of film, has indeed become the most powerful media for fiction in the twentieth century. As it clearly is *stories* that millions of people consume every night watching TV, it should be worthwhile to analyze the specific combination of narrative functions discussed here which is at work in films. The immense popularity of that medium, outrunning even novels by far, may be seen as a clue that the dominance of the perceptual function perhaps meets a universal human bias. It is quite possible that it simply feels more convenient if the evolutionarily elder, more basic system for information gathering is served directly by audio-visual stimulation, without the (maybe richer, maybe not) mediation through language.

Voice and Perception in Nonverbal Narrative

Some narratologists hold that it is not correct to speak of filmic narratives, as the Latin word *narrare* means "declare, enounce, tell," and there is no obligatory voice instance in cinematographic representations. However, some film theorists have tried to determine the implicit discourse level of filmic narrative. For instance, François Jost (1987:15) argued that the picture of a house does not signify "house," but rather "look, there is a house"—that is, a voice-like deictic communication. This sentence, however, is just a factitious verbalization of the autonomous perceptual function (showing). It does not even cover the reflector mode of a subsidiary voice function, which I would rather indicate through "There was a house," or "It is a house," i.e., a clearly declaring kind of sentence. So one question is: how does the demonstrator of events become a declaratory instance (narrator) in the mind of the spectator? Another even more basic question, which shall be treated first, is: how can the observer of events become a demonstrator at all? Jost's suggestion of an implicit deictic speech act implies agency. Correspondingly, the recent debates about whether films have narrators (for surveys see Lothe 2000; Smith 2000) center on narration as agency. I do think that films produce a psycho-poetic mental concept of narrator in the viewer, but I do not believe that this mental construction arises directly from the perceptual instance, because the personified I of the camera primarily is an observer, not a demonstrator. To understand the perceiving instance as an authorial instance of demonstrator already *presupposes* the psycho-poetic effect of a para-social instance rather than being the source of it.

As I said above, the psycho-poetic effect of a narrator, as a subject of attitudes, is only loosely connected to the literary voice instance (that is, only because of its continuous presence), but is not identical with it. So why preclude the theoretical possibility that in films it is the continuous presence of the perceiving instance which occupies the default position of a narrator? In fact, even the literary perceiving instance can sometimes give rise to anthropomorphic mental constructions. But are these anthropomorphic instances communicative instances at the same time?

Vocal calls have a communicative function in all animals, and that is why the linkage of voice with communicative functions, and a thusly implied social agency, is entirely intuitively plausible. Perception, however, as a simulation of individual experience, does not per se imply communicative functions or social agency. Rather the opposite: the illusionary "reality effect" of a differentiated perceptual system is much stronger in cinematographic representations than it could ever be in literary ones. Consequently, the authorial instance of a mediator is much easier to forget in filmic representations. The mental representation of an authorial instance would depend on a rather sophisticated awareness of the artifact character of a film. While the cognitive scope syntax ("This is what someone told me to be true") is continuously busy in verbal communication, there is no intrinsic need to activate this faculty toward experiential simulations, except that the filmic representation overtly deviates from our perceptual habits. Naïve spectators, I would say, do not reflect on the authorial aspect of camera unless they are urged to do so; the unconditioned presupposition that the represented perceptions are shown by someone would rather be a culturally learned mental attitude toward filmic representations.[11] Yet if the supposition of an authorial instance is already conditioned by other means, it should be quite normal to understand the perceptual function as a communicative device of showing. So what other means are there to precondition the psycho-poetic effect of an authorial instance?

Narrative fictions often involve real world references and, by this, interfere with our own general knowledge about the world. The filmic depiction of real world elements thus might entail an "implied hypothesis" of the kind "it's like this, isn't it?" as Murray Smith (2000:164) suggests for his especially clear example of Patrick Keiller's depiction of the city of London (in *London*, 1993). This effect of understanding implied statements about the world may be reduced in more fictitious genres such as for instance romantic love comedies, fantasy films, or horror movies, which often do not refer to historically real times, places, or persons; but even those genres bear implicit statements about "that's what humans are like" and thus give rise to a psycho-poetic effect of the authorial kind.

Another important device to imply authorial instances is the plot. So far, I have treated narrative fictions (literary or filmic ones) as more or less pointless representations of events in a temporal sequence. Yet recent attempts to give a definition of narrative note that it consists in a representation of a *non-contingent* sequence of events (Eibl

2004:255). The non-contingency of narrative sequences is often ensured by simple plot schemata, like, for example, that of "separation and re-union" (see Eibl 2008), "searching and finding," or "punishing the vicious/rewarding the good ones," and we tend to derive meaning from these plots, that is, to infer messages from narrative sequences. The principle of poetical justness is especially clear in this respect: failure signifies wrongness, success signifies rightness. That is, the sheer sequence of events implies hypothetical moral issues. Or, more generally speaking: also the non-contingency of narrations implies hypothetical propositions about the world and thus refers to someone making such a statement.

From interpreting represented perceptions as transmitting a certain kind of knowledge, it is just a small step to other communicative acts such as persuading, suggesting, deducing, arguing, and whatever the genuine functions of verbal communications might be. This is how the common idea of a filmic language could have come into being. Though the metaphor of a language of film does not provide a solid ground for an articulated theory of film, as Gregory Currie (1993) and others have argued very plausibly, it does make sense in that it signifies the com-municative quality of films. The voice instance implied by the metaphor does not literally perform a verbal speech act, which could be analyzed with linguistic instruments, but it expresses the feeling of intentional-ity which also voice-less fictions induce in the recipient, and thus makes it possible to speak of cinematographically represented stories as of narratives.

That noted, we can settle both literary and filmic fiction somewhere on the same continuum of narrative. Both commit declarative acts, the one directly by continuously verbalizing the perceived events, the other indirectly by depicting and arranging them in a way that appeals to our general knowledge about the world. The fact that they make different use of the two systems of information gathering (verbal communica-tion, individual experience) does not determine their capacity to install an authorial narrative instance communicating with the recipient.

Taking film as a representative example for nonverbal narrative, it can be regarded as quite possible that the human species developed narrative skills even before the gift of language. Displaying real world elements in any kind of representational system—be it pictorial, dra-matic, musical, or anything—and intentionally arranging them might correspond to what Abbott (2000:250) described as the "literature ef-fect" of (nonverbal) storytelling: the ritualized staging of mimetic representations as replicable units. And as such ritualizations are another way of reifying information by isolating it from its primary situational contexts, even the linguistic sophistication of third-person narratives seems to be nothing but a later affiliation of an already nonverbal cognitive faculty. What is significant for this faculty of reification is not its particular medial condition but the fact that reified information seems to need a social instance of sender to whom we ascribe a certain message and attitude. Storytelling thus seems to intrinsically be a form of social interaction, and Scalise Sugiyama (2005), though neglecting nonverbal forms of storytelling, seems to be right in assuming that storytelling evolved, in the first instance, to transmit information. Literary voice and perception as observed by Genette represent only the instruments to serve that purpose; they provide apt material for cognitive programs that are designed to extract and restore information, but they at the same time only answer to the need of sharing common information among group members.

Notes

1. This has led to a broad controversy on his notion of "focalization" (among others). For a recent survey on some major discussions see Bortolussi and Dixon 2003:166–178; see also Genette's own responses to early critics in Genette 1988:64–78.

2. Bortolussi and Dixon (2003:172, 174–176) argue that Genette's technical distinction between "voice" and "perception" does not find its equivalent in the mind of most real readers, who tend to synthesize them in their imagination. While I principally join Bortolussi and Dixon's endeavor toward a reader-oriented theory of narrative, my conception of the reader in this theory is not an empirical one, but rather a theoretical one. I posit an "anthropological model reader" (Mellmann 2006:21) that can be understood as a compound of several psychic mechanisms postulated by evolutionary psychology. This conception is not determined by the question of what of the complex cognitive operations is accessible to the introspection of the reader himself, so I cannot agree with Bortolussi and Dixon's conclusion that the analytic distinction of enunciator and perceiver "loses . . . relevance" (172) if not mirrored by interviews with real readers.

3. Terms like that of "perceiving instance"/"enunciating instance," "perceptual situation"/"situation of enunciation," and the like, and what I will introduce in the following passages, were developed in Mellmann 2006:164–204.

4. Chatman's (1990:139–149) distinction between "slant" (in respect of the "reporter" of a story) and "filter" (in respect of "observer" of a story) represents a similar endeavor as mine is here. However, I think he is mischaracterizing Genette when he treats his term *focalization* as just another synonym for *point of*

view, pertaining to both voice and perception. Genette (1980) in fact develops questions of voice and narrator in a separate chapter, while focalization affiliates to the chapter "Mood."

5. I avoid Genette's term *external focalization*, because it makes no sense but in the presence of characters, and then it is rather an antonym to "internal focalization" than a stand-alone type of focalization. Both internal and external focalization show the same (humanlike) degree of scope wideness. "External" focalization can be said to be the default case of this spatio-temporal scope, whereas "internal" denotes the particular case when "the focus coincides with a character, who then becomes the fictive 'subject' of all the perceptions" (Genette 1988:74). But imagine a phrase like: "Leaves tussled by the wind outside the window of Miss Barrett's empty room." It shows the same "situated focus," but there is no character aboard which the focalization could be "external" to. (For further examples see Mellmann, forthcoming.)

6. I emphasize this because it is often said that, after Genette, there are three forms of focalization: "zero," "internal," and "external focalization" (see for instance Martinez and Scheffel 1999:64). First, as I mentioned above, Genette did not aim to give a complete system (wherein three would make a whole), but, in the respective passages (Genette 1980:189–194), built on Stanzel's typology in order to introduce his own concept of focalization. Second, "zero focalization" strictly spoken is no *type of* focalization, but the opposite; unless one takes very wide scopes (like that of "tohubohu/ages") as omniscient views, like Genette (ibid.; 1988:74) himself was prone to do.

7. Genette (1980:162–164) himself integrated this distinction in terms of different grades of "distance"—a concept which then has to be applied to different narrated objects (events, speech, etc.) individually. I hope to give a more economical, systematic solution of the problem. Moreover, I principally avoid the spatial metaphor of "distance" because of its vagueness. Beside its literal meaning of near/far it assembles a great deal of figurative meanings like temporally near/far, directly accessed/mediated, pure/commented, true/stylized, sincere/ironic, positively/negatively evaluated, or even elevated/prosaic. Though all the aspects indicated by these oppositions might bear some relation to the question of mimesis and diegesis, they do not seem very useful for a clear-cut definition.

8. In Niklas Luhmann's system theory, "structural coupling" (after Humberto R. Maturana and Francisco J. Varela) signifies the interrelation between two autonomous ("self-referent") systems which represent "environment" to one another (i.e., do not overlap or interpenetrate), yet can be "irritated" by one another. Applied to our narrative model: The perception system cannot produce attitudes, opinions, or anything that would require more than a sensual system (seeing, hearing, smelling . . .). But it can change its orientation in reaction to an irritation from the voice system. Take for instance the change from external to internal focalization, which can be interpreted as reaction to the proposition "each was surprised"; as if the perceiving instance wanted to focus on what the voice instance had just picked as the central information. The fact that the voice system chose this of all information is due to its own logic: a particular intention of depiction. The voice system itself cannot produce percep-

tions (events), but only words, but it can by this produce a particular selection and appearance of events, which can entail particular focalizations on the part of the perception system again, and so on.

9. See Fludernik (1996; 2003) who observed a dominance of the "teller frame" in pre-modern literature.

10. For a detailed study on emotional triggers in literature see Mellmann 2006.

11. However, once one has taken this attitude, it is again quite natural to follow the social implication of showing. In doing so, we rely on a universal human predisposition, which in developmental psychology is known as the ability for "joint attention." Humans, on a much larger scale than other primates, follow another one's gaze (perhaps the higher percentage of the white in the eye co-evolved together with this endowment), listen if another one seems to be listening, and so on. That is why, once we assume that the filmically mediated perceptions are not our own but those of another, we are still interested in them and are quite prepared to pay to them as much attention as we would to our own perceptions.

CHAPTER SIX

Dreaming and Narrative Theory

RICHARD WALSH

If it is a matter of common experience that there is something uncanny about dreams, this is more than usually true for narratologists. From a narrative point of view, dreaming in itself is both familiar and alien: on the one hand the virtuality of dream experience has long been invoked as the archetypal instance of immersion in a fiction; on the other hand, this same sense of dreaming as hallucinatory experience would seem to disqualify it from consideration as narrative at all.

The ambiguous status of dreaming, as experience or as narrative, is the starting point for the argument of this chapter, which has two stages. The first is to make the case for viewing dreams as narrative; the second is to show that if you accept the narrative view of dreams, there are far-reaching consequences for narrative theory. Of course the need to confront these consequences does not arise if you are unpersuaded by the first stage of the argument, and my fear is that the narrative view of dreams may encounter resistance on the grounds, precisely, that the consequences are unacceptable. But it is a tautological way of thinking that defines its theoretical objects as only those things that fit the theory. I have sought to overcome this resistance (this dream censorship, if you like) by considering the merit of a narrative view of dreams, in the first instance, without regard to consequences.

Daniel Dennett has made a provocative case against the "received view" that dreams are experiences that occur during sleep, from the perspective of an intentional theory of mind; but his argument ultimately focuses upon the hazy boundaries of what counts as experience rather than any sense that dreams are narrative representations. Dennett distinguishes three components of dreaming implied by an experiential model: a (normally unconscious) process of composition; the presentation of the dream (to the experiencing mind); and its recording (in memory, for possible later recollection). The first and third of these, he suggests, may sufficiently account for dreaming without the second, in which case the sense of dream experience is not in fact primary but the retrospective product of a memory trace. That is to say, though he doesn't do so explicitly, the sense of dream experience is a product of narrative representation (1981: 132–137). Dennett's approach conflates an argument against the idea of dreams as (hallucinatory) experience with one against there being any such thing as experiencing a dream. The latter argument seems unwarranted, and indeed disconfirmed by more recent dream research, most obviously by research on lucid (self-aware) dreaming. Dennett accommodates lucid dreams as follows: "Although the composition and recording processes are entirely unconscious, on occasion the composition process inserts traces of itself into the recording via the literary conceit of a dream within a dream" (138). On the face of it this is a plausible move, but in fact it doesn't capture the specific quality of lucid dreams at all: it says, in effect, that you weren't really aware that you were dreaming—you just dreamed that you were. I shall return to lucid dreams, and some of the research results that conflict with Dennett's account, later in this chapter; but for now I want to suggest that a modified version of his argument remains useful for the purpose in hand. Rather than treating the memory trace of a dream globally, as a narrative product only experienced retrospectively (on waking), we can conceive of it as a narrative process, the experience of which is ongoing and recursive for the dreamer. Experiencing a dream, in that case, is experiencing a narrative process: a reciprocal process of creation and reception.

The narrative view of dreams requires a representational discourse: Manfred Jahn, reviewing the status of dreams in response to a question raised by Gerald Prince, comments that "hallucinatory perception, like real perception, cannot be (a) narrative. However, if Freud is right and dreams are the product of a fiction-creating 'dreamwork' device, then they are based on a multimedial mode of composition much like that of film" (Herman, Jahn, and Ryan 2005: 126). The appeal to Freud here is perhaps unnecessary, the notion of a dreamwork being the least specifically Freudian element of his theory of dreams; and if we confine ourselves to the aspect of the dreamwork he labels "secondary revision" (that is to say, the effort to impose order and coherence upon the dream materials), then it doesn't even presuppose the necessary existence of unconscious "dream thoughts" as the obliquely articulated content of the dream (Freud 1976: 628–651). The narrative view of dreams I want to present is broadly based, and while it can accommodate a Freudian interpretation of the source of dream materials, it does not depend upon any such interpretation. The notion of the dreamwork is helpful, and I shall return to it; but the narrative approach is better founded upon an appeal to the work being done on dreams within the context of the cognitive sciences.

The main lines of debate in dream research over the last few decades have been structured around a confrontation between psychological accounts grounded upon mental functions and physiological accounts grounded upon brain chemistry. The debate has been as much about the questions worth asking (and the research worth funding) as the nature of dreaming itself; many points of apparently intractable difference might equally be regarded as complementary, and indeed there has been a significant convergence of views on several key issues in recent years. Representative of the psychological perspective is David Foulkes, for whom dreaming is the operation of reflective consciousness in sleep; a champion of the physiological perspective is J. Allan Hobson, for whom dreaming is best referred, on an activation-synthesis model, to sleeping brain states (Foulkes 1999; Hobson 2002). For the purposes of my argument here, it is worth noting that dreaming according to Foulkes is definitionally representational and narrative rather than experiential, because it corresponds to our waking consciousness of our experience, not to that experience itself (1999: 3). Dreaming is not reducible to stimulus-response because it is creative; it is an aspect of our "reflective ability to think in images" (15). He argues accordingly for a cognitive equivalence between dreams and memories—or, more specifically, "conscious episodic recollection" (145).

Hobson, on the other hand, places much more emphasis upon the dissociation of dreaming from waking consciousness, and particularly the cognitive deficiencies associated with dreaming: "diminished self-awareness, diminished reality testing, poor memory, defective logic . . . inability to maintain directed thought" (2002: 111). These features, he argues, correlate with distinctive states of various regions of the brain during REM and non-REM sleep, as revealed by brain imaging (108–115). Hobson's emphasis upon the chemically distinct conditions of dream consciousness supports his analogy between dreaming and delirium (101), but the analogy does not ultimately resolve into a theory of dreaming as hallucinated experience. This is because the brain activation element of the theory is necessarily complemented by an element of synthesis, which explains how the hallucinatory and emotional effects of brain activation are integrated in a more or less coherent, novel, and personally meaningful way (47). Hobson is cautious about using the word *narration* because he equates it with language and therefore with dream reports, whereas "dreams themselves are experienced more like films. They are multimedia events, including fictitious movement. . . . Thus, we use the term 'narration' advisedly to signal the coherence of dream experience, which is all the more remarkable given the apparent chaos of REM sleep dreaming" (146–147). This is indeed (multimedia) narration, however, and the usage is supported by another descriptive term that Hobson offers as central to the dream-delirium analogy, which is *confabulation*—the psychiatric term for the fabrication of imaginary events as compensation for a loss of memory (101).

Both Foulkes and Hobson define their object of study in terms of the formal features of the process, dreaming, rather than the content analysis of the product, dreams. Both are careful to disentangle the features of dreaming from those of sleep, and REM sleep in particular, and in considering the evolutionary role of dreams, both allow that they may be merely epiphenomenal. REM sleep itself is evidently essential to life in mammals, but for Foulkes dreaming emerges too late, phylogenetically and ontogenetically, to be integral to the basic adaptive functions of sleep, and is probably an incidental by-product of the intersection of two phenomena which are themselves clearly adaptive, consciousness and internally generated cortical activation in sleep (1999: 137–141). For Hobson the brain activity associated with dreaming is adaptive, serving to reorder and update our memory systems, irrespective of dream recall, which may simply result from the circumstantial intrusion of this process into consciousness (2002: 87–88). These considerations do not exclude the possibility that there are cognitive benefits to dreaming, or indeed that any such benefits may be closely related to narrative competence (I'll return to this idea later). Without prejudice to such possibilities, though, the narrative view of dreams can be further consolidated by noting that both Foulkes and Hobson introduce the role of cognitive processing rather late in their accounts. They appear to confine the cognitive dimension of dreaming to reflective consciousness about dream phenomena, or to a synthetic role in the integration of such phenomena. Yet inasmuch as a dream element is recognizable at all (a cigar, say, rather than just a pattern of light), it is not merely phenomenal but perceptual, and already a cognitive product. This is true of all percepts, of course, whatever the phenomenal stimulus, so it doesn't in itself mark any departure from the realm of experience. The essential difference arises with the possibility that the perceptual apparatus may function semiotically, as a representational medium.

I take my semiotic frame of reference from C. S. Peirce, although there is some ambiguity as to whether Peirce himself regarded percepts as fully semiotic, partly because there is a discernible change in his thinking on this point between earlier and later writings, and partly because his usage of the term *percept* appears to be inconsistent (Bergman). He distinguishes the percept from, on the one hand, the "phaneron" or sensory phenomenon, and on the other, the perceptual judgment; the percept as interpreted in perceptual judgment he designates as the "percipuum." However, the percept is known *only* as mediated by perceptual judgment, and Peirce seems sometimes to use the term *percept* to refer to the complex as a whole (Bergman 17–18). In this broader sense the percept is fully semiotic: representational, intentional, and communicative (to the future self). It has been argued that even in the narrow sense the percept should be understood as an iconic sign (Ransdell 1986), but that is incidental here because the internally generated percepts of dreams are necessarily percepts under the interpretation of perceptual judgment, which is in effect the base level of the process Freud called "secondary revision."

One aspect of the iconicity of percepts that merits further comment, though, is their place within the evolutionary hierarchy of signs favored by (for example) Terrence Deacon. In the evolutionary model, the *icon* (of which the percept is the paradigm) is the most primitive kind of sign, characterized by a present relation of similarity to its referent; the *index* still involves a present relation, but is merely associative, as in a conditioned response; whereas only the *symbol* proper, product of a sign system, functions in an absent relation to its referent (Deacon 1997). The hallucinatory percepts of dreams, however, do indeed function in the absence of their referents: they are generated, once we get beyond the initial stimulus of unspecified brain activity, out of the cognitive repertoire of mental imagery upon which the dreamer draws in the sense-making effort Freud called secondary revision, which is—to reiterate—an integral part of the dream formation. The "secondariness" of secondary revision is relative to a Freudian primary process of "dream thought" representation; the distinction is muddied somewhat, though, by Freud's occasional application of the term to a further stage of revision at the point of reporting the dream (1976: 658–659). Similarly, the narrative status of dreams is obscured in much psychological writing on the subject by the standard methodological distinction between the dream experience and the retrospective, narrative dream report (cf. Hobson's reservations above). Nonetheless, all dreams are post-cognitive productions: that is to say, all dreams we are in principle capable of recalling, granted that this is not the sum of measurable brain activity during sleep—but that means all dreams, in the generally accepted sense of the word. In this respect, as Foulkes argues, they are directly comparable to memories, which draw upon the same resources of mental imagery. The apparently qualitative difference in the experiential characteristics of the two—the evanescence of memory versus the perceptual intensity of dreams—can be seen as a difference of degree, not kind: the mental imagery of most dreams is more vivid than that of most memories because any inhibiting awareness of our actual somatic sensory environment is radically attenuated in sleep. As cognitive applications of mental imagery, dreams and memories are discourse; and most of the dreams of most dreamers are narrative, just as episodic memories are narrative, by which I mean simply that they represent discrete temporal experiences: they articulate human time. Accordingly, the element of dream formation that I have until now referred to using Freud's term, "secondary revision," can be redescribed, if with some avowed over-generality, as the process of narration.

The salient difference in kind between a memory and a dream is not that one is true and the other is false: there is such a thing as false memory, and dreams may represent actual experiences, without detracting from the integrity of either mental activity. The difference is that the generative principles of each are antithetical in a crucial respect: the dominant cognitive imperative of memory is its representational adequacy to prior experiential fact, however much that imperative may be co-opted by subjective interests in the particular case; whereas the dominant cognitive imperative of dreaming, however much it may involve representations of prior experience, is the satisfaction of present mental needs (some very obvious and general, such as the expression of desires or the management of anxieties, others rather more obscure or circumstantial). These cognitive drivers, I suggest, are of complementary rhetorical kinds: that is to say, the fundamental distinction between dreams and memories is not between falsehood and truth, still less between illusion and experience, but between fiction and non-fiction.

Perception is a representation of the world: that is, experience, but it is also the foundation of the cognitive narrative faculty, the products of which are available to (episodic) memory. The salient feature of dream

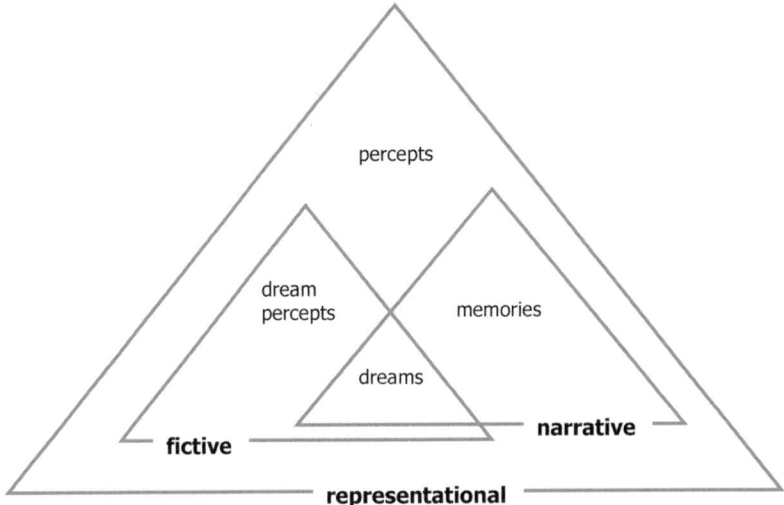

Figure 6.1.

percepts is that they are *fictive* representations, and that of dreams is
that they are fictive *narratives*. These four concepts and three qualities
intersect as in Figure 6.1: percepts are representational, but not fictive
and not narrative; memories are representational and narrative, but not
fictive; dream percepts are representational and fictive, but not narrative; dreams are representational, fictive, and narrative.

Where is the self in this narrative view of dreaming? There appears
to be a tension between the "I" who experiences and participates in
dream events and the "I" who produces the dream. But it is important
to recognize, first of all, that the experiential "I" of dreams, whether as
agent or witness, is itself part of the dream, a product of the process of
dream formation or narration, and a contingent part at that, since there
are dreams that include no self-representation. Second, and in general,
the self is the subject as object: there is a reflexiveness inherent in the
concept that explains why it is impossible for the subject to be entirely
present to itself, for that would exclude any position from which, and to
which, to present itself. Even a minimal sense of self involves reflective
consciousness, which involves some displacement of the subject. Dreams
are by the subject and for the subject, with the same ongoing reciprocity
as waking thought; and they are of course always in some sense about or
of the subject, if not always representationally of the self. The perceptual rather than conceptual nature of self-representation in dreams does

not amount to the primacy of experience, therefore: it is simply inherent in the fact that the medium of dreams is the perceptual system itself.

There is another sense in which reflective consciousness is at play in dreams, which is that normally associated with lucid dreaming. The essence of a lucid dream is the dreamer's *awareness* that it is a dream, a representational use of the perceptual system rather than an attempt to assimilate primary phenomenal data. It may follow from such awareness that it becomes possible to consciously direct the course of the dream, and so lucid dreaming is often understood to include this feature as well. The ability to recognize that one is dreaming can be cultivated, and this has made possible extensive long-term studies of lucid dream experience under laboratory conditions (LaBerge and DeGracia 2000). This research confirms that "lucidity in dreams is not a discrete phenomenon, but that reflective consciousness exists in all dreams and can be measured on a continuum with 'lucidity' and 'non-lucidity' representing two ends of the spectrum" (2000: 269–270). Three components of the lucid dream context may be distinguished, as follows: a metacognitive context, which is reflective consciousness itself; a semantic context, which is the framework of knowledge and belief within which the dream experience is understood; and a goal-option framework within which intentional action becomes possible. Metacognition can occur in all dreams, but whereas in most dreams it is articulated with reference to the dream context itself, in lucid dreams proper it includes waking contexts as well, providing for an awareness of contrasts between the two (274–275). Even the latter form of metacognition is not wholly exclusive to lucid dreams, however: a tacit awareness of such contrasts is a general feature of dream experience, being latent, for example, in the sense of strangeness that accompanies many dreams.

The semantic context of lucid dreaming determines the scope of response on the dreamer's part. Straightforwardly, this amounts to the generic expectations invoked by the recognition, "this is a dream," though it may be assimilated within different generic frames of reference, such as "out-of-body experience" or "astral projection" (275). Such alternative ways of conceptualizing the experience bring with them different expectations, and so tend to inhibit or privilege different kinds of represented action. But this is also true, of course, for different senses of what "this is a dream" might imply. Most narratologists, I'm sure, are aware that they tend to have narratologists' dreams; and lucid dream researchers are similarly influenced by their own preconceptions. The kinds of intentional action emphasized in lucid dream research are clearly oriented toward the context of the laboratory-based study of lucid dreams: they frame the general goal-option context in terms of voluntary choices of action within the dream environment, with particular emphasis upon deliberate metacognitive checking of state of consciousness (i.e., techniques for consolidating and sustaining lucidity), and the recall and performance of pre-planned actions in order to further experimental objectives (276, 282). These latter actions include, for example, making emphatic upward glances, a signal designed to exploit the laboratory apparatus for monitoring REM sleep as a channel of communication; and counting to ten between upward glances, in order to make a verifiable estimate of the passage of time within the dream. (The effectiveness of these strategies is unclear: certainly the eye movements during REM sleep do not generally map well onto dream content.)

It might seem that, in principle, the onset of lucidity in a dream should make anything possible. In fact it has proved far easier, at least within the context of these experimental assumptions, to exercise control over the represented self—even beyond the limitations of natural laws—than to consciously affect the dream environment. It is perhaps unsurprising that the creative freedom of the lucid dreamer should be so circumscribed, however, for several reasons. The first, as I have already suggested, is that the semantic context within which dreaming is conceptualized has parameters of its own—for example, a tendency to discriminate sharply between the experiential self of the dream and the dream environment. The second (which is in part a pragmatic justification for the first) is that the precondition of lucidity is reflexivity, and it is therefore best sustained by focusing upon the dream self as the representational embodiment of that reflexivity. Third, the precondition of any narrative dream whatever is some degree of coherence and continuity, a cognitive effort which necessarily draws upon the cognitive contexts of waking life. And finally, there is a curious double relation between lucidity and immersion which is worth closer consideration.

Metacognitive awareness of the dream state is justifiably associated with a degree of detachment from the perceptual-cognitive experience. Indeed, the need to control and contain overwhelming experiences of fear or anxiety, for example, is one of the commonest reasons for the onset of lucidity: it's all right, it's only a dream. And although lucid dream experiments have proved that "some sensory experiences are well modelled by the brain in the absence of primary sensory input" and can be voluntarily induced, the sensation of pain in particular is much less accessible (296). In this respect, lucidity appears to be a defense against the risks of immersion in a dream, allowing the dreamer to hold it at arm's length and keep out of harm's way. On the other hand, lucidity strongly correlates with immersion in the sense of a participatory involvement in the dream environment, rather than detached observation of it—or in narratological terms, homodiegesis rather than heterodiegesis. Similarly, the perceptual environment of dreams in general may range from minimal realization, in which sensory qualities are mostly absent or attenuated, up to typical sensory perception and even beyond, to vibrantly psychedelic experiences: lucid dreams are typically at the higher end of the scale, appearing more perceptually vivid than other dreams, and accompanied correspondingly by relatively intense brain activation. A reasonable inference from the association between lucidity and psychedelic dreams would be that lucid dreams, far from being more superficial and detached, often bring to perception deeper neurological processes than usual (285–256). These conflicting views of the relation between lucidity and immersion suggest that the latter is a rather catch-all concept, and that it is important to distinguish between different senses of the term.

Immersion within a simulation, whether a physical environment, a technologically virtual analogue, or a mental model (a thought experiment, say), is an experiential matter in the sense that it provides for agency, action, and reaction, within the limits defined by the parameters of the simulation. Specifically narrative immersion is generally understood as a special case of mental simulation, provided for by mimetic representation and necessarily constrained to a passive, receptive stance by the determinate nature of narrative. Note that this account of narrative immersion concerns the consumption of narrative: narrative creativity, on the same basis, is understood as a prior, authorial run through the simulation, of which the narrative itself is the product or trace (see Ryan 2001: 110–114). Dreams, it is worth insisting, are creative: but it is not plausible to define that creativity as the tracing of a path through the preconceived parameters of a simulation. Dreams necessarily unfold within the terms of some situational premises, without which they would have no coherence or stability at all, but they are remarkable precisely for the fluidity with which these parameters can change in the course of the ongoing dreamwork, and in accordance with ideas emergent out of the representational particulars of that activity. Simulations and narratives are in a crucial sense antithetical, as ludologists insist: simulation is a top-down concept, a modeling of the logic and general laws defining an environment, whereas narrative representation unfolds as a bottom-up process, via particulars. Because dreams use the perceptual system as their medium of articulation, they lend themselves to the assumption that they conform to, or even are paradigmatic of, an experiential concept of immersion in a virtual environment. But the perceptual system in this context is discursive and generative—its semiotic capacity harnessed creatively—so that the sense of immersion here is not experiential, but itself a semiotic product, and therefore in no way incompatible with the foregrounded awareness of artifice that constitutes lucidity in dreams.

If there are good grounds for regarding dreams as narrative fictions, however, the implications of doing so are far-reaching for narrative theory. Dreams elude many basic narratological assumptions, perhaps because of the peculiarity of their circumstances, or perhaps because these very circumstances bring them closer to the nub of the matter than other kinds of narrative. At the very least, narrative theory should be able to accommodate dreams; and it might do well to learn from them.

Fictionality

The fictionality of dreams in itself resists explanation in conventional narratological terms. Fictionality is treated, by both the pretense model of fictive discourse in speech act theory and the fictional worlds model of fictive reference, as fundamentally a problem of truthfulness. In the first case fiction is distinguished from seriously asserted narrative; in the second, it is distinguished from reference to the actual world. Yet it seems bizarre, on the one hand, to conceive of a dreamer pretending the dreamwork, or of dreaming as pretending to remember; and on the other hand the contingency of dream representations—their ad hoc fluidity in response to the demands of the moment, however those may be understood—exposes rather starkly the cumbersome redundancy of a fictional worlds account of fictive reference. Dreams suggest a view of fictive communication that is not subordinate to directly assertive communication at all, nor anchored by the assumption of a global referential ground, but rather accountable to generative principles of relevance or salience. I have characterized the difference between dreams and memories in terms of rhetorical orientation: while memories spring from and are accountable to a criterion of representational adequacy to experiential fact, dreams have the same recursive relationship to the representation of subjective significance—to desires, anxieties, values; in short, to the realm of meaning. In the dream case then, the generative principle of relevance is a criterion of accountability within the sphere of discourse; accountability to the expressive and cognitive needs of the dreamer as they emerge and evolve in terms of the narrative process in train. The dream case is the fictive case: with the appropriate substitution of terms the principle of relevance operates in the same way for fictive communication in general, just as it encompasses, via the rhetorical antithesis I have outlined, the non-fictive, assertive communication against which fiction is commonly defined. In the non-fictive case, the theoretical framework for such a rhetorical, pragmatic model of communication has been extensively elaborated in terms of relevance theory (Sperber and Wilson 1995). Its implications for fiction, though, have not been widely recognized. A serious consideration of dreaming forces this theoretical issue, I think, because it insists upon a direct relation between narrative fictionality and imagination, and the elemental status of the latter as a mental faculty.

Narrativity

It will be clear that while a narrative view of dreams provides for a principled account of the (rhetorical) distinctiveness of their fictionality, it also insists upon the pervasiveness of narrativity as a feature of cognitive sense making. In particular, dreams foreground the sense of narrative representation as a process, and lend themselves to description in terms of, for example, Paul Ricoeur's concept of mimesis as configuration, a model which does indeed tend to encourage the equation of narrativity with panfictionality, as Ricoeur himself acknowledges (1984–1988: 1:267n1). But the ubiquity of narrative artifice is a consideration of a different order from the distinct pragmatic features of fictionality. Dreams exemplify most emphatically the way in which the general configuring activity of mimesis in Ricoeur's sense is performed, in the fictive case, only upon what the creative mind proposes to itself. It is not so much the application of a cognitive narrative faculty in order to make sense of certain particulars, as it is the conjuring of such particulars in the exercise of such a faculty. The raw improvisatory quality of dreams strongly suggests a view of fictionality as characterized by just such a reciprocal interplay of narrative particulars and general narrative competence: as the serious-playful exercise of the narrative faculty, which is to say both its use and its development. Indeed such a notion of fiction as a kind of cognitive exercise also has the merit of offering, without assuming, a broadly plausible adaptive rationale for the phenomenon of dreaming: narrative creativity in the absence of empirical constraints enhances our capacity to assimilate novel phenomena.

Story and Discourse

The distinction between story and discourse, or *fabula* and *sjuzet*, commonly rests upon a view of story as event-sequence, despite the fact that there is nothing storylike about events in themselves. If story cannot be any kind of (narrative) representation, which is discourse, neither can it be something intrinsic in the phenomenal world. This has been pointed out by Marie-Laure Ryan, who offers instead a view of story as a mental image, a cognitive construct (Herman, Jahn, and Ryan 2005: 347). The distinction between story and discourse, then, rests only upon a distinction between material and mental representations. But as the case of dreams makes forcefully apparent, the media of mental representations, whether perceptual or linguistic, are as semiotic as material signs, and the representations themselves are as specific. Mental representation is indeed discourse, and entirely compatible with the narratological sense of the term. To insist upon a story-discourse distinction on such a basis, then, is to make it something quite other than the distinction we were taught to think of in terms of the "what" and "how" of narrative representation. Dreaming, as an irreducibly cognitive instance of the fictive paradigm, insists that narrative discourse cannot be referred to an underlying, conceptually prior story or fabula. The recursive nature of the dreamwork as a creative process, on the narrative view, suggests a better concept of story as a product of the generative-interpretative feedback loop, as something contingent and progressive, at close range to the process itself. On such an account, story does not ground discourse, but arises from it, as an ongoing narrative interpretant.

The Narrator

The concept of the narrator as a distinct narrative agent is also undermined by a narrative view of dreams. It is perhaps plausible to understand some dreams as homodiegetic narration, that is to say as narrated by a self within the dream, but any such self is necessarily a representational product of dreaming consciousness like any other. Narration thus conceived, then, is subordinate to representation, and in no way consti-tutes a separate element of narrative transmission. More fundamentally, for those dreams without self-representation, there is no case for argu-ing that the dreamer makes sense of the dream (either in process or ret-rospectively) as the report of a distinct heterodiegetic narrator. Dreams are directly fictive and experienced directly, not framed as the discourse of a distinct agent for whom the events are known rather than imagined. Imagination, indeed, is precisely the relevant concept: it does not need to be redeemed by a dissociative framework providing for suspension of disbelief, willing or otherwise. Issues of belief and disbelief are not germane to the mind's capacity for narrative elaboration in itself, and of limited importance to the affective salience of such narratives. The fact that dreams may be lucid in varying degrees, and still be dreams, requires us to think of fictive narration as an act that can be directly owned by the creative imagination without being conflated with delu-sion or (self) deception. Dream lucidity is not an elaborately contrived metafictional game, but a possibility latent in the ordinary conditions of dream consciousness; not a reflexive framing of the dreamwork (as a dream within a dream), but simply one of the available cognitive con-texts within which it takes place.

Voice

Pursuing the communicative framework of dreams a little further, there is much to be gained from reflecting upon the concept of voice in this context—both because the perceptual medium of dreams helps to off-set the linguistic bias of the term *voice*, and because the interiority of dreams helps tease apart entangled senses of agency and selfhood in the concept. The dreaming subject is the agent of a narrative act that may itself include representation of the self as agent, and as a discursive agent (in the broadest sense), as well as representation of the discursive agency of others. Retroactively, the dream discourse also implies and constructs selfhood around the subject position being established in the act of narration, the dream situating the dreamer in the process of its own construction and reception—because dreams can change you. The dreamer's internally dialogic relation to the dream, as both its producer and its consumer, foregrounds the dialogism of narrative discourse in general, but it also draws attention to the insufficiently recognized fact that such discourse does not only involve multiple voices, it involves multiple senses of voice. Voice, in narrative theory, may refer to a representational act (a narrative instance—the Genettian sense), a (discursive) object of representation (a represented idiom), or a representational subject position (in the perspectival or ideological sense the term acquires in, for example, feminist or Bakhtinian theory). The interplay between these crucially distinct features of narrative representation accounts for many of its subtlest effects, and theoretical discussion of such issues as free indirect discourse and focalization is much impaired by their conflation under a single term.

Medium

The medium of dream narrative is the cognitive-perceptual apparatus itself; narrative, on this view, is a cognitive sense-making faculty that proceeds from the outset in representational terms. Accordingly, there is no meaningful sense in which narrative can be thought of as medium-independent: no event, however minimal, is structured as such except in cognitive terms, and the same applies a fortiori for sequences of events. The possibility of remediation, or the transposition of narratives between media, is no grounds for attributing any abstract deep structure to narrative, since the sense of narrative sameness upon which that possibility depends requires cognitive articulation, and so is itself within the domain of mediation. The base level of mediation, then, is not the transmission of some already otherwise encoded meaning, but the inaugural articulation of meaning that is semiosis itself. Moreover, the semiotic nature of mental imagery carries inherent within it the possibility of metacognitive awareness, just as lucidity is latent in all dreams: the implication would seem to be that recursiveness is integral to sense making, and therefore that the formulaic reciprocity between the two views of narrative cognition formulated by David Herman—making sense of stories and stories as sense making (2003: 12–14)—is in fact irreducible to a hierarchical relation. Neither view can provide for a foundational concept of narrative within a cognitive paradigm. If the cognitive-perceptual apparatus itself is representational, narrative cognition is always the medium-bound articulation of meaning (that is to say, concurrently the creation and expression of structures of meaning), the only ground for which must be pragmatic efficacy within a given context. In the fictive case for which dreams stand as exemplar, then, the rhetorical direction of fit between meaning and ground is reversed:

such contextual criteria of accountability are no longer the final court of narrative, but its occasion, subject to the possibilities of imaginative elaboration and exploration, and qualified or refigured by that process.

Narrative Creativity

In dreams, narrative creativity and reception go hand in hand: "authorship" and "readership" are symbiotic. The narration proceeds in a continuously reciprocal relation with the dreamer's own evaluative response to it, because this response is the only constraint upon it, and entirely defines the parameters of its accountability. The specific line of narrative development is a reflexive negotiation between the current dream state at any given moment and the cognitive context of the dreamer's narrative competence. But isn't this the case for all fictions? Dreams present in naked form the interplay of the particular and the general that supplies the fundamental logic of all narrative creativity, and not just in an external global sense, but intimately throughout the creative process. Authorship is itself a kind of readership, a process of discovery, in the particulars of a conceit, of the "right" narrative development, step by step, in the context of a general framework of narrative understanding. Such authorial discernment is successful to the extent that the narrative understanding it draws upon is collective; to the extent that the author belongs to a community of readership. The communication model of narrative transmission, on this view, is seriously misleading: the meaning of a story is not conceived by the author, ab initio, to be transmitted in narrative form to a reader who attempts to decode it. Author and reader approach the emergent possibilities of a narrative from the same side—the author a little in advance—and share in the achievement of an understanding that is itself, from start to finish, narrative in form.

Affective Response

Our understanding of affective response or emotional involvement with fictions is also called to account by a consideration of the directly fictive nature of dream narration, together with the reciprocity of creation and reception exemplified by dreams. The affective power of dreams is not consequent upon an illusion of reality, nor upon an assumption of discursive truth, because it is an integral quality of the unfolding fictive representation itself as it is assimilated and generated by the dreamer. One phenomenon supporting this view is the persistence of immersion within lucid dreams, and the fact that lucidity, while certainly constituting a redirection of the dreamer's cognitive attention, by no means vitiates the emotional valency of dream representations. Or again, the discursive rather than experiential nature of the affective quality of dreams can be inferred from the common circumstance of a dream in which the mood is incongruous with the apparent narrative content. This discrepancy is highlighted whenever a dream report—a summary of the events of the dream narrative—does not satisfy the dreamer's own sense of its emotional force. It suggests that such affective qualities are not products of a discrete stage in the dreamer's reception of the dream, but integral to the representational qualities of the dream discourse itself, and the feedback loop of narrative creativity and reception sustaining that discourse. In other words, emotional valency, or affective value in general, is inherent in the process of (narrative) representation, rather than a secondary response to the products of representation. If this is so, the various ways in which narrative theorists and philosophers have sought to reconcile emotional involvement with fictionality are misconceived and redundant. Affective response does not depend upon suspension of disbelief or any equivalent framing of our engagement with fictions because it is inherent in, and continuous with, the semiotic process of representation, and need not wait upon an assessment of the reality or truth of its products.

It is hard to see how dreaming can be understood as anything other than a narrative process; but if this is so, dreams present a challenge to narrative theory on several fronts. I don't think this challenge can be minimized by regarding dreams as marginal phenomena, not only because they are a near universal feature of human consciousness and intimately tied up with our common intuitions about the narrative imagination, but also because they are inescapably key instances of (fictive) narrative cognition. The main thrust of my argument here has been to bring that cognitive frame of reference to bear upon narrative theory, but I hope it's clear that there is also a great deal to be gained from orienting our attention the other way. Narrative theory can bring a great deal to our understanding of cognition, provided only that it opens itself to the questions posed by that still enigmatic domain of inquiry.

PART II

A SELECTION OF NEW APPROACHES

Cross-Cultural Mind-Reading; or, Coming to Terms with the Ethnic Mother in Maxine Hong Kingston's *The Woman Warrior*

KLARINA PRIBORKIN

The Chinese mother and American-born daughter in Maxine Hong Kingston's memoir *The Woman Warrior* repeatedly misinterpret each other's intentions. This occurs not only because of the psychological complications typical of mothers and daughters in most nuclear families, but also due to specific socio-cultural misunderstandings. The initial hostility between the narrator-daughter who endeavors to assimilate into the American majority culture and her mother, Brave Orchid, who insists on a more traditionally Chinese way of life, slowly evolves into the daughter's understanding of her mother's customs and storytelling practices. In the course of Maxine's maturation and self-construction, she develops various imaginative and cognitive techniques that enable her to reinterpret her mother's character by retelling the stories of her own and her mother's lives. In order to come to terms with her mother, the American-born daughter creates a fictional mother figure from some of the "facts" she knows about her mother's past life in China. The woman who emerges from this creative process is an independent doctor and powerful exorcist, who battles against and triumphs over a powerful ghost. This empowered image does not match the mother's self-perception, not to speak of the way the daughter perceives her mother on a daily basis. Nevertheless, it does allow the daughter to identify with her actual mother by reconstructing the socio-cultural setting of her life in China. Through her own acts of storytelling, the daughter reconstructs her ancestral heritage and culture; this enables her to perceive her mother's life-choices and actions from a new and not necessarily Western perspective.

My approach here takes into account how distinct narrative strategies map onto cognitive needs, producing a story in which the rhetoric itself mirrors the cognitive work of the narrator in solving a problem of the American-born daughter coming to terms with her ethnic mother. Combining the cognitive and multicultural insights, I will trace the mental and emotional processes of the daughter's reconciliation with her mother and her ethnic heritage through storytelling. While the multicultural approach enables me to examine the dynamics of multi-generational assimilation processes in immigrant families, the cognitive perspective allows a closer look at the unspoken intentions behind the charged interactions between the mother and daughter in the memoir. By analyzing the mother's and daughter's respective assumptions about each other, it is possible to identify the reasons for both the pitfalls and successes of their inter-generational, cross-cultural communication and to draw some conclusions regarding the possibility for mutuality and cooperation between minority and majority cultures in America.

Cross-Cultural Misunderstandings: My Mother Is a Martian to Me

Cognitive psychologists and neuroscientists have hypothesized that humans possess the inherent ability to represent mental states of others by means of a conceptual system, commonly termed *Theory of Mind,* or "mind reading" (Gallese 2001; 42, Gallese and Goldman 1998). Our capacity to make assumptions and predictions regarding the mental states of others is also referred to in the research as mind-reading.[1] In their article "Social Cognition, Language Acquisition and the Development of the Theory of Mind," Garfield, Peterson, and Perry argue against regarding mind-reading as strongly modular and suggest, instead, that the development of mind-reading abilities depends both on successful language acquisition and social interaction. If making assumptions regarding other people's internal world is at least in part a social skill, we must recognize that different cultures posit different sensibilities and interpret gestures, facial expressions, bodily movements, silences, etc., differently. Since people from different cultures do not possess a common system of symbolic representation, they may easily misinterpret each other's mental states.

Lisa Zunshine claims, in *Why We Read Fiction,* that people universally pay a great deal of attention to such paralinguistic signs as gestures and bodily movements, automatically assuming that they are indicative of the underlying mental states of others (6). Yet, although *the attempt* to decode the meaning of these paralinguistic signs is universal, the *actual interpretation* of such signs is culture-specific. Expectations regarding specific external signals become a part of each culture, allowing efficient communication and facilitating communal bonding. Yet, in attempts at cross-cultural communication, what appears obvious and natural to one cultural group may be completely inexplicable to another. In interactions in which the shared system of universal signs is complicated by additional sets of signs, the interpretation of kinesthetic and paralinguistic signals may be fraught with difficulties and eventuate in failure.

The cross-cultural misunderstandings between Maxine and her mother, Brave Orchid, arise from Maxine's growing involvement in the American milieu. Once she starts attending an American public school, Maxine is gradually drawn away from her domestic ethnic environment, and her mother's worldview becomes foreign to her. Like the other children of Chinese immigrants, Maxine is expected to suspend her Chinese identity in order to adapt herself to America, a process which results in self-censorship. The daughter starts perceiving herself and her parents from an American perspective, subconsciously measuring her behavior according to the yardstick of American appropriateness. Maxine is embarrassed by her mother's behavior: "The immigrants I know have loud voices, unmodulated to American tones . . . I have not been able to stop my mother's screams in public libraries or over telephones" (*WW* 11). A stranger to her ancestral ethnic culture, Maxine is situated in a cognitive position that resembles that of her American readers.

Mutuality cannot be taken for granted in cross-cultural settings; even a mother and daughter may fail to understand each other, in spite of their repeated attempts to establish a constructive relationship. Cultural misunderstandings create such alienation that Maxine often fails to predict her mother's intentions. She imagines her mother doing awful and unpredictable things that would embarrass her. For instance, when Brave Orchid, distressed by the mistaken delivery of medicine to her house, commands Maxine to go to the pharmacy and make them "rectify their crime," Maxine feels sick because she thinks that her mother wants to "make [her] swing stinky censers around the counter, at the druggist, at the customers. Throw dog blood on the druggist" (*WW* 170). The mother, however, only asks her to "get reparation candy." Later Maxine explains that it is a custom in Chinese pharmacies to give raisins along with the medicine as a wish for good health. The gap between the mother's intentions and the daughter's original interpretation of her request testifies to the lack of shared cultural references, and undermines their confidence in their cognitive capacity to understand each other.

The frequent misunderstandings which occur in *The Woman Warrior* initiate and feed the daughter's growing hostility toward her mother. For instance, in the fourth chapter of the memoir, the mother is described opening the front and back doors of the house, mumbling words.

> "What do you say when you open the door like that?" Her children used to ask when they were younger.
> "Nothing. Nothing," she would answer.
> "Is it the spirits, Mother? Do you talk to spirits? Do you ask them in or ask them out?"
> "It's nothing" she said. She never explained anything that was really important. They no longer asked. (*WW* 121)[2]

Maxine's mother, assuming her children's ability to learn traditions by observation as she herself did, refuses to provide verbal explanations for her ceremonies, even when the children ask for them explicitly. However, since her behavior is devoid of its socio-cultural context, her customs simply appear odd and inexplicable to her children. What is perceived as obvious to the mother, born and raised in China, is foreign to her American-born children and to the Western reader. Brave Orchid relies on her children's mind-reading abilities, but since the generations do not share a common sign system, the mother's signals prove insufficient. Yet the children's inability to understand their mother's behavior without words is experienced by the mother as insulting. To have to spell things out explicitly strikes her as demeaning.[3]

The children lack what Endel Tulving refers to as semantic memory, something that enables us to have culturally shared, general knowledge not tied to learning experience. Such knowledge is obtained from our cultural environment, and is stored in our memory as objective, reliable information that we tend not to doubt (*Organization of Memory* 385–386). Yet the children's semantic memory, shaped by an American rather than a Chinese environment, makes it impossible for them to experience and store their mother's behavior as self-evident. Since Brave Orchid's American children could not attain a cultural knowledge of Chinese behavioral norms through constant observation of their environment, they require a different approach that is tied to a specific learning experience, referred to in the literature as "episodic." A verbal explication of their mother's customs would have enabled the children to store them in their memories as episodes that took place in a specific time and space, rather than as culturally accepted phenomena. Whereas the mother's verbal representation of her rituals could have enabled the children to develop such episodic memory, she refuses to explain, expecting her children to remember Chinese traditions as a generally accepted truth rather than as her particular ethnic representation of reality. It pains her that these expectations are not fulfilled.

In *China Men*, Kingston explains the difference between the natural learning of culture and the immigrant's conscious learning. The latter makes them aware of the differences between their home and the majority culture. In "Cross-Cultural Wordplay," Rufus Cook summarizes Kingston's stand: "If you are 'an authentic Chinese,' then you aren't supposed to have to learn the language or the proverbial stories: you are supposed to know them instinctively 'without being taught, [to be] born talking them'" (Cook 137). While the native Chinese store their parents' customs and stories as semantic memories, the Chinese-Americans, who early in life are made aware of the discrepancies between their ethnic and majority cultures, remember their parents' traditions as episodic learning experiences peculiar to their home environment.

The above mentioned problems of cross-cultural communication are gradually resolved as the daughter learns to use her imagination and narrative skills in order to come to terms with her mother's communication preferences. By reconstructing her mother's socio-cultural context, the daughter learns to understand the mother's motives and intentions. This allows her to retell her mother's stories from a new perspective, employing the narrative modes of free indirect discourse and thought report. These narrative strategies of thought representation, as Alan Palmer refers to them in *Fictional Minds*, indicate the daughter's evolving ability to read her mother's mind; Maxine is learning to understand the meanings implicit behind her mother's words. Cross-cultural mind-reading between the mother and daughter is achieved through a succession of cognitive processes that produce an empowering narrative in which these women achieve productive dialogue.

The narrator/daughter constructs her mother as a fictional character by imagining her mental states in relation to her socio-cultural setting. The mother's medical diploma, the old photographs of Medical College graduates as well as fragments of stories about her life and education in China, encourage the daughter to imagine her mother's life before she left China. This process, according to Palmer, is crucial for the construction of a fictional mind, since an "action cannot be separated from the milieu in which it is carried out" (158). It is only after imagining the circumstances of her mother's life in China that the daughter can make inferences about her mother's mental states and understand her intentions. Drawing not only on her mother's stories, but also on her own historical research, the daughter interprets her mother's cultural environment in terms of Western modes of thought—logic and palpable evidence—as well as through Brave Orchid's representations of her past, which the daughter had often misunderstood.

Maxine becomes capable of understanding the meaning of her mother's life-choices only by comparing Brave Orchid's reality to her own American environment and translating the mother's decisions into their more-or-less equivalents in the American framework. By reinterpreting the events of her mother's autobiographical stories within the Western contexts of feminism, individualism, and self-sufficiency, the daughter makes her mother's character accessible and understandable both to herself and to her readers. In the chapter titled "Shaman," for example, the mother is shown to be an independent, self-sufficient woman—a scholar, a doctor, even a ghost exorcist. This image is appealing to the daughter; although this story about the mother takes place in China, in Kingston's book it stands for American values of self-empowerment. Maxine's recreation of her mother's environment evokes the daughter's admiration for the woman who was able to leave her family's house, rebel against traditional female roles, and succeed in her studies. Brave Orchid is revealed as a progressive, somewhat radical, professional woman who can serve as a role model for her American-born daughter. In the context of the mother's life in China, leaving home for college can be interpreted as a heroic, feminist, even rebellious act, which the narrator/daughter can employ to identify with her mother.

The daughter opens her narration of the mother's autobiographical stories with Brave Orchid's education in the Medical College, rather than with her mother's childhood, adolescence, or even marriage to her father. She thus finds something in her own present experience that can match her mother's past experience. Since the mother's medical training fits the daughter's Western environment, Maxine can use it as a starting point for understanding her mother's mental states. In comparison to Brave Orchid's stories about the strange customs of her native village or about ghosts, the stories about the Medical College are easily accessible to Maxine. The daughter perceives her mother and the other female students as quite similar to herself; they are "new women, scientists who changed the rituals" (*WW* 75); they too pursue a common dream of independence and self-empowerment. Maxine can share this dream "about a carefree life" that allows women to have "a job and a room of their own" (*WW* 62). This reference to Virginia Woolf's feminism further demonstrates how the daughter enfolds Western ideas into the narrative she creates of her mother's life in China, thus emphasizing the goals and dispositions that the two women share.

As Maxine familiarizes herself and the reader with her mother's former socio-cultural environment, she imagines additional details which help her to understand her mother's current mental states. Since the mother regularly provides her daughter with only minimal details she feels necessary for her educational purpose, the daughter must fill in the gaps. Maxine's imaginative reconstruction appears as what Alan Palmer refers to as "thought report," a form of thought representation in narrative that is used to describe emotions and behavior by conveying motivation and intention (13). This flexible category is best used for "linking individual and mental functioning to its social context"; providing background, contextual information for the character's feelings, and mediating between the character's setting and the reader (54, 76).

Kingston uses a thought report, for instance, when the daughter recognizes the enormous change in her mother's life when she leaves her home for college: "Free from families, my mother would live for two years without servitude. She would not have to run errands for my father's tyrant mother with the bound feet or thread needles for old ladies, but neither would there be slaves and nieces to wait on her" (*WW* 62). Such words as *servitude* and *tyrant mother* in Maxine's description demonstrate her imaginative attempt to understand her mother's feelings about her environment. She assumes her mother was not happy in her role as a daughter-in-law, but realizes that this social status also had its positive sides. Although the narrator/daughter's representation of her mother's life seems to be quite authentic, her interpretations of her mother's feelings regarding her social role are based on Maxine's Western perception of the Chinese familial framework and, thus, do not necessarily match the mother's actual feelings. By means of thought report, the narrator/daughter introduces her own emotions about her mother's environment *as if* they were her mother's. This shows that the daughter substitutes or interchanges her own mental states with her mother's, either because she attributes her own feelings to the fictional representation of her mother, or because she is convinced that her mother's mental states would resemble her own in the described socio-cultural circumstances.

The mother/daughter identification is apparent not only through the daughter's report of her mother's thoughts, but also through free indirect speech employed by the narrator/daughter in order to intertwine her own narration of the present with her mother's language and subjectivity. According to Palmer, free indirect thought "combines the subjectivity and language of the character . . . with the presentation of the narrator," enabling the author to explore inner speech as well as other areas of the character's mind (54). Brave Orchid's thoughts are relayed in free indirect speech: "Her American children could not sit for very long. They did not understand sitting; they had wandering feet" (*WW* 113). By presenting the mother's worldview and discourse in third-person narration, the author combines the mother's and the narrator/daughter's voices, thus making it difficult for the reader to distinguish between them. Kingston experiments with the reader's ability to navigate what Lisa Zunshine refers to as "the multiple levels of intentionality" (28); that is, the text tells us what the daughter is thinking that her mother is thinking. While the narrator/daughter succeeds in reading the mind of her mother's fictional character in the novel, on the autobiographical level the daughter may still be misinterpreting her mother's intentions due to both cultural barriers and generational gaps.

Since the narrator/daughter only partially succeeds in reading her mother's fictional mind through thought representation, she turns to other narrative strategies in order to identify with her mother. Maxine's construction of her mother's character may be a necessary stage in the daughter's process of coming to terms with her, but this feeling of similarity is insufficient to allow the daughter to establish efficient communication with her autobiographical mother. As Winnicott suggests, in order to "use the other,"[4] or in Jessica Benjamin's terms, to "creatively benefit from another person," one has to enter into exchange with this other, putting this other outside of oneself. At first, the other is not experienced as real, external, or independent; it is experienced as a part of our mind, but there is a point in our development when it is necessary to acknowledge the other as separate in order to be recognized back (Benjamin 37).

According to Benjamin's Theory of Intersubjectivity, human beings construct their identity through recognition by another independent subject: "recognition is that response from the other which makes meaningful the feelings, intentions and actions of the self. It allows the self to realize its agency and authorship in a tangible way. But such recognition can only come from an other whom we, in turn, recognize as a person in his or her own right" (Bonds 12). Recognition is reflexive; it includes not only the other's response to our existence, but also our own reaction to this response. It is important for us to recognize ourselves and our mental states in the other, since it enhances our sense of effective agency. Therefore, when the daughter attributes her own mental states to her mother, convincing herself and her readers that she has managed to understand her mother's thoughts and narrate them, she actually reaffirms herself and her mental representational abilities. In this way, she comes to assert her own sense of authority.

However, the daughter must also learn to place the mother outside of herself, distinguish her from her own mental experience, and acknowledge that the mother is an independent person who is not subject to her daughter's control. The discrepancy between the mother's fictional character and Brave Orchid's own perception of herself, apparent in the dialogues between them, eventually leads Maxine to accept that her mother is a separate individual who also needs to be recognized as such. Unlike Maxine, who describes her mother as a woman warrior and a capable exorcist in the beginning of the chapter, Brave Orchid perceives herself somewhat differently. Although she depicts herself as "the one with the big muscles," she attributes her physical strength to the circumstances and necessities of her American life, where "a human being works her life away" (*WW* 104). Instead, interestingly, Brave Orchid's self-definition revolves around motherhood. She compares herself to "a mother cat hunting for its kittens. She has to find them fast because in a few hours she will forget how to count or that she had any kittens at all" (*WW* 105).

When Brave Orchid reveals to Maxine her constant concern about her children, the daughter acknowledges her mother's feelings and reassures her that she has succeeded as a mother, because she taught her "children to look after themselves" (*WW* 106). By recognizing her mother's efforts to do her best at raising her children, Maxine responds to her mother's appeal, thus establishing a positive atmosphere of mutuality, allowing her mother to recognize her daughter's independence. Although the daughter's initial identification with the mother is achieved through the narrator/daughter's cognitive and imaginative narrative construction of her mother's character, Maxine must also attempt reconciliation with her mother by engaging in a dialogue with her.[5]

According to Peter T. F. Raggatt's "Multiplicity and Conflict in the Dialogical Self," dialogical interaction of opposing voices in the author's mind is a necessary precondition for the construction of life narrative. Basing his argument on Mikhail Bakhtin's idea that "'the mind' is not a contained center but rather a product of dialogical relations" (17), Raggatt argues that a life story is constructed through a *conversation of narrators* in the author's head and therefore "could never be encompassed by a *monologue*" (15–16). Kingston indeed creates her life story by giving voice to herself through the dialogue with her own mother; their opposing voices produce the necessary drama and catharsis for the emergence of two separate fictional selves. The dialogue, therefore, serves for Kingston as a narrative strategy for simultaneous self-creation and reconciliation with the mother. Moreover, this mother/daughter verbal exchange coordinates the women's expectations and allows them to accept each other's self-definitions, thus mutually acknowledging each other. Kingston employs dialogue as a narrative technique for mother/daughter interaction in order to enable the establishment of an intersubjective relationship based on mutual trust.

Initiating Cross-Cultural Interaction: Mother's Narrative Strategies in Action

Although it may be suggested that the narrator rebels against her mother's authority by writing down her oral narrative and inventing alternative versions to the original talk-story, a more positive interpretation would see the daughter's memoir continuing the mother's tradition of storytelling. By using the techniques of her mother's oral narration to construct her written narrative, the daughter asserts the possibility of cross-cultural and cross-generational communication through artistic expression. Kingston's narrative, like her mother's, projects multiple versions and meanings. In fact the very ambiguity and the multiplicity of truths which the mother advocates allow the daughter to invent her own versions of the traditional stories. As Kingston explains in her interview with Islas and Yalom, the oral stories change from "telling to telling. [They] change according to the needs of the listener, according to the needs of the day . . . so the story can be different from day to day . . . [while] writing is static" (31). Kingston says that she would like "the words to change on the page every time, but they can't. So the way [she] tried to solve this problem was to keep ambiguity in the writing all the time" (Islas and Yalom 31). By maintaining the ambiguity of the text, by presenting multiple interpretations of the same story, the narrator/daughter imitates her mother's fluid narrative style that allows the storyteller to change the narrative while telling the story. This notion of multiple truths challenges the settled monological perceptions of Western culture, thus impelling Kingston's readers to look at events from the perspective of people who were marginalized by the dominant culture as ethnic others.

Although Brave Orchid's fluid narrative technique at first prevents Maxine from understanding her mother, the adult narrator/daughter adopts her mother's narrative style, along with its element of ambiguity and unreliability, which she rejected when she was younger. For instance, in the first chapter, the daughter retells her mother's story about an aunt who was intentionally forgotten by her relatives as a punishment for adultery and the aunt's consequent infanticide and suicide. The mother explains to Maxine that the family has denied the aunt's memory to such a degree as to "say that [Maxine's] father has all brothers because it is as if [this aunt] had never been born" (*WW* 3). In order to imagine her aunt's story in a way that will be meaningful to her, the daughter invents two opposing versions. While in the first story the aunt is portrayed as a victim of rape, in the second version she appears as a sexually liberated woman. Although the narrator admits that "imagining her [aunt] free with sex does not fit" (*WW* 8) her socio-cultural environment, the invention of the second version of the aunt's story enables the daughter/narrator to identify with her aunt as a sexually empowered woman. By retelling the mother's story in this way, Maxine makes an effort to reconnect to her dead aunt and to her mother's way of telling stories. Moreover, the multiple interpretations that the narrator/daughter presents to the reader signify the daughter's assimilation of the mother's ambiguous storytelling strategies.

Another vivid example of the narrator's freedom to the point of inconsistency is found in the fourth chapter, in which Maxine imagines how her mother and aunt confront the aunt's bigamous husband. The third-person narrator relates the events in great detail, including the character's thoughts and reactions in free indirect thought, thus creating an illusion of reliability. However, the fifth chapter begins with the narrator's confession: "What my brother actually said was, 'I drove Mom and Second Aunt to Los Angeles to see Aunt's husband who's got the other wife'" (*WW* 163). When the brother is further questioned about the trip, he provides only the basic details of the plot. However, even Maxine's account of the brother's account of the story turns out to be fictionalized; the narrator confesses again, "In fact it wasn't me my brother told about going to Los Angeles; one of my sisters told me what he'd told her" (*WW* 163). When the reader realizes that the narrator's original account of the aunt's trip to Los Angeles is three times removed from the real events, it becomes difficult for the reader to trust the narrator; yet it also allows us to experience Maxine's frustration regarding her mother's unreliable storytelling. We are shown how hard the daughter has to work in order to enter her mother's reality.

The inconsistencies in the narrator's storytelling technique undermine her reliability, producing a challenge to what Lisa Zunshine refers to as the "reader's metarepresentation abilities" (*Why We Read Fiction*, 47). According to Zunshine, metarepresentation, sometimes referred to as "a representation of a representation," allows us to store certain information/representations "under advisement," and thus carry out inferences on information that we know is incorrect (50). The readers can keep track of metarepresentations by means of "source tags" that allow us to track the mind behind the represented sentiment (89–90). Without these source tags, which signify the source of information we receive, it would be impossible to differentiate between reliable and unreliable accounts of narrated events. In *The Woman Warrior*, the source tags the author provides enable us to navigate between the different versions of the same story narrated alternately by the mother and daughter. In the story of the "no name" woman in the first chapter, for instance, the narrator signifies the sources of the information provided by means of such source tags as "my mother said" in relation to Brave Orchid's version, as well as "maybe" and "perhaps" when the narrator/daughter presents her own contrasting interpretations of the mother's original story.[6]

While in the first chapter of the memoir, the borders between fiction and reality are not completely obscure, in the fourth chapter, the narrator manages to "cheat" her readers by eliminating the source tags from her imaginative account of what occurred in Los Angeles. The detailed representation of the characters' minds misleads the readers. Since there is no indication in the text that signifies the fictional nature of the story, we tend to take it for granted that the narrator has been present when the relayed events took place, and therefore her account of the events is both reliable and autobiographical. The narrator, however, first creates an illusion of autobiographical narration and then overthrows the reader's impression by revealing her own unreliability. By manipulating our metarepresentation abilities, the daughter replicates in the reader her own frustration with her mother's stories.

We, too, often fail to recognize which stories are autobiographical and which fictional, since the narrator plays with the different levels of fiction, blurring the boundary between the imaginary and the autobiographical accounts of her life. By making readers experience the daughter's difficulties in understanding her mother through their own frustrations while reading the memoir, Kingston constructs a sophisticated textual mechanism that enables the readers to identify with the emotional states of the narrator/daughter and her mother. The text impels the readers to re-experience the daughter's cognitive states through the narrative structure of the memoir, rather than merely through the narrator's representation of the character's behavior on the thematic level. The reader and the daughter undergo similar cognitive processes in relation to the mother's storytelling, processes that begin in misunderstanding and end in the possibility of dialogue, thus making possible both the daughter's artistic expression and the reader's comprehension of it.

Negotiating Otherness, or Individualism Reconsidered

The possibility for constructive dialogue between a Chinese mother and her American-born daughter, established though the daughter's artistic expression in *The Woman Warrior*, opens up a possibility for cross-cultural understanding and mutuality. The mother/daughter relationship as presented in Kingston's memoir may enable the readers to re-conceptualize diversity and otherness. Since Maxine's attempts to achieve subjectivity through both identification with and separation from the mother have only limited success, she realizes that it is by co-operating with the mother, rather than resisting her, that she can construct her personal and ethnic selves. The daughter has to come to terms with her heritage and her anger toward her mother, thereby accepting their imperfect relationship, as the foundation for her self-construction. When Maxine realizes that her own life story is interdependent with the story of her mother, she has to redefine her self-definition strategies.

The transformation in the daughter's perception of subjectivity is apparent in the mother/daughter interactions throughout the memoir. Initially, in order to differentiate herself, Maxine upholds her individuality by means of a long monologue which leaves no room for her mother's response. Almost every sentence begins with the shouted first-person pronoun: "*I* can get into colleges. *I'm* smart. *I* can do all kinds of things. *I* know how to get A's . . . *I* can make a living and take care of myself" (*WW* 201; emphasis added). As the monologue progresses, Maxine emphasizes her separation from her mother by means of the pronoun *you*. Addressing her mother, the daughter accuses her: "*You* lie with stories

. . . *You* cannot stop me from talking. *You* tried to cut off my tongue, but it did not work" (*WW* 202; emphasis added). As a young woman, Maxine perceives her individuality in Western terms in which differentiation from and rebellion against one's family and social environment are legitimate ways of self-construction.[7]

Although the Western, individualistic self needs the other in order to define itself in opposition, the other also becomes a symbol of rejection, something one fears and even disdains. The individualistic self does not acknowledge its dependence on the other, but rather diminishes the other in order to uphold its own superiority. As Michelle Wallace suggests in "The Search for the 'Good Enough' Mammy: Multiculturalism, Popular Culture, and Psychoanalysis," the racial and social binaries that define people of color in opposition to white people perpetuate "monological identities" that prevent efficient cross-cultural interaction. It is therefore through a renewed perception of our identity as relational rather than individualistic that a change in our perception of diversity can occur.

The glorification of individualism leads to self-definition in opposition rather than in relation, and therefore generates the daughter's interpersonal and intercultural estrangement from her mother. It is through the dialogical and the relational, rather than through monological and aggressive interaction with the mother, that the daughter's subjectivity finally consolidates in a narrative. This is because in order to tell her own life, the daughter has to narrate her mother's story as well as the stories of her female ancestors (the No Name Woman, Fa Mu Lan, and her aunt Moon Orchid). In *How Our Lives Become Stories*, Paul Eakin suggests that *The Woman Warrior* belongs to the category of narratives that depict relational lives, a term Eakin uses to describe "the story of a relational model of identity, developed collaboratively with others, often family members" (57). Kingston's memoir projects a relational model of identity construction because the narrator/daughter constructs her self through the stories of others.

The concept of relational identity allows perceiving the ethnic other not as different, exotic, or inexplicable, but rather as an equal subject that has the power to acknowledge another individual as an independent self. According to Charles Taylor's "Politics of Recognition," "nonrecognition or misrecognition can inflict harm, can be a form of oppression, imprisoning someone in [a] false, distorted, and reduced mode of being" (75). Taylor further suggests that "recognition is not just a courtesy we owe to people. It is a vital human need" (76). Thus when a majority culture intentionally withholds recognition from minorities, it not only harms the people's self-esteem but also makes them internalize a distorted "picture of their own inferiority" that the majority culture inflicts upon them. According to Taylor, "We become full human agents, capable of understanding ourselves, and hence of defining our identity, through our acquisition of rich human languages of expression" (79). Since these modes of expression are learned through exchanges with others, by employing her mother's fluid narrative techniques in her memoir the narrator/daughter suggests a possibility for the construction of an individuality and unique narrative voice in connection rather than in opposition.

The final reassurance of the daughter's symbolic homecoming and reconciliation with the mother takes place though the mother and daughter's mutual recognition of each other as storytellers. When Maxine tells her mother that she, too, is telling stories, the mother tells Maxine a story that serves as her blessing for the daughter's artistic endeavors. As Judith Oster puts it, "For the first time in the book Brave Orchid seems to be recognizing what Maxine is doing, and showing in her subtle storytelling manner that it has value to her too" (243). The daughter's acknowledgment of herself as a storyteller is another milestone in her own self-definition. Earlier in the memoir, Maxine portrays her writing as a protective shawl of success that she wraps around herself when she visits her family. Now, after the mother has acknowledged her daughter's artistry, the mother herself wraps this shawl of protection around her daughter. Brave Orchid grants Maxine the great power of storytelling that the two women are now willing and able to share.

The final story the narrator/daughter tells in the memoir is both connected to and yet separate from her mother's story. On the one hand, it is important to Maxine to show how her own creativity flows from her mother's storytelling, yet it is equally essential for her to tell her own, separate story that defines her Chinese-American self. The story of Ts'ai Yen, a poetess who was captured by a barbarian chieftain, encapsulates the poignancy of the mother's experience of American reality as well as the narrator/daughter's struggle to voice her bicultural identity. After twelve years in captivity, Ts'ai Yen succeeds in communicating with the barbarian tribe by singing a tune that matches the high sound of their flutes. The high notes of the barbarian music disturb Ts'ai Yen so much that she answers back with a song in her own language: "a song so high and clear, it matche[s] the flutes. Ts'ai Yen sang about China and her family there. Her words seemed to be Chinese, but the barbarians understood their sadness and anger. Sometimes they thought they could catch barbarian phrases about forever wandering" (*WW* 209).

Ts'ai Yen's creative self-assertion inspires Maxine, because it allows the poet not only to voice herself, but also to create an understanding and mutuality between conflicting cultures. She finds a way to make a difference through collaboration rather than opposition. Instead of separating herself from the tribe, she finds a way to relate to them on the basis of a genuine feeling. Ts'ai Yen's success shows that people from different cultures can understand each other's emotions, if only they establish enough common references to enable them to communicate. When Ts'ai Yen is ransomed, she brings the songs of the barbarians to her homeland; they translate well, because they generate a genuine emotion that transcends cultural differences and appeals to common human experience.

The poetess and the barbarians, as well as the mother and daughter in the memoir, create a common language through artistic dialogue. The poetess symbolizes both the mother and the daughter; while she embodies the mother's experience of exile and estrangement in America, she also stands for the possibility of cross-cultural communication conducted by means of the daughter's writing. Brave Orchid and Maxine finally meet within the framework of storytelling and in the character of Ts'ai Yen, who shows the way to cross-cultural mind-reading. If, after twelve years in captivity, Ts'ai Yen becomes capable of communicating with the barbarians by collaborating with them in musical performance, Maxine gradually learns to negotiate between the Chinese sensibilities of her mother and her own American perspective. Just as Ts'ai Yen develops a sensibility to the barbarians' music, the daughter comes to understand her mother when she manages to attune herself to the mother's unique narration techniques.[8]

Conclusion

The arguments presented in this chapter lead to the conclusion that in spite of the initial psychological and cultural misunderstandings, a Chinese-born mother and her American-born daughter can eventually achieve cross-cultural communication. Although the universal cognitive ability of mind-reading is hampered by cross-cultural misunderstandings, it is restored, in this case, through the act of storytelling. The processes of mother/daughter reconciliation involve both psychological and cognitive developments that become apparent in the daughter's retelling of her mother's autobiographical and fictional stories. The kind of tacit communication that allows one person to understand another's intentions is established through the daughter's construction of her mother as a fictional character, and its readjustment to her real mother by means of dialogue. Since direct display of emotions is usually unacceptable in Chinese culture, Brave Orchid at first resists direct communication. Eventually, however, she recognizes her daughter's need for the sort of straightforward manifestation of feelings typical of Western communication. The dialogue between Brave Orchid and Maxine prepares them for the final stage in their reconciliation process—communication and recognition through storytelling. The storytelling exchange that occurs at the end of the memoir implies the possibility of cross-cultural mind-reading, not only between the mother and daughter, but between national minorities and majorities as well. In this case, the mixed genre of this memoir also serves as a channel for cross-cultural translation between Chinese and American culture, between a Chinese-American author and her American audience.

Notes

1. Zunshine, *Why We Read Fiction*, 6.
2. In *Articulate Silences*, King-Kok Cheung explains that "the practice of recycling victuals (in this case Seagram's 7) is a common one. Many Chinese regularly observe religious holidays by setting dishes in front of the altars of dead ancestors or of gods. Once the symbolic worship is over, the food and drinks are either consumed by the worshipers or saved for use in another round of worship . . . A child who grows up as a member of the Chinese *majority* is unlikely to be nonplussed by a routine followed by many a family" (92).
3. Brave Orchid explains neither rituals nor superstitions. As King-Kok Cheung suggests, the "mother withholds a verbal explanation . . . because she believes in—and therefore is wary of—forms of 'speech acts.'" As she warns Maxine: "Be careful what you say. It comes true" (204). "Words have a magic potency and no Cantonese equivalent of 'Touch wood' can fully neutralize inauspicious remarks or gestures" (Cheung 92).
4. Winnicott, *Playing and Reality*, 103.
5. The development of Maxine's empathy and compassion toward her mother can also be attributed to her maturation. In "A Complete Theory of Empathy Must Consider Stage Changes," Michael Commons and Chester Wolfsont present a sequential, hierarchical stage model of empathy, "beginning with 'the infant's automatic empathy' and ending with the advanced adult's 'construction of empathetic reality'" (30). Although we are born with the basic capacity to empathize with others, as we grow up our cognitive and empathetic

capabilities evolve, enabling us to improve our social interaction. While Commons and Wolfsont discuss thirteen stages of empathy development, the early (sensory-motor, nomical, and sentential stages) are not as relevant to the memoir analysis as the later stages that begin with the "pre-operational" (storied empathy). At this stage Maxine is able to empathize with the heroines in her mother's stories, while still confusing real and imaginary. In "White Tigers" she identifies with the woman warrior so much that she retells her mother's story, casting herself as the fearless heroine. As the memoir progresses, Maxine's "storied empathy" evolves into the "concrete" stage, which involves understanding the "other's motives in terms of one's own motives in a similar situation" (31). Returning to the example already discussed in this chapter, Maxine manages to understand her mother's feelings when leaving home to study in medical college by comparing Brave Orchid's reality in China to her own, American environment. Finding similarities between her own life-choices and her mother's situation, the daughter understands her mother's intentions in the context of her socio-cultural framework. A more detailed discussion of Commons and Wolfsont's empathy model in relation to Kingston's memoir could serve as an interesting basis for further research.

6. The source tags also play a role in drawing the reader's attention to the agents of speech and authority in the memoir. In *The Woman Warrior*, speech stands for power, authority, and sanity, while silence represents victimization and madness. As Maxine says to the silent girl whom she terrorizes at school: "If you don't talk you can't have a personality" (180). The unnamed aunt in the first chapter, for instance, is largely silent. The narrator/daughter tells the various versions of her story, yet the only sentence Maxine imagines her aunt to utter is: "I think I'm pregnant" (7). The source tag "she told the man" makes us realize how little the aunt actually says in this dramatic situation. This reticence also symbolizes her inability to resist the patriarchal oppression. The aunt behaves according to the traditional Chinese ideology that privileges silence over speech, especially among children and women (Cheung 82, 83). In *The Woman Warrior*, the narrator/daughter resists this victimization and silencing, gradually asserting her subjectivity and authority through speech.

7. According to Taylor's "The Politics of Recognition," the concept of *individualized* identity emerged in the eighteenth century. The notion that human beings are naturally "endowed with a moral sense, an intuitive feeling for what is right and wrong" inspired the idea that an individual can reach self-empowerment and moral ascendancy through introspection without any external social or divine intervention (77).

8. The cut reeds that the barbarians used in battle as arrows were also used as a musical instrument. When the poet learns to associate the high notes of the reed pipe with music rather than war, she also hears the sadness of the barbarians expressed in their music.

CHAPTER EIGHT

Theory of Mind and Michael Fried's *Absorption and Theatricality*: Notes toward Cognitive Historicism

LISA ZUNSHINE

It is warm outside. Spring blossoms brush against the house. Leaning over the windowsill, propping his right hand with his left, a young man is blowing bubbles. Just now a particularly large bubble is trembling at the tip of his blowpipe.[1] The man is holding his breath. The world is standing still.

The Soap Bubble is one of Jean-Baptiste-Simeon Chardin's "paintings of games and amusements" done in the 1730s (Fried 51). His subjects build card castles, sketch, and play knucklebones. They are so completely absorbed in what they do that they are unaware of being watched, and they draw us in precisely with their peculiar obliviousness to our presence, their utter lack of theatricality.

Absorptive paintings are anti-theatrical and as such both irresistible and difficult to create. This is the argument advanced by Michael Fried in *Absorption and Theatricality: Painting and Beholder in the Age of Diderot*. Following the development of French genre painting from the 1750s to the early 1780s, Fried shows how artists tried to minimize the self-awareness of art—grounded in the "primordial convention that [art is] made to be beheld" (157)—by depicting persons not aware of the presence of the beholder. He also shows how quickly the established methods of representing absorption would become stale and how desperately the artists would cast about for new ways to convince their audiences that the people in paintings did not care about their gaze.

Published in 1980, *Absorption and Theatricality* won academic prizes and stirred up controversies. Today it continues to reach beyond the disciplinary boundaries of art history, influencing debates in literary criticism, cultural studies, and performance theory. My goal in this chapter is to expand its reach yet further—into cognitive science. I suggest that

Figure 8.1. *Soap Bubbles,* Jean-Baptiste-Simeon Chardin, National Gallery of Art

to grasp fully the brilliance of Fried's argument and the significance of his insights for contemporary cultural criticism, we need to consider what he says in the context of recent cognitive-evolutionary research on Theory of Mind.

As I will demonstrate shortly, studies in Theory of Mind confirm our intuitions about the performative nature of all human communication. As such they provide a broader theoretical framework for Fried's articulation of the difficulties faced by artists who wished to minimize the theatricality of their pieces. Fried's discussion of these difficulties is particularly illuminating for literary critics interested in bringing together cognitive science and cultural historicism because he is deeply invested in historicizing, yet he also wants to understand the psychological dynamics behind the historically specific concerns of artists and critics. Today we can use insights from cognitive science to mediate the relationship between psychology and history in our cultural and literary analyses, which is why I consider Fried's approach cognitive, ahead of its time, and want to explore the implications of his argument for cognitive literary theory.

From Bubbles to Blindness: Struggling to Ensure Absorption

And so Chardin's canvases of the 1730s as well as Chardin's, Greuze's, Van Loo's, and Vien's works of the 1750s depicted people so caught up in praying, playing, sketching, learning difficult lessons, blowing bubbles, grieving, rejoicing, listening raptly to charismatic speakers, or simply sleeping, as not to be aware of being watched. By "negating the beholder's presence" these paintings resisted the pervasive theatricality of representational art. And by resisting theatricality, they became more spectacular. As Fried puts it, "Only by establishing the fiction of [the beholder's] absence or nonexistence could his actual placement before and his enthrallment by the painting be secured" (103). The sight of people so absorbed in what they are doing that they are unable to put on any special body postures or facial expressions—the absence of performance—seemed mesmerizing.

By the early 1760s, however, the subject matter wore itself thin. It became increasingly difficult for artists to use the established contexts of absorption (such as reading, praying, sleeping) to convincingly exclude the beholder from the picture. And so "deliberate and extraordinary measures came to be required in order to persuade contemporary audiences of the absorption of a figure or group of figures in the world of the painting" (Fried 67). One such measure involved ratcheting up the drama: Greuze's *Le Fils ingrat* (1777) and *Le Fils puni* (1778) depict a family so distraught over the rebellion of the ungrateful son and the resulting early death of his father that it is obvious that none of them would be able to gather their wits enough to look about themselves and realize that they are being observed.

Another measure involved opening a painting "to a number of points of view other than that of the beholder standing before the canvas" (159): In David's *Belisaire* (1781), the "off-center perspective [places] the beholder to one side of the painting, away from [the central] figure of Belisarius" (156), thus implying that the perspective and viewing convenience of the beholder are simply not part of the characters' worldview.

Yet another way to create the illusion of absorption was to make the titular character blind, as in Vincent's *Belisaire* (1777), David's *Belisaire* (1781, 1785), Peyron's *Belisaire* (1779), and David's *Homere endormi* and *Homere recitant* (both 1794): The blind protagonist is by default unaware of the beholder.

Fried calls these measures "extreme." The intensification of drama, the experimentation with different perspectives, and the introduction of blind historical and mythical figures all seem to testify that by the 1770s, "the everyday as such was in an important sense lost to pictorial representation" (61). The absorptive charm of mundane activities of listening, watching, and daydreaming was broken. In fact, Fried argues that if we follow "the evolution of David's art between 1780 and 1814," we can trace in it "a drastic loss of conviction in [both] action and expression as resources for ambitious painting, that is, in the very possibility that either could be represented other than as theatrical" (176). In other words, "the persuasive representation of absorption" may have remained a "positive desideratum" (13) for artists, but, at least within the context of that specific period in French art history, the means of achieving that absorption and thus escaping the theatricality of representational art seemed to have been exhausted.

Moreover, for some critics, such as Rousseau, those means had always been futile. Unlike Diderot, who actively considered the ways of transcending theatricality both in drama and painting, Rousseau "not only [argued] that the theater is beyond redemption" but also strongly implied "that there is no aspect of social life that is not comprised within the dangerous, because readily theatricalized and theatricalizing, realm of the spectacular" (168). In other words, one cannot reduce the self-conscious quality of the whole painting by reducing the self-consciousness of its subjects.

To clarify this latter point, consider a hypothetical question, which Rousseau and Diderot would answer somewhat differently. Let's say that the work of art itself cannot be cleansed of intentionality and, hence, theatricality, for it was *made with the intention to be looked at*. Perhaps, then, its intention-less subjects—that is, the persons caught at the moment when they are not aware of being looked at—can diminish the overall theatricality of the piece to some extent. Rousseau would say no: the allegedly beholder-free subjects make the whole piece even more insidiously theatrical. Diderot might say yes, if only to reflect his fascination with the process of hiding, ignoring, or diminishing the beholder.

And what would Fried himself say? On the one hand, he would agree with Rousseau that theatricality pervades our art and social life. In his earlier essay, "Art and Objecthood" (1967), Fried called his readers' "attention to the utter pervasiveness—the virtual universality—of the sensibility or mode of being . . . corrupted or perverted by theater" (168). On the other hand, "Art and Objecthood" focuses on works of modernist art that "defeat theater" by their quality of "presentness," that is, by their apparent ability to just *be* there independently from the perspective of the beholder. If, as Fried puts it in the famous last sentence of that essay, "presentness is grace" (168), then this grace is attainable or at least imaginable—a position that aligns Fried somewhat more with Diderot than with Rousseau.

For a cognitive cultural critic, the issue of escape from theatricality, as formulated by Fried, captures an important cognitive paradox underlying much of our culture. To get at the root of this paradox, we need to turn to the concept of Theory of Mind. For it seems that both the impulse to transcend the theatricality of representations (and indeed of our whole social life) and the nagging suspicion that no such transcendence is possible might be grounded in the workings of our evolved cognitive adaptations for mind-reading.

What are these adaptations?

Theory of Mind and Two Underlying Assumptions

Theory of Mind, also known as mind-reading, is a term used by cognitive psychologists and philosophers of mind to describe our ability to explain behavior in terms of underlying thoughts, feelings, desires, and intentions (e.g., we see somebody reaching for a cup of water, and we assume that she is thirsty).[2] We attribute states of mind to ourselves and others all the time. Our attributions are frequently incorrect (the person who reached for the cup of water might have done it for reasons other than being thirsty). Still, making them is the default way by which we construct and navigate our social environment. When Theory of Mind is impaired, as it is in varying degrees in the case of autism and schizophrenia, communication breaks down.

Note that the words *theory* in Theory of Mind and *reading* in mind-reading are potentially misleading because they seem to imply that we attribute states of mind intentionally and consciously. In fact, it might be difficult for us to appreciate at this point just how much mind-reading takes place on a level inaccessible to our consciousness. For it seems that while our perceptual systems "eagerly" register the information about people's bodies and their facial expressions, they do not necessarily make all that information available to us for our conscious interpretation. Think of the intriguing functioning of the so-called "mirror neurons." Studies of imitation in monkeys and humans have discovered a "neural mirror system that demonstrates an internal correlation between the representations of perceptual and motor functionalities" (Borenstein and Ruppin 229). What this means is that "an action is understood when its observation causes the motor system of the observer to 'resonate.'" So when you observe someone else grasping a cup, the "same population of neurons that control[s] the execution of grasping movements becomes active in [your own] motor areas" (Rizzolatti et al. 2001, 662). At least on some level, your brain does not seem to distinguish between you doing something and a person that you observe doing it.

In other words, our neural circuits are powerfully attuned to the presence, behavior, and emotional display of others. This attunement begins early (since some form of it is already present in newborn infants) and takes numerous nuanced forms as we grow into our environment. We are intensely aware of the body language and facial expressions of other people, even if the full extent and significance of such awareness escape us.[3]

Let me now spell out two assumptions underlying the present argument. First, I think of our cognitive adaptations for mind-reading as promiscuous, voracious, and proactive, their very condition of being a constant stimulation delivered either by direct interactions with other people or by imaginary approximations of such interactions. To amplify this point, it is useful to compare our adaptations for mind-reading with our adaptations for seeing. Because our species evolved to take in so much information about our environment visually, we simply cannot help seeing once we open our eyes in the morning,[4] and the range of cultural practices grounded in the particularities of our system of visual adaptations is truly staggering. Similarly, as cognitive evolutionary psychologist Jesse M. Bering observes, after a certain age, people "cannot turn off their mind-reading skills even if they want to. All human actions are forevermore perceived to be the products of unobservable mental states, and every behavior, therefore, is subject to intense sociocognitive scrutiny" (12). This means that although we are a far way off from grasping the full extent to which our lives are structured by our adaptations for mind-reading, we should be prepared to find that the cultural effect of those adaptations may prove just as profound and far-ranging as that of being able to see.[5]

The second assumption is a paradox. We perceive people's observable behavior as both a highly informative and at the same time an unreliable source of information about their minds. This double perspective is fundamental and inescapable, and it informs all of our social life and cultural representations.

To begin to appreciate the power of this double perspective, consider the reason we remain suspicious of each other's body language. When I am speaking to you, you count on my registering information conveyed by your face, movements, and appearance. That is, you can't know what particular grin or shrug or tattoo I will notice and consider significant at a given moment (indeed, I don't know either). Our evolutionary past ensures, however, that you will intuitively expect me to "read" your body as indicative of your thoughts, desires, and intentions. Moreover, the same evolutionary past ensures that I intuitively know that you expect me to read your body in this fashion. This means that I have to constantly negotiate between trusting this or that bodily sign of yours more than another. Were I to put this negotiation in words—which will sound funny because we do *not* consciously articulate it to ourselves in such a fashion—it might go like this: "Did she smile because she liked what I said or because she wanted me to think that she liked what I said, or because she was thinking of how well she handled an argument yesterday, or was she thinking of something altogether unrelated?"

In other words, paradoxical as it may seem, we treat with caution the information about the person's state of mind inferred from our observation of her behavior and body language precisely because we can't help treating them as a highly valuable source of information about her mind—*and we both know it*. Because the body is *the* text that we read throughout our evolution as a social species, we are now stuck, for better or for worse, with cognitive adaptations that forcefully focus our attention on that particular text. (Nor would we want to completely distrust the body—our quick and far-from-perfect reading of each other is what gets us through the day.)

What all this adds up to is that we are in a bind. We have the hungry Theory of Mind that needs constant input in the form of observable behavior indicative of unobservable mental states. And we have the body that our Theory of Mind evolved to focus on in order to get that input. And that body—the object of our Theory of Mind's obsessive attention—is a privileged and, as such, potentially misleading source of information about the person's mental state.

Note how at this point the research on Theory of Mind complements our own discipline's insight about the body as a site of performance. Because we are drawn to each other's bodies in our quest to figure out each other's thoughts and intentions, we end up *performing* our bodies (not always consciously or successfully) to shape other people's perceptions of our mental states. A particular body thus can be viewed only as a time-and-place-specific cultural construction—that is, as an attempt to influence others into perceiving it in a certain way. As Ellen Spolsky (Chapter Two, this volume) puts it,

> The clues to which we sensibly learn to be attentive cannot be relied on absolutely because bodies themselves, the bodies that are evolved to give external expression to internal states, learn to produce these clues within contexts differentiated by cultural categories such as gender, age, social class, and occupation. Not only our interpretations of them but the evolved physical expressions themselves are enriched and/or distorted by social overlays, making both misinterpretation and deliberate deception possible.

Cognitive evolutionary research thus lends strong support to theorists in cultural studies who seek to expand the meaning of performativity, such as Joseph Roach, who argues that performance, "though it frequently makes references to theatricality as the most fecund metaphor for the social dimensions of social production, embraces a much wider range of human behaviors. Such behaviors may include what Michel de Certeau calls 'the practice of everyday life,' in which the role of spectator expands into that of participant" (46). Indeed, work on Theory of Mind indicates that our everyday mind-reading turns each of us into a performer and a spectator, whether we are aware of it or not.

Let us consider another, closely related, implication of the studies on Theory of Mind. They encourage us to think of a broad variety of cultural institutions and social practices as both reflecting our overarching need to attribute minds *and* remaining subject to the instabilities inherent to our mind-reading processes. For example, our social infrastructure seems to be chock-full of devices designed to bypass our fakeable, performable, constructable body in reading the person's mind. We use various tokens, legal documents, credit and medical histories, recommendation letters, gossip, blood and hair samples, and polygraph tests to avoid the situation when we have to make an important decision based on the information provided solely by the person's immediate observable behavior.

Some of these devices succeed better than others, and none is perfect. We may not yet be living in the future depicted in *Gattaca* (1997), whose protagonist (played by Ethan Hawke) fakes his blood and hair samples to deceive others about his intentions, but that sci-fi moment does capture an important sociocognitive feature of our world: There is a constant arms race going on between cultural institutions trying to claim some aspects of the body as essential, unfakeable, and intentionality-free, and individuals finding ways to perform even those seemingly unperformable aspects of the body.

Fictions of Embodied Transparency

As one example of such an arms race, consider a peculiar representational tradition—I call it a tradition of embodied transparency—of putting protagonists in situations in which their bodies spontaneously reveal their true feelings, often against their wills. Manifesting itself differently in different genres and individual works, moments of embodied transparency are carefully foregrounded within larger narratives. In each case an author builds up a specific context in which brief access to a character's mental state via her body language stands out sharply against the relative opacity of other characters, or of the same character a moment ago.

Every moment of transparency is thus entirely relative and context-dependent, but the wish to create and behold such moments seems to be perennial, grounded in our evolutionary history as a social species. Representations of embodied transparency must be immensely flattering to our Theory of Mind adaptations, which evolved to read minds through bodies but have to constantly contend with the possibility of misreading and the resulting social failure. The pleasure derived from moments of embodied transparency is thus largely a social pleasure—a titillating illusion of superior social discernment and power.

Elsewhere I consider examples of embodied transparency in novels (such as Austen's *Pride and Prejudice*), in nineteenth-century genre paintings (such as "proposal compositions" discussed by Stephen Kern in *Eyes of Love*), and in twentieth-century mock-documentaries (such as *The Office*).[6] In each case, I demonstrate that to be effective, moments of embodied transparency have to be spontaneous, unexpected, and short. They also have to look unconventional in the larger context of their genre. That is (returning to my earlier argument about the arms race), writers, artists, and movie directors have to keep inventing *new* ways of forcing the body into a state of transparency because as soon as one way of doing it emerges as an established convention, it loses credibility. The mind retreats further, leaving the body as a front going through the expected motions of "revealing" the "true" states of mind. The double perspective of the body comes back with a vengeance.

The issue of unconventionality will bear directly upon our discussion of eighteenth-century paintings of absorption later in this chapter. But before returning to Fried, let us consider two other case studies of eighteenth-century embodied transparency: narrativized, but not exactly fictional. As segues to *Absorption and Theatricality*, these case studies will illustrate my point about the centrality of brevity and spontaneity in constructing convincing representations of direct access to people's mental states.

The first case study comes from poet and playwright Johanna Baillie's *Plays on the Passions*, published in 1798; specifically the "Introductory Discourse" to *Plays on the Passions*, as discussed recently by cognitive literary critic Alan Richardson. For Baillie, as Richardson points out,

> reading human emotions and intentions through their embodied manifestations is an innately driven, experientially developed, species-wide human practice. [Thus] Baillie argues that public executions drew large crowds precisely because of this universal fascination with emotional expression. Few spectators "can get near enough to distinguish the expression of a face, or the minuter parts of a criminal's behavior" under such unusually intense emotional pressure, yet even "from a considerable distance will they remark whether he steps firmly; whether the motions of his body denote agitation or calmness." ("Facial Expression Theory," in press)

Baillie's focus, of course, is not executions but theater: that "grand and favorite amusement of every nation into which it has been introduced." She believes (to quote Richardson again) that "in the right hands and under the right circumstances" theater can "provide a more intimate look" at a variety of "nonverbal emotional behaviors" ("Facial Expression Theory"). What interests me, however, is precisely the fleeting description of the greed with which onlookers take in the impromptu spectacle of the condemned man's body language. I agree with Richardson that Baillie sees public executions as feeding the "universal fascination with emotional expression"—feeding our Theory of Mind, we can say now.[7] At the same time, I think that there is something unique about this particular crowd-drawing occasion. For a public execution does more than merely provide spectators with a show of strong emotions on the part of the criminal: it also promises a privileged access to his feelings. As a real-life social event and as a secondhand description of this event, a public execution represents a striking instance of embodied transparency.

As a condemned criminal walks to his death, his mind is pried open against his will. Whatever behavior he may display as he approaches the gallows—"agitation or calmness"—spectators *know*, or think that they can largely guess, what lies beneath: terror, despair, fear, and perhaps an improbable hope for a miraculous reprieve. The extremity of the occasion narrows down drastically the possible range of the man's mental states, while the transience of this moment of transparency—the fateful walk will end very, very soon—amplifies its value. Hence the onlookers gaze and gaze[8] even when the great distance precludes them from discerning the exact facial expression of the criminal: the awareness of the inimitable value of this soon-to-be-over spectacle of emotional access keeps them riveted to the spot. Some of that raw value trickles down as they comment on what they see (or think they see), and as they retell the event or re-imagine it in service of a very different cultural project, such as, for example, Baillie's discourse on passions and theater.

My second case study also originates in the late-eighteenth-century theatrical discourse.[9] In 1807, Henry Siddons (son of Sarah Siddons and himself an actor) published *Practical Illustrations of Rhetorical Gesture and Action*, a translation from German of Johann Jacob Engel's *Ideen zu einer Mimik* (1785), revised to reflect the conventions of the English stage. At one point, to illustrate what he calls "the communicative power of gesture" (36), Siddons treats the reader to the following tableau:

When a person sits at the theater, after having seen a play acted three or four times, his mind naturally becomes vacant and inactive. If among the spectators he chances to recognize a youth, to whom the same is new, this object affords him, and many others, a more entertaining fund of observation than all that is going forward on the stage.

This novice of an auditor, carried away by the illusion, imitates all he sees, even to the actions of the players, though in a mode less decisive. Without knowing what is going to be said, he is serious, or contented, according to the tone which the performers happen to take. His eyes become a mirror, faithfully reflecting the varying gestures of the several personages concerned.

Ill humour, irony, anger, curiosity, contempt, in a word, all the passions of the author are repeated in the lines of his countenance. This imitative picture is only interrupted whilst his proper sentiments, crossing exterior objects, seek for modes of expressing themselves. (35–36)

What interests me in this scene is the implicit contrast between the "reality" of emotions as they are portrayed onstage and as they are mirrored by the unsophisticated observer. For note that nobody in this tableau apparently experiences the real feelings of "ill humour, irony, anger, curiosity [or] contempt." The actors put on a show of those emotions, but who knows what they really feel? The "youth" unselfconsciously mimics their body language, but does it mean that he is *really* angry or contemptuous at this point? I doubt it. However much I may fear and hate a psychopathic murderer from a movie, those feelings are *nothing* compared to what I would experience were I to encounter such a person in real life. In this respect, the body of Siddons's impressionable "youth" is as unreliable an index to his true feelings as the acting bodies on stage are to theirs.

However, this weak version of ill humor, irony, or anger is not *all* that animates our young man. He feels something else—and very deeply, too—and that something else is plainly written all over his body. *It is his engagement with what he sees onstage.* The smile of contempt that momentarily curls his lips as he watches the actress stare down the double-dealing villain thus expresses not so much any actual contempt on his part but rather his deep involvement with the performance: his complete surrender to the power of the actors.

If we focus on this particular aspect of the young man's feelings, it means that, at least for the duration of this episode, his body language reflects his state of mind more accurately than the body language of the performers reflects their state of mind. He is completely taken by what happens onstage, and because he is not faking that state of deep emotional engagement for the benefit of the observer (for he does not know that he is being observed), his unpremeditated show of feelings becomes more engrossing for the theatergoer than the official show of feelings put on by the actors.

Siddons's voyeuristic tableau thus plays with our double view of the body as the best and the worst source of information about the person's mind by teasing us with a vision of a highly readable body in the setting (theater) that thrives on cultivating the gap between the body and the mind. Moreover, this specific setting also ensures that the moment will not last: the "youth" is in thrall now, but this spell will be broken at any second. As with the context of public execution, which renders both plausible and short-lived the moment of the man's embodied transparency, the context of theatrical spectatorship renders transparency both possible and transient.

I don't think that either Baillie or Siddons consciously set out to construct what I now call contexts of embodied transparency. Instead, they wanted to make specific points about the power of theater, and the representations of bodies rendered briefly and radically readable allowed them to advance these points. One may thus speculate that a moment of embodied transparency can make a rhetorical point more compelling and vivid. Writers may intuitively cultivate such moments to increase the imaginative charge of their arguments, their immediacy and spontaneity.

Using Theory of Mind to Negotiate between Psychology and History

Let us now turn to *Absorption and Theatricality*. What happens if we approach Fried's arguments from the point of view of cognitive theory of mind-reading? First, we notice that the attempts to eschew theatricality in eighteenth-century French genre painting are on a par with other representational endeavors (such as Baillie's execution scene and Siddons's voyeuristic tableau) to render the body transparent. For what is the state of "absorption" if not the carefully constructed moment when the observable body language provides the direct access to the person's state of mind? Chardin's young man balancing a soap bubble on the end of his pipe is completely absorbed by what he does. As a result, his mental state is as transparent as that of an enthralled youth from Siddons's imagined theater and that of a criminal approaching the gallows. Absorption signifies transparency.

Hence the important point that I revisit several times in my chapter.

Research on Theory of Mind begins to explain some of the intrinsic pull of the absorptive paintings. Fried notes that "absorption emerges as good in and of itself, without regard to its occasion" (51). A cognitive literary critic such as myself will agree with this and speculate that representations of absorption may feel "good in and of [themselves]" because they flatter our mind-reading adaptations. Such representations regale us with something that we hold at premium in our everyday life and never get much of (i.e., moments of perfect access to other people's minds), and they intensify our pleasure by constructing plausible social contexts for these fleeting mind-reading feasts.

Here and elsewhere in his argument, Fried demonstrates his interest in the interplay between psychology and history. Let us see how the cognitive perspective may anchor his—and our—intuitions about the relationship between the two.

Fried begins with a strong assertion of the historical limits of his argument. "This study is exclusively concerned with developments in France," he tells us on the first page. Then again on page two: "I am convinced that there took place in French painting starting around the middle of the century a unique and very largely autonomous evolution; and it is the task of comprehending that evolution as nearly as possible in its own terms—of laying bare the issues crucially at stake in it—that is undertaken in the pages that follow."

By insisting that the French absorptive paintings should be considered on their "own terms," Fried distances himself from two interpretive traditions. First, he disagrees with those art historians who think that by focusing on the human body in action, Chardin and others were taking an ideological stand against Rococo's indifference to historical figures and heroic endeavors. As Fried puts it, authors of absorptive paintings were not really interested in upholding "the doctrines of the hierarchy of genres and the supremacy of history painting as they were held by anti-Rococo critics and theorists." In his view, the artists' interest in representation of absorption was not ideological or primarily concerned with the subject matter but was rather "determined by other, ontologically prior concerns and imperatives." And these had to do, among other things, with the relationship, "at once literal and fictive, between painting and the beholder" (75–76).

I suggest that such "ontologically prior concerns"—particularly when framed in terms of the relationship "between painting and the beholder"—are ultimately bound up with the cognition of mind-reading. The absorptive painting titillates us with the illusion of embodied transparency. Our responses to this powerful illusion certainly draw on an idiosyncratic mix of personal ideologies and aesthetics, but the socio-cognitive—the drive to read minds and the anxiety about misreading minds—is inextricably there, heightening and structuring our interest in the painting.

Here is the second interpretive tradition that Fried wants to keep at arm's length. He is wary of our tendency to assume that if two works deal with the same subject one must have influenced the other. Thus, when he looks at the 1770s–1780s paintings by Vincent, David, and Peyron, featuring the blind general Belisarius receiving alms, Fried has to address the role of the much earlier canvas on the same theme: Luciana Borzone's *Belisarius Receiving Alms* (1620). It may seem obvious that the 1620 painting had a lasting influence on the late eighteenth-century *Belisarius*es, yet, as Fried puts it, "the notion of influence is what I wish to see beyond" (145). In his view, the *Belisarius*es of the 1770s–1780s were informed by the increasingly desperate attempts of the artists to eradicate the beholder (and hence create the illusion of complete absorption), and it would be anachronistic to project this desperation back onto the 1620 canvas.

Consider one important detail shared by the seventeenth-century *Belisarius* and its eighteenth-century counterparts. Both feature a younger officer who is looking at the general while remaining invisible to him. Fried's careful analysis demonstrates, however, that in Vincent's and David's *Belisarius*es, the posture of this observer implies a more aggressive endeavor to ensure the subjects' engrossment in the world of the painting. For example, in stark contrast to the merely pensive officer of Borzone's painting, the officer of Vincent's painting "gazes anxiously, almost mistrustfully, at the sightless eyes of the great general" (152). For Fried, this intense gaze is strongly indicative of the artist's near-desperate attempt to render both men completely absorbed in the present moment and thus oblivious to the presence of the beholder. The general cannot perform for the beholder because he is blind, and the officer cannot perform for the beholder because he is too preoccupied by figuring out what the blind man is up to.

The officer's mistrust is actually somewhat ill warranted. Does he think that Belisarius is faking blindness? Why should he doubt the old general? It is precisely because the officer's attitude is not entirely psychologically convincing that we infer that it must serve other representational needs. Specifically, it increases his absorption in what is going on. And it is this forced absorption, as Fried argues, that renders the whole painting a pointedly 1770s project, whose goals and mood are strikingly different from those of its alleged 1620 predecessor.

To see why Fried's insistence on seeing beyond the "notion of influence" is crucial from the cognitive perspective, let us broaden the context of his discussion and think of how we generally construct our narratives of influence. For example, imagine a cultural historian who has just read Siddons's description of the enthralled youth in *Practical Illustrations* and is now looking at the young man of Chardin's *The Soap Bubble*. The suggestive parallels between the representations of absorbed bodies may prompt this historian to look for evidence of influence of one on the other, inferring, perhaps, that the French tradition of absorptive painting informed Siddons's sensibility by the way of Engel's earlier *Ideen zu einer Mimik*. Such an argument would sound quite plausible; in fact, it would be typical of claims we routinely make in our cultural analyses. Looking at the two representational traditions side by side seems to call for some sort of narrative of influence.

But, let us say, we do not want to insist that one directly influenced the other. What alternative do we have? We may have to come up with an argument that would demonstrate that the idea of emotional transparency as a representational desideratum was "in the air"—that is, present in a variety of cultural discourses at the time—and as such got picked by Chardin, Engel, Siddons, and Baillie.

And it is not that this historicized account will be wrong. It could be quite insightful. However, it will always remain a "just so" story—a result of our earnest wish to explain what it was about this or that historical moment that made artists and writers feel that representations of embodied transparency would be particularly desirable and valuable right then.

That is, unless we posit a cognitive foundation for our enduring interest in visual and verbal representations of embodied transparency, a "notion of influence" or a historicized "just so" story (or some combination of the two) is what we will have to fall back onto again and again. By contrast, see what happens if we establish, once and for all, that we remain perennially fascinated by socially rich representational contexts that construct bodies as transparent and thus flatter and titillate our Theory of Mind. If we do so, we free ourselves from the obligation to endlessly explain one such construction through another, or to make historical contexts carry more weight than they can bear.

This is to say that usually we make them carry *all* the weight, and they don't have to. For if we factor in the cognitive aspect, we can attend to specific historical contexts of cultural representations and speak of influences only when we have compelling evidence for such influences and contexts, and not because we simply have no other ways of explaining their powerful appeal. The cognitive perspective thus makes possible a more balanced and responsible historicizing than we are currently pressed into.

This is why I suggest that Fried's study was "cognitive" before its time and that now, with the advent of research on Theory of Mind, it can be properly appreciated as such. By resisting the "notion of influence" and the primacy of the considerations of "hierarchy of genres" and "subject matter," Fried articulated the need for alternative conceptual frameworks, which would address the "ontologically prior" relationship between painting and the beholder. With their double view of the body, studies in Theory of Mind go right to the heart of this relationship. They suggest that absorptive paintings are riveting because they present us with an illusion of direct unmediated access to the subjects' mental states. Sociocognitive satisfaction thus underlies aesthetic pleasure. It does not *define* this pleasure: too many culture-specific and personally idiosyncratic factors are at play in each case. In fact, as Fried demonstrates, a number of eighteenth-century critics found various faults with absorptive pieces, which means that a rich visual illusion of privileged mind-access does not directly translate into aesthetic pleasure for everybody. Still, at least to some important degree, it makes this pleasure possible.

How to Construct Brevity, Spontaneity, and Unconventionality

I suggested earlier that, to be convincing, contexts of embodied transparency have to strike observers and readers as transient and unexpected. Let me restate this. By presenting the body as faithfully reflecting the mind, such contexts attempt to transcend the double position of the body as a highly privileged yet unreliable source of information about the person's mental state. However, this transcendence—this moment of truth—is always suspect for our mind-reading and hence body-performing species. This is why contexts of embodied transparency require painstaking planning and foregrounding on the part of the author, yet have to strike beholders as brief, spontaneous, and unconventional.

Now consider Fried's observation that the scenes of absorption demanded both intricate plotting on the part of the artist *and* a very peculiar handling of time. It seems to me that Fried addresses here the same cognitive dynamics as I do above, only that he articulates them through the idiom of a specific genre at a specific historical moment. Here, for example, is the elaborate background narrative that had to precede the "chance" occurrence captured by Greuze's *La Piete filiale* (1763). This painting features a paralyzed old man surrounded by his family at the precise moment when they all respond emotionally to his interaction with his benevolent son-in-law. Here is Diderot's description of that painting:

> The moment . . . chosen by the artist is special. By chance it happened that, on that particular day, it was his son-in-law who brought the old man some food, and the latter, moved, showed his gratitude in such an animated and earnest way that it interrupted the occupations and attracted the attention of the whole family. (Quoted in Fried, 55)

As Fried observes,

> Diderot's statement is the most forthright assertion of the primacy of considerations of absorption that we have so far encountered. He seems almost to be saying that Greuze was compelled first to paralyze the old man and then to orchestrate an entire sequence of ostensibly chance events in order to arrive in the end at the sort of emotionally charged, highly moralized, and dramatically unified situation that alone was capable of embodying with sufficient perspicuousness the absorptive states of suspension of activity and fixing of attention that painter and critic alike regarded as paramount. (56)

The cognitive perspective strongly supports both Diderot's and Fried's intuition that the painters had to go to great lengths to build their scenes of absorption. A successful representation of embodied transparency requires a convincing background narrative and something akin to a "mania for plotting" (55) in an artist (a charge sometimes leveled against Greuze).

If you disagree and think it is easy to come up with a context in which a character is forced to embody her true feelings, try it. Chances are that you will settle on either of these two scenarios: a violent surprise—a subject is shocked by some news and his body immediately shows it—or physical torture; as Walter Benn Michaels puts it in his discussion of *American Psycho*, "You can be confident that the girl screaming when you shoot her with a nail gun is not performing (in the sense of faking) her pain" (70). But these two scenarios, surprise and torture, don't begin to cover the variety of contexts that authors invented over the years to render bodies transparent.[10] A plausible instance of embodied transparency and, moreover, one that does not strike your audience as tedious (you can rely on surprises and tortures only for so long) requires a careful combination of generic conventions, familiar cultural realities, and specific plot turns. It's a lot of work all around.

And we have already seen how quickly the established methods of creating absorptive contexts start feeling stale. As Fried argues, by the early 1800s, David felt that none of those methods were effective anymore and that neither expressive body language nor carefully thought-through narrative could salvage the represented body from the grasp of theatricality.[11]

What David and other artists perceived as a specific representational crisis is actually an expression of a broader cognitive challenge involved in constructing contexts of embodied transparency. We can call it a long-term challenge because it indicates the impossibility of relying on the same type of narrative construction for long. This is to say that there is a relatively short historical window of opportunity within which the audiences would buy the idea of the complete unselfconscious absorption of the sleeping, card-castle-building, and bubble-blowing subjects. We can see why after a short while what used to be a fresh and convincing visual narrative of absorption would ossify into a convention. That is, we can imagine a hypothetical portrait of a woman not merely blowing bubbles but *blowing bubbles*—engaging in an activity that is *supposed* to be absorptive—*performing* unselfconsciousness for the beholder and thus completely defeating the original purpose of the endeavor.

Then there is also what we can call a short-term challenge. An ideal context of transparency must come with the blueprint for its self-destruction. In fact, we can say that such a context is convincing in direct proportion to its fragility. Think back to Siddons's theatrical vignette. It captures the moment of transparency that cannot last because the particular emotional state of the "transparent" subject cannot last. At any point now the young man will shift his attention or grow self-conscious and thus lose that focused single-mindedness which is now written all over his body. In other words, it is because we are aware of the impending and inevitable loss of transparency that we are more poignantly attuned to its presence. ("Presentness is grace"—Fried again.)

There seem to be two different routes for arriving at the realization that transparency is convincing in proportion to its transience: by observing specific contexts of transparency, or by thinking through the implications of the work on Theory of Mind. I used the second route: taking as a starting point the view of the body as the best and the worst source of information about the mind, one can predict that all contexts in which the body seems to tell the truth have to be short-lived. This is to say that the inherent instability of moments of embodied transparency can be influenced by a broad variety of historically specific factors, but such subversive factors will *always* be available in one form or another: their perennial availability is determined by the nature of the phenomenon that such representations grapple with. Because they attempt to circumvent our double view of the body by imagining a context in which the body is completely readable—a barely sustainable state for our mind-reading species—they must remain unstable and vulnerable to subversion.

Hence I believe that Fried describes the same dynamic of transience, only that he arrives at it via the first route. Working closely with modernist as well as eighteenth-century art, he becomes aware of the peculiar handling of time in sculptures and paintings that sought to "defeat theater . . . by virtue of their presentness and instantaneousness" ("Art and Objecthood," 167). And so, as we are reading what he says about the construction of time by absorptive paintings (see the long quote below), we can see how well his insights mesh with those made possible by research in Theory of Mind.

Here is Fried on Chardin's bubble-blowing, card-castle-building, knucklebones-playing characters:

> Chardin's paintings of games and amusements, in fact, all his genre
> paintings, are also remarkable for their uncanny power to suggest the
> actual duration of the absorptive states and activities they represent.
> Some such power necessarily characterizes all persuasive depictions
> of absorption, none of which would *be* persuasive if it did not at least
> convey the idea that the state or activity in question was sustained for
> a certain length of time. But Chardin's genre paintings, like Vermeer's
> before him, go much further than that. By a technical feat that almost
> defies analysis—though one writer has remarked helpfully on Chardin's
> characteristic choice of "natural pause in the action which, we feel, will
> recommence a moment later"—they come close to translating literal
> duration, the actual passage of time as one stands before the canvas,
> into a purely pictorial effect: as if the very stability and unchangingness

of the painted image are perceived by the beholder not as material properties that could not be otherwise but as manifestations of an absorptive state—the image of absorption in itself, so to speak—that only happens to subsist. The result, paradoxically, is that stability and unchangingness are endowed to an astonishing degree with the power to conjure an illusion of imminent or gradual or even fairly abrupt change. (*Absorption*, 50)

To put it in cognitive terms, such canvases make us believe that we have the direct access to these people's minds *now* by making us expect to lose this access at any second. They effectively reinforce our anxious suspicion that other people's minds are never transparent by presenting *this* moment of transparency as an exception, an accident, a fluke. By doing so, they make us value this fluke—they encourage us to seize the moment and to look and look and look at it while it lasts. Or as Fried puts it, amplifying the view of Diderot and his contemporaries: A painting has "first to attract . . . and then to arrest . . . and finally to enthrall . . . the beholder, that is, a painting [has] to call to someone, bring him to a halt in front of itself, and hold him there as if spellbound and unable to move" (92).

To clarify: research on Theory of Mind does more than simply support Fried's insight. Paintings featuring subjects deeply absorbed in what they do are certainly not the only ones that have the power to bring the beholder "to a halt . . . and hold him there as if spellbound"—other works of art have that power too (or at least aspire to it). In explaining how they do it, we need to draw on cognitive, historical, aesthetic, and other factors; both the content and the combination of those factors would be different in each case. The work on Theory of Mind thus brings to light some of the cognitive factors that go into capturing the attention of the beholder in the case of absorptive paintings. To put it differently, that work suggests that there is a specific cognitive pattern—that is, our dual view of the body—that such paintings actively seek to exploit by their careful background plotting and their peculiar handling of time.

Theory of Mind and Sentimentalism

No discussion of eighteenth-century cultural representations of people gripped by strong emotions (e.g., the distraught family in *Le Fils ingrat* and *Le Fils puni*; the profoundly moved patriarch in *La Piete filiale*) can avoid the issue of the period's sentimentalism. Here, again, Fried's treatment of this issue renders his approach cognitive before its time, especially in the context of his larger view of the relationship between psychology and history. Looking at Greuze's *La Piete filiale*, Fried insists that correlating the effect that this painting must have had on its contemporaries with what we call eighteenth-century "sentimentalism, emotionalism, and moralism" does not really explain as much as we think it does when we evoke all these "isms." As he puts it,

> For a long time now it has been traditional, almost obligatory, to remark that we, the modern public, no longer find it in ourselves to be moved by the sentimentality, emotionalism, and moralism of much of Greuze's production. But the truth is that we take those qualities at face value, as if they and nothing more were at stake in his pictures; and that we therefore fail to grasp what his sentimentalism, emotionalism, and moralism, as well as his alleged mania for plotting, are in the service of, pictorially speaking—viz., a more urgent and extreme evocation of absorption than can be found in the work of Chardin, Van Loo, Vien, or any other French painter of that time. (*Absorption*, 55)

Translating what Fried says into the language of cognitive theory, we can say that perhaps when we talk about sentimentalism to describe effects of certain representational methods on the audience we confuse ways with means. Perhaps what we call sentimentalism is really a way—one of many—to smuggle transparency into the representation. That is, eighteenth-century writers and artists faced the same challenge that writers and artists always face: they needed to construct convincing representational contexts for forcing the body to reveal the mind. The fact that we now group some of their methods under the unflattering rubric "sentimentalism" shows again how quickly those methods become outmoded with frequent use and how vulnerable they are to parody and subversion.

As a related example of such a subverted context of transparency, consider the eighteenth-century sentimental novel, with its loving attention to blushing, crying, panting, fainting bodies. In Samuel Richardson's *Pamela* (1739), such bodily displays still stand for real feelings (see Mulan, 1–61), but in his next novel, *Clarissa* (1747–1748), they are already consciously faked for the benefit of naïve observers.[12] The term *sentimental* itself undergoes a change between 1740 and 1820. Originally neutral, "characterized by sentiment," or positive, "characterized by or exhibiting refined and elevated feeling," it acquires a pejorative mean-ing of "addicted to indulgence in superficial emotion."[13] In other words, keep looking at the emoting body in hopes that it will keep providing direct access to the person's mental states, and you will soon be treated to "superficial emotion," performed for your viewing pleasure.

But perhaps the negative connotations of the late-eighteenth-century term *sentimentalism* show something else, too. Think of how many nov-els, movies, and songs produced within any recent decade can be easily characterized as sentimental, not in the pejorative sense of the word, but in the earlier eighteenth-century sense: as "characterized by sentiment." That kind of sentimentalism is here to stay because what it does, again and again, is correlate body with mind in convincing social contexts—and we can never get enough of such correlation in our representations of the world.[14]

Now think of the effects of claiming that sentimentalism is an eighteenth-century phenomenon and that "we, the modern public, no longer find it in ourselves to be moved" by *La Piete filiale* the way Greuze's contemporaries did. On the one hand, common sense suggests that this claim is correct. Surely, in the 1760s, they must have indeed responded to *La Piete filiale* in some ways different from the ways we respond to it now, just as audiences in the 1960s must have responded to "Green, Green Grass of Home" differently than we respond to it now.

On the other hand, one practical effect of this claim is that sentimen-talism begins to seem safely *contained*—sealed off as a relic of a long-gone epoch associated with a very specific list of texts and works of art. And, so contained, sentimentalism becomes usable again. That is, what-ever writers and artists do now can be sentimental, but it cannot add up to "sentimentalism," for we have been done with that for more than two hundred years, haven't we?

And such containment and recycling are of course necessary given that authors are always in need of new ways to render the body con-vincingly transparent. The rubric *sentimentalism* covers a broad variety of representational methods, many of which can never really go out of use. In fact, we can say that when one method of forcing the body into transparency is declared passé and appended with a proper condescend-ing "ism," it is an indication that this method is now on the way to being recycled in a different guise and reinvented by a new genre or group of artists.

And so when Fried suggests that sentimentalism and emotionalism and mania for plotting were but the ways to the means—that they were "in the service of" bringing about convincing representations of absorption—a cognitive literary critic such as myself can't agree more. I agree because I see absorption in terms of transparency (that is, the absorbed person is transparent to the viewer), and because I see transparency in terms of the cognitive paradox underlying our everyday social functioning. Since we cannot help reading bodies for states of mind, and since we can never be sure that the states of mind that we are reading into the bodies are correct, we continue to be fascinated by representations that create illusions of privileged mind-access in complex social environments. Embodied transparency thus remains what Spolsky (Chapter Two, this volume) would call a "representationally hungry" problem: although "especially talented writers and artists" repeatedly turn their attention to this problem, the sociocognitive need that drives it can never be satisfied.

Notes

1. As Michael Fried describes it, "The transparent, slightly distended globe at the tip of his blowpipe seems almost to swell and tremble before our eyes" (*Absorption and Theatricality*, 51).
2. This section of the chapter first appeared in my essay "Theory of Mind and Fictions of Embodied Transparency" (*Narrative*, 2008), and I am grateful to Jim Phelan, the editor of *Narrative*, for letting me reproduce it here.
3. For an important related discussion of mirror neurons, see Spolsky, Chapter Two in this volume.
4. Unless, of course, our visual system is severely damaged.
5. A note on the imaginary approximations of real life mind-reading: So important is the mind-reading ability for our species, and so ready is our Theory of Mind to jump into action at each hint of intentionality, that at least on some level we do not distinguish between attributing states of mind to real people and attributing them to fictional characters. Figuring out what the fictional character is thinking as she is complimenting the protagonist on his reading choices feels *at least on some level* almost as important as figuring out what a real woman is thinking as she looks us in the eye and holds forth on how she enjoyed reading the same book that we currently have in our hands. Hence the pleasure afforded by following various minds in fictional narratives is to a significant degree a social pleasure—an illusive but satisfying confirmation that we remain competent players in the social game that is our life.
6. Zunshine, "Theory of Mind and Fictions of Embodied Transparency."
7. What Spolsky (Chapter Two, this volume) calls "the terminology of feeding" seems to be becoming crucial for discussion of the cognitive structures underlying human consumption and production of culture.
8. Also, this situation is quite peculiar in an ethical sense. On the one hand,

gaping at an individual who is known to be in such dire distress should be ethically questionable; see, for example, Spolsky's discussion in this volume (Chapter Two) for various cultural taboos on staring. On the other hand, we may speculate that by committing a heinous crime and by being presently known as having committing that heinous crime, the criminal has removed himself beyond the pale of such ethical considerations, and thus has become fair game for intrusive observation by strangers.

9. For a detailed discussion of this example, see Zunshine, "Lying Bodies."

10. As Diderot himself put it (he was speaking of theater, but what he said applies equally well to paintings and fiction), the scenes "of violent passion are not those that reveal superior talent in the declaiming actor nor exquisite taste in the applauding spectator" (quoted in Fried, 117).

11. Of course, new ways of rendering the body transparent arise all the time, but one needs a broader perspective, made possible by research on Theory of Mind, to recognize them, in their untold variety, as adding up to the enduring representational tradition.

12. See Zunshine, "Richardson's *Clarissa*."

13. *OED*; second online edition, 1989.

14. Incidentally, by "convincing" I do not mean "realistic"; for example, an otherworldly setup of a science fiction story can be completely socially convincing but not realistic in the conventional sense of the word.

CHAPTER NINE

Garden Paths and Ineffable Effects: Abandoning Representation in Literature and Film

H. PORTER ABBOTT

In Chapter Eight and elsewhere Lisa Zunshine has made a powerful case that fiction is no mere epiphenomenon of natural selection but a critical instrument in the maintenance and improvement of a key survival skill: reading the minds of others. In *How the Mind Works*, Stephen Pinker developed a contrasting take on the subject: "Fictional narratives supply us with a mental catalogue of the fatal conundrums we might face someday and the outcomes of strategies we could deploy in them" (Pinker, 543). There are still other quite plausible accounts of the evolutionary utility of fiction, but where such theorizing runs into difficulties is when literature fails to perform what scientists say it is good for, notably in the alien terrain of some modernist and much postmodernist fiction. Tellingly, the response to such reader-resistant texts by Pinker, E. O. Wilson, Ellen Dissanayake, and others has been to write them off as various forms of perverse indulgence—in effect, evolutionary misbehavior. In this they echo the outraged hosts whose disapproving response to works of the avant-garde has often been a key part of an intended effect.[1]

I believe humanists err in their response to such a critique when they argue that, strange as they are, such texts "capture in their own ways insights that Pinker and other cognitive scientists have been offering."[2] Such a response implicitly bases the value of apparent misbehavior in the arts on its contribution to theorizing in other fields. It also sidesteps the tricky question of what such texts do for the evolutionary well-being of the common reader, which may well be nothing. But even if no evolutionary value can be found for the perversity of these texts, this does not necessarily mean that their perversity is of no value. In this chapter, I focus on a species of textual derangement that yields a kind of experiential knowledge rarely, if ever, available to us in the ordinary course of our lives and that is worth having. What it conveys is different from the "how true" effect of mimetic fiction—"what oft was thought but ne'er so well expressed" (Pope 61)—or the rich science that Wordsworth argued was what poetry delivered when it was at its best: "the primary laws of our nature . . . not standing upon external testimony, but carried alive into the heart by passion" (Wordsworth 239–251). Readers have always known something like what Wordsworth claimed: that we lose ourselves in fiction, getting inside it, enjoying what Richard Gerrig has called "the phenomenology of being transported" (238). In narrative texts, especially, we feel and think with the characters in a process variously called projection, identification, empathetic understanding, or, currently, the exercise of our capacity to read minds. This is a rightly acclaimed distinction of fiction: it provides representations of human behavior that come alive in ways that the representations of analytical discourse do not.

In this chapter, however, I focus on three closely related ways of not doing what fiction is so acclaimed for. My demonstration texts each provide examples of a deliberate failure to meet common syntactical or narrative expectations, for which the principal template is the garden path of a garden-path sentence (this will come clear below). Moreover, in their egregious failure, they replace an art of representation with an art of cognitive states that are inaccessible or understandably avoided in the ordinary course of life. As such, though they would seem to serve no apparent evolutionary purpose, indeed may even interfere with evolutionary success, they generate real knowledge of who we are. *Wisdom* may be a better term than *knowledge*, since what is often involved is knowing what it feels like not to know.

Nouns Straining at the Syntactical Leash

To get to my subject, let's first clarify what a garden-path sentence is. Here's one:

Fat people eat accumulates.

This is a perfectly grammatical sentence. But by inviting us at the outset to apply an inapplicable syntactical template, it reads like nonsense. We are led down a garden path (hence the term) that dead-ends, and it is only by a recursive effort (going back and finding the right syntactical path) that we find its meaning. On first reading, customary usage will invariably read the word *fat* as an adjective modifying *people*. But doing so will then require *accumulates* to be read as a noun (what fat people eat). The only way to restore *accumulates* to its proper status as a verb is to read *fat* as a noun. Then it makes sense. Garden-path sentences usually rely on words like *fat* that can have more than one grammatical status (or meaning), but also, and often crucially, on a little bit of grammatical discourtesy as well. The writer has left out a couple of helper words that any editor would note: "*The* fat *that* people eat accumulates." Here are a few more:

> The old man the boat.
> The player kicked the ball kicked the ball.
> Mary gave the child the dog bit a band-aid.
> All women who like a man who paints like Monet.[3]

The term "garden-path sentence" was coined by the linguist Mitchell Marcus in 1980 as central evidence for "the incremental parsing of natural language" or "discourse-as-process." The argument is that, to some extent, we understand a sentence incrementally, that is, word by word. This is still an area of some contention, but garden-path sentences do foreground a challenge in the field of computational linguistics and the construction of computerized translation programs, which generally rely on whole sentence reading. At the same time it underscores how much the disambiguation of sentences can be assisted by an understanding of context—something that it is very hard to train a computer to do. On your computer, you often see this inability to disambiguate even mild garden-path effects when your word-processing program, trying to be helpful, warns you that there is a problem with your grammar when in fact there isn't.

There can be a pleasure in reading garden-path sentences, and one might make the same kind of evolutionary argument for this pleasure that Zunshine makes for the way novels "test the functioning of our cognitive adaptations for mind-reading while keeping us pleasantly aware that the 'test' is proceeding quite smoothly" (Zunshine 2006, 18). With garden-path sentences, what is pleasurably honed is our ability to disambiguate challenging sentences. They fine-tune our language motors. This is no doubt why jokes often deploy the garden-path effect:

> Time flies like an arrow, fruit flies like a banana.
> Take my wife, please!

It is fun to play like this with language, and I think the pleasure is sharpened by the way such a sentence tempts linguistic chaos in the split second before we parse the sentence correctly and restore linguistic order. When you get the joke, chaos is brought under control by your own acuity. It is a way to achieve the excitement of living on the edge. The risk is not as great as taking a curve at high speed on a motorcycle, but losing control of one's language can be hard on the mind. How painful it is anyone knows who has struggled with a language poorly understood amid a group of native speakers.

In one's own language, garden-path effects usually arise from the relative incompetence of a would-be communicator or the challenge of meeting economies of space, as in news headlines ("Officer shoots man with knife"). But sometimes it can be inflicted intentionally by a writer of certified competence. Here's a sentence, one of many, by a canonical modernist:

Any little thing is a change that is if nothing is wasted in that cellar. (Stein 1998, 350)

In this instance, the garden-path effect takes hold in the phrase "that is if" at which point readers must abandon the expectation of some kind of modifier of "change" (e.g., "Any little thing is a change that is welcome"). Instead they must retroactively insert a different grammatical understanding of "that is" that leaves "change" unmodified and introduces a contingency that in the world of the sentence would keep any little thing from being a change. The grammatical discourtesy in this case is the absence of two commas that would signal the sentence's altered course ("Any little thing is a change, that is, if nothing is wasted . . ."). What makes this sentence different from other garden-path sentences is that the syntactical correction does not deliver the relief that ordinarily comes with "getting" the meaning. And this is because there is insufficient information through which we can establish full confidence in our reading. The nouns and the verbs are referentially impoverished, with context providing only a modicum of help. The sentence comes from the section of Gertrude Stein's *Tender Buttons* titled "Rooms," and a cellar is a room, so it belongs. But the deictic modifier ("that") points to a specific cellar that we know only through this sentence. Context also helps somewhat in figuring what qualifies as "any little thing." "Any" is a repeated modifier of things in the neighboring paragraphs ("any smile," "any coat") amid an abundance of other things, though some not so little ("a can," "a cape," "a hill," "a curtain," "a package and a filter and even a funnel"), and some not so concrete ("a measure," "a success," "a religion"). We also learn that "when there is a shower any little thing is water" (348). But all this does not help (me, at least) to understand what it is in the cellar that might be "wasted," nor how the presence of waste would keep any little thing from being (making?) "a change."

So, whatever pleasure we may take in the sentence's grammatical and lexical swerves and indeterminacies, it is not the same pleasure as that of a joke. The pleasure of the joke depends on the small triumph of getting it. But here there is nothing to get without serious under- or over-reading. Instead, for the mind that refuses premature closure, the pleasure of the text is necessarily threaded with a feeling of anxiety or frustration in the face of the indeterminable.

Stein has certainly taken her lumps for all this linguistic havoc, even from sympathetic critics. David Lodge claimed she went too far in violating "the very essence of her medium, language." It is all "exhilarating," he wrote, but "the treatment is so drastic that it kills the patient" (Lodge 1977, 154). Wendy Steiner made pretty much the same criticism when she blamed Stein's botched syntax for robbing her of even a minimal chance to make a significant modernist contribution (156). By contrast, those who defend Stein stress the way she liberated language from the "real world" constraints of reference, practical usage, and even the order of time and space. In Neil Schmitz's words, "the denoted world collapses," nouns float free from any "clarifying knowledge of the nature of things, and . . . nothing can be named and then classified, given as real" (cited in Perloff, 101). This comports pretty much with Stein's own view that in "Tender Buttons" she was "doing nothing but using losing refusing and pleasing and betraying and caressing nouns" (Stein 1971, 138). In rough agreement, Brian McHale described her style as "words disengaged from syntax" (153).

But I don't believe it is possible to disengage from syntax—that is, when sentence or sentence-like expectations are cued. Certainly there are moments in this text when Stein makes rhythm and sound so predominate that the reader is truly released from the prison house of syntactical constraints.

Alas a dirty word, alas a dirty third alas a dirty third, alas a dirty bird. (341)

A no, a no since, a no since when, a no since when since, a no since when since a no since when since, a no since, a no since when since, a no since, a no, a no since a no since, a no since, a no since. (344)

This way of writing dominates a poem like Stein's "Susie Asado." But not "Tender Buttons." And when sentence structure begins to take shape, as it does almost always in "Tender Buttons," the quality of Stein's lexical swerves takes on a different coloration. Commonly the syntax of her sentences can be quite clear, with no garden-pathing:

A can containing a curtain is a solid sentimental usage. (348).

Yet it is the syntax that elicits a feeling of "wrongness." Without the perception of sentence structure, Stein's semantic disorder is much more purely a lexical romp. With it, you feel a syntactical resistance as the nouns pull on your expectations.

Almost very likely there is no seduction, almost very likely there is no stream, certainly very likely the height is penetrated, certainly certainly the target is cleaned. (349–50)

Stein's irresolvable garden-path sentences, then, are simply another hard, sharp turn of the screw in this gallery of disturbing effects. In these sentences, the conflict between syntax and semantics is compounded by a conflict in the syntax itself.

Come to season that is there any extreme use in feather and cotton. (313)

Releasing the oldest auction that is the pleasing some still renewing. (349)

There is no avoiding an intimation of chaos when cued syntactical expectations are redirected, and, even more, the discomfort when, in the end, the syntactical uncertainties fail to resolve. In short, our cognitive grooves are too deep. Stein must have been aware of this and in fact depended on it. However much her writing may resemble the chaotic speech of aphasics suffering from Jakobson's "contiguity disorder" and "contextual deficiency" (Lodge 1977, 148), Stein herself was no aphasic. This is an important point. From her earliest years, she was, like her readers, irreversibly stamped with linguistic competence. So she must have known what she was inflicting on her readers and depended on it as an integral part of the effect she wanted. If syntax can be considered a restraining (paternal) hand, then that hand is felt almost everywhere in the reader's transaction with this text. It is there in the sensation of its resistance. In fact, the feelings of liberation and play are inseparable from the feeling of syntactical constraint. To put this another way, the sensation of wildness in the sudden flights of the nouns is enabled by what constrains them—we only know their weird excess by the syntactical tether that wants to hold them back.

"Tender Buttons" is a strange text. Well beyond paraphrase, it can only be known by entering into it. And once in it, you know it, whether you want to or not. This may be what Stein meant in "Composition as Explanation," when she wrote paradoxically of "everybody in their entering the modern composition" that "they do enter it, if they do not enter it they are not so to speak in it they are out of it and so they do enter it" (Stein 1998, 521). I have begun with "Tender Buttons" because it creates the unparseable garden-path effect in its rawest form. In so doing it goes much further down the line than Oscar Wilde ever did in fulfilling the standard that "all art is quite useless." It is like an evolutionary dead-end, where strategic adaptation is displaced by an art that succeeds at the expense of our survival skills. In this it accords with Stein's aestheticism and the value of "the art that lasts": such an art "is an end in itself and in that respect it is opposed to the business of living which is relation and necessity." The masterpiece, she wrote, "has nothing to do with human nature or with identity, it has to do with the human mind" (Stein 1971, 358). And this is my focus: she brings us into a space in our minds that we rarely visit. We may enjoy this or we may detest it, but it lets us feel the enormous power of syntax, to know something of how deeply our minds are invested in it, and the vertigo we experience when its power is challenged.

A Poetics of Syntactical Parsimony

The demonstration sentences in my second set are at once similar to Stein's and quite different.

> Here without having to close the eye sees her afar. Motionless in the snow under the snow. The buttonhook trembles from its nail as if a night like any other. (Beckett 1996, 68)

These are three consecutive sentences from Samuel Beckett's haunting late work, *Ill Seen Ill Said*, each one of which, as you've probably noticed, is a garden-path sentence. In the first, the garden path terminates in the phrase "to close the eye sees," in which the noun *eye* must be converted from an object to a subject in order for the sentence to make grammatical sense. The garden path in the second would have the snow under the snow, rather than her standing in the snow and under the snow that falls on her. In both of these sentences, the "discourtesy" that enables the garden-path effect is a missing comma. The moment of hesitation in the third sentence comes in the phrase "as if a night," which runs astray for want of a verb: "as if [it were] a night like any other." Technically, the verb has been there all the time, since, once we get the sentence right, we feel its presence even without seeing it in print. But less is more. Beckett's syntactical parsimony guarantees that we also cannot *not* be aware of the verb's absence and feel the mute impedance that comes with its visual suppression.

Once you "get" these sentences (and, unlike Stein's, they can be gotten), they come together with an austere and eerily precise beauty. Eerie, because the hesitations of a first reading remain a part of the whole effect, however adept readers may get at hearing and grasping the "proper" syntax of these sentences. Precise, because in this way more is said than grammatical correctness will allow. And much of this more, as the title of the work implies, is that its difficulties of seeing and saying are a single condition: perception as expression and expression as perception. Infected throughout its system, each of the work's sentences is an instance in miniature of this "illness." In each, the reader feels the struggle of seeing as saying and saying as seeing. This is no mere case of what Ivor Winters called "imitative form," in which the form of the work imitates what the work is about. "Here form *is* content, content *is* form," as the young Beckett wrote of *Finnegans Wake*. "His writing is not *about* something; *it is that something itself*" (Beckett 1929, 14). The difficulty of this text, then, is no mere signifier of difficulty but, like the difficulty of Stein, inseparable from its condition as art. And, as in Stein, the syntax of this art is critical to its success.

Beckett is a "sentence man," as Hugh Kenner originally observed, and as such the work of his maturity stands in marked contrast to the lush verbal art of Joyce the "word man." From this perspective, the verbal pyrotechnics of Beckett's apprenticeship in the 1930s were a legacy of the master, which, through constant formal exploration and exile in a foreign language, was refined to an art of syntax, "the local order of language" (Kenner 1990, 293), which only grew more austere and subtle as he aged. In a powerful extension of Kenner's insight, Ann Banfield has analyzed how Beckett increasingly relied on words whose meaning is almost entirely a register of their syntactical function. Unlike the nouns, verbs, adjectives, and adverbs so rich in semantic content that Joyce gathered together and cross-germinated, Beckett's work depended more and more on the "nonproductive" words, the pronouns and determiners that "lack highly specified semantic content, having only cognitive syntactic features." Joseph Emonds called these, collectively, "the syntacticon," and in Beckett's late work they predominate along with semantically "'light' nouns, verbs, adjectives, and prepositions," words like "be" (is, are), "one," "self," "thing," "have," "go," "let," "say," "other," "same," "mere," "such" (Banfield, 16–17). Citing Emonds, Banfield notes that "the philosophical vocabulary is drawn from the syntacticon," which in Beckett's hands is the instrument of a negative quest for the beyond of language, a goal he first wrote of in his essay on Proust and, Banfield argues, never gave up:

> "What is the wrong word?" (56) is *Ill Seen Ill Said*'s refrain. For the ill-seen haze, not direct sensation's language of precision but one itself vague, semantically light, "miscalled" (63). "Any other would do as ill" (93). One is not mined in the riches of the dictionary, with its adjectives of infinite shades and qualities and nouns filled to the brim with meaning, but out of the poverties of the syntacticon. For Beckett's minimalism is a semantic lightness. "Proust does not share the superstition that form is nothing and content everything" [Beckett 1931, 67], Beckett observes. So the late style is extracted from the lexical formatives with the least semantic content. (20)

One paradox of Banfield's take on Beckett's style is that his pursuit of an art in which "the object is perceived as particular and unique and not merely the member of a family" (Beckett 1931, 11) required letting go of much of language's semantic specificity. Thus, in *Ill Seen Ill Said*, "'this old so dying woman' is never named (apart from this single definite noun phrase) but only 'pronamed' by a function word . . . with no meaning but the syntactic features of feminine gender, third person, and singular number" (19).

I think Banfield is right. However, it is not only the semantic vagueness of his language that works for Beckett but also the contortions of his syntax. This is my focus. Banfield's paradox that Beckett's precision requires semantic lightness overlies another: that it also requires the serial impedances of his syntactic parsimony. And this again brings us into literary terrain that is neither representation of the world nor postmodern world-building. If I am right, Beckett's project is to return us to a feature of the world we are always in, one that is neither fiction nor fact, neither "figment" nor its "counterpoison," an object of neither the "vile jelly" of "the eye of flesh" nor that "other eye" in "the madhouse of the skull and nowhere else" (58), and (fleshly or other) neither clearly the reader's eye, the author's, or the subject's. It is a deep "confusion" of "Things and imaginings. As of always" (58).

> Weeping over as weeping will see now the buttonhook larger than life (57)

Though there was enough of Joyce the punster left in Beckett to enjoy the impoverishment of "weeping willow" in "weeping will," the energy in this sentence is syntactical. It lies in the opposing tugs of "will see" and "will, see." "Will see," if it could work syntactically, would become an implicit "she will see," and the sentence would resolve into a narrative in the third person (e.g., no longer weeping, she will now see the buttonhook). But this would make the tautological "weeping over as weeping" the subject modifier. Disambiguated, the sentence requires the verb *see* to take the imperative mode with "weeping over as weeping will" as the modifier (e.g., weeping at an end as weeping always does come to an end, see now the buttonhook). But then who is addressed? Who is being asked to see? There are three possibilities—the reader, the narrator, the woman—each of which is curiously encumbered. If it is the reader, then it is the reader who has been in tears and thus unable to see the buttonhook until finished weeping. Likewise for the narrator, who is also by this construction addressing himself. If the woman, then the narrator is no longer describing her and her actions but giving her directions as well.

In my view, all of these options are meant to apply. By abusing our hard-earned evolutionary capabilities in this way, forestalling any imposed clarity, Beckett makes us feel what happens when the borders separating the imagined, the real, the seen, the said, and the seer are not just erased but erased *and* maintained. On the one hand, the distinctions of the true and the false, the self and the other, are constructions embedded in language. On the other hand, they seem, for all their "constructed" nature, to correspond to distinctions that have a nonlinguistic existence. So Beckett seeks not to abolish them, but at one and the same time to abolish and promote them. Again, the exquisite lessness of Beckett's syntax is a moreness of sensation as he pursues his project of driving language ever "worstward," seeking at every turn to "fail better."

It brought him to *Worstward Ho*, the final text of the "second trilogy," in which the quality of music, always present in Beckett's verbal art, is accentuated almost to the point of displacing the dramas of syntactical meaning.

> Stooped as loving memory some old gravestones stoop (Beckett 1996, 115)

An old woman in a graveyard is seen "stooped as [in] loving memory some old gravestones stoop." Or is it "stooped as, loving memory, some old gravestones stoop"? Or is it "stooped as loving memory [is stooped], some old gravestones stoop"? Again, I would argue for all three—the memory of one loved, loving memory, and the burden of loving memory—three states of mind that are at once different and susceptible to melding. Yet at the same time, in the same sentence, Beckett has created a perfectly symmetrical cataleptic alexandrine, bookended with the strong chiasmatic chiming of *stooped* and *stoop*. The sentence moves toward a pure evocation of rhythmical sound, a music that threatens to assert its own priority, absorbing all of the interest. Yet it doesn't, and that's the important point. This is the case even when the effects are so compacted as to approach a kind of nonsensical hilarity:

> Not that as it is it is not bad (99)
> Whenever said said said missaid (109)
> So far far far from wrong (110)

It has often been observed that if you repeat a word often enough it loses its meaning and becomes a moment of pure sound. Beckett loves to test the limits of this effect, but, uniquely, he does so without relieving readers of their syntactical obligations. In each of these brisk passages, modest words drawn from the syntacticon are rung like bells, but only by observing the rules of the syntactical game (whenever said [is] said, said [is] missaid). And even when you do, uncertainties can remain (is it *said* that is "missaid" or is *said* meant to be read as *missaid?*). In short, the struggle for meaning and the release from it are experienced simultaneously, but you can only know this by going into it.

Reuven Tsur proposed in "Two Critical Attitudes: Quest for Certitude and Negative Capability" that readers and critics can be ranged on a scale between those determined to find interpretive certitude at all costs and those who adopt what Keats called "negative capability" or a capacity "of being in uncertainties, Mysteries, doubts, without any irritable reaching after fact and reason" (Keats cited in Tsur, 776). The latter attitude, he argued, depends on a willingness to shift "mental sets" without settling on any one as final and requires a high degree of tolerance for "uncertainty and ignorance" (787). Tsur's distinction between these two critical attitudes complements a similar distinction between works themselves—those that invite certitude and those that require openness to a plurality of meaning. The latter certainly applies to much of the work of both Stein and Beckett, and indeed Tsur's prime example of a work demanding negative capability is Beckett's *Waiting for Godot*. But what I have been focusing on in Stein and Beckett requires, I believe, an attitude that goes one step beyond the far end of Tsur's scale—it does, that is, if I am right in thinking that the critical attitudes Tsur describes are still attitudes toward representations. As such, negative capability is an openness to the meaning of what is represented, but it does not deny that representation is going on. The reading of Beckett (by Günther Anders) that Tsur features to demonstrate this attitude can therefore assert that part of the meaning of Estragon's game of "shoe off, shoe on" is that we, the audience, in our everyday lives, are doing "nothing but a playing of games, clownlike without real consequences" (Anders cited in Tsur, 786). But part of the function of the impedances that I have been examining in both Stein's and Beckett's works is to foreclose such interpretive moves, taking their creations outside the realm of representation altogether.

Critical attitudes of whatever degree of subtlety, including Tsur's version of negative capability, tend to draw even the most bizarre artistic departures back into the realm of *aboutness*. But in Stein's definition of "Master-Pieces," she distinguished between a "thing to see" which belongs to the world of identity and human nature and a "thing to be" which is the rare work of great art (363). This is a leading principle of Stein's aestheticism much as it was a leading principle in the young Beckett's description of Joyce's *Work in Progress* that I noted above: "His writing is not *about* something; *it is that something itself.*" This in turn may account in part for why a significant number of modernist and postmodernist authors tend to avoid interpretive commentary on their own work. Indeed it often seems that there is a kind of battle going on in which the successive experimental departures that characterize modernist and postmodernist oeuvres are a form of evasive activity. Yet as fast as writers generate an art of *isness* through modes of unfamiliarity, the academy reconstructs an art of *aboutness* through modes of familiarity. What I have sought to do here, however, is to avoid any such aboutness and rather to triangulate as best I can the immediate cognitive/affective states in readers that, if their reading is done right, complete that "something in itself" that is served by the text. At the same time, I have argued that we have in the work of these two authors an art that achieves its effects through a deliberate and constant abuse of what has been undoubtedly our finest instrument in the work of species survival. Finally, whether this can be called aestheticism or not, I hope it is clear that the states of mind generated by this art, as it stretches understanding beyond the limits of language, have an importance that goes beyond an art pursued for its own sake.

Wars of the Worlds

In an ingenious essay, "Speak Friend and Enter," Manfred Jahn has expanded the concept of garden-pathing to include, among other discourse forms, narrative. His narrative examples are James Thurber's "The Secret Life of Walter Mitty" and Ursula Le Guin's "Mazes," both of which start out by leading the reader down a narrative garden-path in one kind of world, only to require that world's recursive reconstruction as another. As with garden-path sentences, there is a range in the way the effect takes hold and is exposed: quickly in "Walter Mitty," gradually in "Mazes." In works like Agatha Christie's *The Murder of Roger Ackroyd* (1926), Ian McEwan's *Atonement* (2002), and M. Night Shyalaman's film *The Sixth Sense* (1999), exposure of the garden path comes quite late in the narrative and with little warning, so that these works almost demand rereading in order to understand the extent to which the world the reader/viewer had accepted as the actual world of a traditional mystery or novel or film was in fact another kind of world. This need to go back and immerse oneself in the world of the narrative can be a significant difference between the garden paths of narrative and grammar. In *Ackroyd* and *Atonement*, the entire narrative world is discovered to be one in which the narration itself is a complex act of concealment and distortion.[4]

In *The Sixth Sense*, the garden path requires an even more radical revaluation in that it requires a revised understanding of the way the world itself works. A psychiatrist, Malcolm Crowe, discovers late in the film that he had not survived an attack at the film's outset. Ever since that early point in the film, he, along with most first viewers, had assumed he had survived the attack. But in the actual world of the film he does not survive the attack and from that point on has been a ghost, visible and audible only to the audience and a clairvoyant child, Cole, whom he has been trying to help.

The Sixth Sense is an instance of what Edward Branigan has called breaking the frame of "world knowledge" by "passing through to a world that has new laws" (288 n. 64). But unlike other instances Branigan cites (*Through the Looking Glass*, *Dracula*), the break in *The Sixth Sense* returns us to the world we thought we knew, but now reconstructed. In its world–re-creating radicalism it makes vivid what is true in the other examples above: that garden-pathing in fictional narrative is a matter of "worlds" or more precisely, "storyworlds." These worlds are what correspond to the syntactical alternatives in a garden-path sentence. And just as in the garden-path sentence a grammatical crisis is brought on by the pressure of grammatical impossibility, in a garden-path narrative an ontological crisis is brought on by the pressure of an ontological impossibility—an impossible world—which must be absorbed into the actual world of the story for the narrative to be coherent. For *The Sixth Sense* to maintain its narrative coherence, the world in which Dr. Crowe appears to be alive must give way to a different world in which he has died, an enlarged world governed by different rules. Much has been written in the last twenty years about the ways in which narrative is not just a matter of linear action but of worldmaking (Ryan, Gerrig, Hernadi, Doležel, Herman), but few narrative moves make us so vividly aware of this as a garden path with its necessary cognitive recalibration of worlds.[5]

Central to Jahn's expansion of garden-path effects is his adoption of Ray Jackendoff's concept of "preference-rules": that hierarchy of probabilities or "non-necessary conditions" packaged in frames or scripts that give them their efficiency and "cognitive power." The frame of a restaurant or the script of a waiter's behavior contain certain data that are always true ("necessary conditions"), but they are also packed with assumptions of varying degrees of probability ("non-necessary conditions").[6] Michael Haneke's award-winning 2005 film, *Caché*, for example, starts with a short garden path: a street-level view of apartments that most first viewers take for a conventional establishing shot in the film's storyworld. But the credits come and go, and still the scene persists. The camera remains fixed and nothing happens except for the passing of four pedestrians and a bicyclist. After two full minutes have passed, we hear the sound of intermittent comments of indeterminate import by people who cannot be located anywhere in the scene ("Well?" / "Nothing." / "Where was it?" / "Out front in a plastic bag." / "What's wrong?"). Finally, after just under three minutes, the camera shifts to a different angle in the same locale to follow the TV culture journalist, Georges Laurent (Daniel Auteuil), as he emerges from the apartment, crosses the street, points in the direction from which the first shot was taken, says to his wife, Anne (Juliette Binoche, now visible in the doorway), "He must have been there," and goes back into the apartment. The camera then reverts to the fixed scene it began with, there is more dialogue, and, eventually, a fast-forward that puts it beyond doubt that what we had been watching is a tape, recorded at some earlier time.

Here, what corresponds to the expectations generated at the level of the sentence by syntax are worldmaking expectations generated at the level of narrative by the cognitive frames and scripts (including, crucially, those belonging to the special language of film) that are cued as the narrative begins. Any sustained street shot includes a high probability of pedestrians and a bicyclist. Voices without perceivable speakers are less probable, yet still possible. Such voices loud enough to sound as if they are quite near are less probable still, but conceivably coming from people just outside our range of vision. The cognitive tension increases with more improbabilities: the speakers fail to appear, they continue to make mysterious references, and, perhaps most disturbing, the scene itself grows unconventionally tedious with its lack of significant action or camera movement. Perceptive film veterans may make the "naturalizing" leap before Georges appears, grasping that the scene itself is the subject of the discourse. For some the appearance of Georges going out front on the heels of Anne's answer to his question ("C'etait où?") will do it. As for the stragglers in the audience, still crushed under the weight of improbabilities, liberation comes with impossibility: that the actual world cannot go on fast-forward.

As Jahn writes, garden-path effects challenge our overriding preference for "maximum cognitive payoff": "What [leads] one astray is one's anticipation of good sense, and this is itself a consequence of a human understander's unwillingness to consider nonsense, of a general horror of semantic emptiness, of a craving to make satisfactory sense of a discourse, in short, of preferring what makes sense or even most sense" (177). Correlatively, a minor pleasure of Haneke's opening strategy comes when the viewer understands how the opening belongs to a coherent filmic universe. As in the garden-path sentence, the pleasure depends on there being an initial resistance to sense.[7] In this film, it is an increasingly vexed pleasure as the device recurs in different variations, each with its small payoff of surprise and recovery. A close approximation of the first tape turns out to be in the filmic present. A sequence seemingly in the filmic present, first through the window of a moving car, then down the hallway of a seedy apartment building to a certain door, turns out to be another tape in the filmic past. Later, the same sequence down the hallway turns out to be in the filmic present when Georges moves into the frame to knock on the same door. A scene of Georges anchoring a discussion of Rimbaud turns out to be a tape that Georges is editing. In most of these and other garden paths, there is the play of probability, improbability, impossibility, and resolution that delivers the small pleasure of successful parsing. Thematically the device works very well, since much of the film is about the perils of certainty. The palpable tension of the film is in large part a product of Georges's adamant refusal to acknowledge the guilt he feels for lies he told as a child, lies with terrible consequences for Majid, the Algerian orphan his parents had loved and might have adopted. Georges's persistent denial, dissembling, and belligerence are clearly meant to parallel the way France as a nation has failed to address its own collective feelings of suppressed guilt. Haneke's garden-path editing, then, re-creates this theme of the perils of certainty as a condition of viewer response by repeatedly reminding us how easily we can go wrong in our assessment of what it is we are seeing.

"Every path leads to nothing," Haneke has said in reference to *Caché*, "The truth is always hidden" (Haneke 2006). Well, yes and no. On the one hand, there is extraordinary truth in Auteuil's performance of a cerebral TV celebrity, harried out of the blue by anonymous, incriminating tapes and pictures. It is a highly skilled performance, and for it Auteuil was awarded Best Male Actor in the 2005 European Film Academy Awards. In fact, across the board the acting in this film is exceptionally convincing, and it gives the film much of its power. To this degree the film is clear. On the other hand, what is not clear is the question of guilt and how to deal with it. Terrible things were done, both in French-Algerian relations (Majid's parents died in a historically documented massacre in 1961 when a rally called by the FLN was broken up by the police and two hundred French Algerians were thrown into the Seine to drown) and in Georges's relations with Majid when his lies led to Majid's expulsion from his family. But Georges was only six at the time, and by keeping us close to him as he struggles with his literal and figurative nightmares, Haneke also keeps us from any easy judgments.

Similarly, with regard to Haneke's garden-path filming, one can say that it is not entirely true that "every path leads to nothing." Repeatedly, as the examples above show, we do find the right path. We make our mistakes and we successfully correct them. Where we lose our way lies not, with one exception, in navigating the issue of film present or taped past, but in figuring out who is doing the taping and sending the pictures. The chief suspects, Majid and his son, involve major contradictions. Moreover, they are thoroughly convincing in their denial of having anything to do with it. This is a drama of attribution that we share with the Laurents and that could conceivably go on forever. Compounding this mystery is the final shot in the film, which is also the major exception to our ability to resolve the question: is it in the filmic present or is it another surveillance tape (perhaps even another installment in the harassment of Georges)? It serves as a bookend with the opening shot, but with this difference: where we were allowed to recognize the opening shot as a garden path, we are not allowed to determine whether this one is. It is a shot of school steps at the end of classes with young students doing all the probable things students do at such a time, milling about, chatting in groups, getting in cars. But it is also, like the opening shot, a fixed, "unprofessional" piece of camera work, stretching through the credits, and seeming to go on forever. In this it violates both what would normally be the preference rules for conventional film and the garden-path expectations to which we have become accustomed.

Compounding this difficulty, the Laurents' son, Pierrot, and Majid's unnamed son meet in the upper left-hand corner of the scene (something missed by many first viewers), then descend the steps in animated and seemingly amicable discussion, and finally part. This is the first we learn that these two boys even know each other. How long have they known each other? Are they friends? Are they perhaps accomplices? Have they been involved together in the plot against Pierrot's father? Or is Majid's son cultivating Pierrot in order to visit upon the son further vengeance for the sins of his father? There is simply no way to answer these questions. Haneke has said of an earlier film that it "should not come to an end on the screen, but engage the spectators and find its place in their cognitive and emotive framework" (Wheatley), and viewers at Cannes did their best to oblige. According to Mark Lawson, "this sequence proved to be a sort of Rorschach montage in which viewers saw what they expected or wished to see. Different observers believed that they had detected evidence of a death, a divorce or a dispute over paternity. A critic who is a conspiracy theorist saw a clue to the involvement of the French secret service" (Lawson).

All of this frenetic activity is interpretation by supplementation (creating meaningful coherence through strategic alterations of the text as given). And though this may be a sign of intellectual vigor, it may also be the better part of wisdom not to give in to it, that is, to accept our ignorance. At the level of representation, there is much powerful material in the film that we do grasp, but there is also much that we don't. We don't know what the boys are up to, we don't know whether Anne Laurent is having an affair with one of the Laurents' friends, we don't know much at all about Majid or what else may have contributed to the grotesque and ingenious vengefulness of his suicide. After all, to accept that we don't know what we can't know, that there will always be unconnected dots, does in fact harmonize with the central theme of the film referred to above: the danger of premature judgment, of assuming knowledge of what is hidden.

But there is one more twist to this film and to what constitutes a full response to it. This brings us back to the question of who is making the tapes and sending the pictures. It also brings us back to my main theme. So far, we have been dealing with questions of what is knowable in this narrative and what remains a mystery. And certainly the question of who is doing the taping would seem a preeminent mystery. Watching the first tape, Georges and Anne see Georges as he leaves for work in the morning. He walks right toward the camera, passing to its right. "How come I didn't see him?" he asks. Maybe the camera was in a car? No, it doesn't look like it was filmed through a window. "It'll remain a mystery." But actually our problem is not that we know too little about the taping, but that we know too much. For we have here an agent who influences the course of events;[8] who is aware of intimate details from a past known only to one, possibly two, characters besides Georges; who has a capacity to tape very close up without being seen; who can catch a key event at a time impossible to predict and from a spot that is clearly visible during the scene's enactment; who can deliver a tape through a closed security gate; and who can post drawings that not only depict intimate details of what has happened in the distant past but also strongly suggest what will happen in the future (Majid's suicide). In short, we have been garden-pathed not in our serial assumptions about *who* is doing the taping but in the assumption that it is being done by someone at all, that is, by a human agent who belongs to the film's storyworld. Rather, given the set of details we have to work with, such a participant must be something else altogether. A supernatural agent capable of intervening in a storyworld sufficiently enlarged to accommodate the supernatural? An extra-diegetic agent capable of making metaleptic forays into the film's storyworld? We simply will never know.[9]

This, I think, is an altogether different kind of mystery from those belonging to the storyworld in which the rest of the action takes place. In such a world, the preference rules consistent with that world would require any taping to be done by a human agent, however hidden. When Haneke violates this rule, he creates a different kind of mystery from the one Georges had in mind when he said "It'll remain a mystery." It is an ontologically disorienting mystery, a garden-path sequence that can't be "gotten" the way the opening sequence and its successors can. Even while the tapes and pictures and the manner of their delivery are doing their efficient and chilling work in opening up the psychological and ethical dimensions of this film, they also introduce an eerie extra-dimensional element that runs crosswise to the grain of this otherwise powerfully realistic film. In this they give new meaning to the title of the film, *Caché*. The often ingenious hiding of other kinds of information in this film is, like the ethical issues it raises, basically an extension of the realistic tradition in film. When Haneke calls on his viewers to take over the job of completing the film as they wrestle with its ethical issues, he is asking them to join with him in what is basically a humanistic project.[10] But with regard to the agent behind the tapes and pictures, the viewer cannot proceed in the same way without seriously underreading. To respond fully to this aspect of the film with its manifest impossibilities is to experience a jamming of cognitive schemata that leaves the viewer stranded between worlds.

I have chosen *Caché* for my demonstration text in this section because it includes narrative versions of both kinds of garden path: those that can be parsed and those that can't, the latter a narrative version of the jammed garden-path effect that I dealt with in the sentences of Stein and Beckett. It is not hard to find much more thoroughgoing narrative versions of unparseable garden paths in postmodern literature. There are many in a novel like Alain Robbe-Grillet's *Dans le labyrinth* (1959), including the oft-cited scene in which the minute description of a nineteenth-century etching, "The Defeat of Reichenfels," almost imperceptibly transforms into a continuation of the narrative. Critical response to *Dans le labyrinth* has not uncommonly attributed this and other narrative impossibilities to narrator exhaustion. But in Brian McHale's words, "This is an 'expensive' reading, in the sense that it requires us to smooth over a good many difficulties and to repress the text's own resistance to being read this way" (14). Similarly, readers of J. M. Coetzee's *In the Heart of the Country* (1977) have at times sought to "naturalize" its wealth of impossibilities as instances of narrator insanity.[11] In the terms I have been using, exhaustion and insanity are default "possibilities" that allow the reader to re-frame the narrative as symptom and to relocate its action from a fictional actual world of the narrative to a fictional mental world of the narrator. And though, as McHale says, both interpretive moves require strategic reductions of the texts involved, it is certainly understandable why readers should want to make them. Narratives, like sentences, can be intolerable when they can't be parsed. Our biological and cultural evolution did not give us the cognitive engineering to accommodate them with ease. But if you are going to do justice to a novel like *In the Heart of the Country*, you cannot opt "for psychological realism," as Coetzee himself put it: that is, for "a depiction of one person's inner consciousness." In his words, he's playing "a different kind of game, an anti-realistic kind of game" (cited in Richardson 2006, 138).

One argument throughout this essay has been that such a game, whether in sentences or narratives, involves a relocation of the action from the world of the text to the mind of its beholder. It becomes a drama of cognition and at the same time a much more intimate transaction between creator and audience. Of course, it is not news that much of twentieth-century experimental literature and film abandoned not only traditional conventions of representation but representation itself. What I have tried to show is that the development of a cognitive approach would fill a special need in dealing with texts that have relocated the central action to what happens in the mind of the reader or viewer: from what it is *about* to what it cognitively *is*. My other argument brings me back to where I began and to my quarrel, not with the abundance of good work on art and evolution, but with its tendency at times to equate what we are evolved to do with what we should do. It is an old quarrel, but in an age when authorities can tell us how the mind works, it is at least humbling to know what it feels like when you are pushed against the limits of that mind. I think this is one of the objects of this perverse art. I think this is a good thing: to have us look over the edge and, in the process, to feel in our bones an intimation of how much there may be that we were not designed ever to know.

Notes

1. Pinker 2002, Wilson 1998, Dissanayake 1992. One could make the case that the production of, and market for, such texts is a case of privileging one survival trait—the "novelty drive" (Miller)—at the expense of others: cultural solidarity, ritual, the preservation and transmission of knowledge. That such texts draw on our penchant for novelty is indisputable. But this does not necessarily mean that they can be written off as novelty for novelty's sake, nor the elitist racket Pinker suggests (1997, 522; 2002, 407).

2. James Phelan, cited in Zunshine (*Why We Read Fiction*, 43). In Zunshine's own words, such texts "offer valuable insights into the workings of our consciousness" (42). Patrick Colm Hogan, also responding to Pinker's critique, argues that "[u]nderstanding a postmodern novel is an exercise of the same order as interpreting life itself from the fragmentary evidences of quotidian experience" (*Cognitive Science, Literature, and the Arts*, 87). See Abbott, "Immersions in the Cognitive Sublime."

3. All these were taken or adapted from www.site.uottawa.ca/~kbarker/ garden-path.html. A quick internet search will provide other sites with good examples.

4. One problem with writing about garden paths in recent narratives is that, of necessity, one "spoils" the garden-path surprise for those who have not read them yet by revealing what is exposed and how it is done. I have tried in this chapter to keep such spoiling to the minimum needed.

5. Current work on the worldmaking character of narrative was anticipated by Nelson Goodman's groundbreaking and still useful *Ways of Worldmaking* in 1978. In 1987, Brian McHale's *Postmodernist Fiction* identified the obsessive worldmaking (and unmaking) of postmodernist narrative as a way of bringing into the foreground the arsenal of sleights of hand and unexamined assumptions that traditional realistic narrative has relied on in its worldmaking.

6. Jahn accepts the common practice of using "frames" to refer to "*states and situations* (seeing a room, making a promise)" and "scripts" to refer to "*stereotypical action sequences* (playing a football game, going to a birthday party)" (174).

7. The makers of *The Sixth Sense* took considerable pains to anticipate the viewer's search for inconsistencies in the revised world. At one point Cole lists the rules that govern ghostly behavior, including "They don't know they're dead" and "They see what they want to see," which explains why Malcolm, up to the last scene, fails to recognize all the evidence that he is a ghost. In this regard, a considerably more ragged example of garden-pathing in film is the 2003 horror mystery *Identity*, in which a collection of travelers, stranded at a motel in a rainstorm, are mysteriously murdered, one after another. Late in the film, we discover that they are each a separate personality of a convict afflicted with multiple personality disorder. In order to escape his imminent execution, he has

been frantically contriving to have them kill each other off with the exception of his one decent personality (John Cusack).

8. It is worth bearing in mind that the chain of events set in motion and sustained by this hidden daemon, including the humiliation of Majid's being jailed with his son, is quite possibly the trigger that brings on Majid's suicide.

9. Catherine Wheatley argues that "the person who is really 'sending' the tapes can only be the director himself" (Wheatley), but in my view this involves simply another metaleptic move that is insufficiently tagged as such.

10. In an interview with Richard Porton, Haneke stresses that *Caché* is a "realistic film" (Porton, 50). *Hyperrealistic* might be a better term, since much of the focus of the film is on the degree to which reality has been obscured by its simulacra.

11. I use the term *naturalize* here in the sense Jonathan Culler uses it in chapter 7 of *Structuralist Poetics*, that is, as an "operation" on the text performed by readers in order "to recognize the common world which serves as a point of reference" (135). Naturalizing by inferring an insane narrator is basically a way of preserving incoherence as a kind of coherence.

CHAPTER TEN

Consciousness, Ethics, and Narrative: Reading Literature in an Age of Torture

PATRICK COLM HOGAN

There is a scene in the third act of *King Lear* that may be particularly resonant to American readers today. Cornwall has received intelligence reporting that the armies of France have landed in England and that the former king and the Earl of Gloucester are collaborating in this invasion. The informer is Gloucester's own son, Edmund. Cornwall advises Edmund to leave. The interrogation, he explains, is "not fit for your beholding" (III.vii.9–10). Edmund departs with Goneril, who has urged that Gloucester be blinded. Cornwall has Gloucester arrested and binds him in a chair. After Regan insults him, Cornwall threatens him and demands information. Finally, he pulls out Gloucester's eyes. Before he finishes the blinding, Gloucester finds himself opposed by one of the servants present at the interrogation. They fight. The servant is run through from behind by Regan. When Gloucester is fully blind, Cornwall has the remaining servants throw him out onto the heath. Cornwall retreats to nurse a wound that he received from the traitorous servant. The remaining servants determine that they will provide aid to Gloucester.

Cornwall's methods are not too distant from our own. Humiliation, as we find in Regan's taunting of Gloucester, is an important part of contemporary torture by the United States (see "Report" 262). So are sensory deprivation (see Millett 100–101 and citations), often total darkness (see "Report" 262), and disorientation (see McCoy 51, 90)—just what Cornwall achieves by blinding Gloucester and throwing him onto the heath, where he can only "smell/His way" (III.vii.91–92).

Torture, here as elsewhere, faces us with issues of ethics, consciousness, and narrative. There are issues of ethics because we need to determine just what sort of action is right in these conditions—what is right for the informant, for the government official, for the potential torturer, for the observer. There are issues of consciousness. It is consciousness—both that of the torturer and that of the victim—that allows there to be any such thing as "severe pain or suffering," both "physical" and "mental," which is "intentionally inflicted" (to draw on the definition of torture given in the "Convention against Torture"). Finally, there are issues of narrative. As my invocation of *King Lear* already suggests, and as such stories as the "ticking bomb scenario" indicate more clearly, our ethical evaluations are bound up with the ways in which we emplot our experiences and actions.

More exactly, this scene—like the contemporary governmental policies it might call to mind—suggests questions about conflicting impulses, the relation of such impulses to ethical adjudication and emplotment, and the place of consciousness in both. Of course, Shakespeare leaves us with no doubt who is right in this scenario. It is the rebellious servant. He is the one hero. Moreover, I believe Shakespeare was right to do this. But, before joining Shakespeare in this conclusion, we need to think through the issues he raises.

The entire scene is framed in terms of conflict and doubt. At the start, when he first divulges his father's rebellion, Edmund protests that, in this act, "nature . . . gives way to loyalty" (III.v.4), and he must "repent to be just" (III.v.11). Of course, we do not trust Edmund. We know how easily he puts on the feelings that are expected of him and that serve his purposes. Yet it makes sense that Cornwall sends him away before the interrogation—and not only because a son's torture of his own father would be unsympathetically received by others. Perhaps Edmund's anger over his father's preference for Edgar is not purely a matter of greed for money and power. Perhaps it also suggests an attachment and a desire for affection. In any event, the reference to "nature" is apt. This could be the "nature" of a son's feelings for his father or simply anyone's feelings toward a helpless old man. Either way, witnessing Gloucester's suffering would almost certainly generate a conflict in Edmund's feelings. He phrases it as a conflict between "nature" and "loyalty." We know perfectly well that loyalty is irrelevant. Edmund might try to justify his actions by appealing to reasons of state, the precarious condition of England, the clear and present danger represented in a paradigmatic scenario of putatively justified torture. But, in fact, the conflict would be only with his own self-interest in wealth and power.

The other end of the scene involves a similar conflict. When the first servant tries to stop Cornwall, his two comrades stand by, though the three together could almost certainly have overcome Cornwall and Regan and saved Gloucester's remaining eye. At first, we do not know what their feelings are in the matter. But when Cornwall and Regan leave, they express their compassion for Gloucester and condemn their master and mistress. It is clear that they were suffering conflict. We do not know how they would have accounted for this conflict. On the one side, there was "nature"—their human sympathy with the suffering old man. On the other side, one assumes, there was their own self-interest, their fear of Cornwall and Regan, concern about their employment and security. Perhaps there was also some sense that the nation was imperiled, that extraordinary times permit extraordinary actions. This idea can have an effect even if it is not a conviction, but a mere possibility.

Indeed, even the servant who rebelled must have suffered some conflict up to the point where he tried to stop the torture. When he intervenes, he says that he would now mock Cornwall in precisely the way Regan mocked Gloucester (III.vii.77–78). Thus, he goes back to the first moment in the short sequence that initiated the torture. The point is astute. As Fiske and colleagues explain, referring to Abu Ghraib, torture "starts with small, apparently trivial actions (in this case, insulting epithets), followed by more serious actions (humiliation and abuse)," as torturers "overcome their hesitancy and learn by doing." Thus the rebellious servant explains his rebellion, and its emotional motivation, by referring back to the first, almost trivial act, which initiated the torture and culminated in the blinding. But if Regan's mockery already provided him with a motive for rebellion, why did he intervene only now? He explains this as well, telling Cornwall, "I have served you ever since I was a child;/But better service have I never done you/Than now to bid you hold" (III.vii.74–76). There are two ways of understanding this. The first and perhaps more obvious is, roughly, ethical. The servant sees this rebellion as itself an act of loyalty, an act required by his duty—to Cornwall, or perhaps to England. But something else is hinted at by the reference to his childhood. The servant has a filial relation to Cornwall. There is a hint of personal attachment in this intervention, the sort of personal responsibility we feel for the acts of people with whom we are closely linked. It is crucial that both are emotional. Though both feelings now turn him toward rebellion, their initial force was inhibitory. At first, both loyalty and attachment to Cornwall worked against yet a third emotion—compassion for Gloucester. What happened, then, to change this situation, and how is this change related to the ethical aspect of his actions? The ethical aspect of the question demands our attention because, in the end, the servant's emotional motivations take on a moral coloring. His compassion for Gloucester and his loyalty to Cornwall—even in opposing him—are both moral attitudes, or at least attitudes that are consistent with his moral judgments.

To understand these cases, we need to consider how our minds operate when making decisions in the face of motivational conflict. We also need to understand the place of specifically ethical decision in this operation.

Ambivalence, Ethics, and the Human Brain

Let's begin with a relatively simple, but I believe broadly accurate, model of human action. On the one hand, we have what might be called "generative principles of action." These are patterns in our motivational or, in a broad sense of the term, emotional response to experiences. The experiences at issue may be internal or external, thus bodily conditions (such as hunger) or external objects or events (for example, the approach of a wild animal). Moreover, they may be perceptual or imaginative, since research shows that our imagination of internal or external experience has neurological effects that mimic actual perception or action of the same sorts.[1] These generative principles may be thought of as primarily "limbic." In other words, they have their source in the largely subcortical areas of the brain that form the neural substrate of feelings such as anger, fear, attachment, and so forth.

In connection with this, I should note that I do not accept the standard appraisal account of emotion in which emotions result from our evaluation of conditions relative to goals. Rather, in my view, emotion results from a much more proximate triggering of emotion systems or emotional memories (i.e., unselfconscious episodic memories which, when activated, cause us to experience relevant emotions). Appraisal may lead to our experience of emotions. However, the emotional effects of appraisal do not result from inferential calculations per se. Rather, they result from the perceptions, imagery, and memories that accompany appraisal. Suppose, for example, that I get bad news from my doctor. The news leads me to appraise the likelihood that I will be in good health or that I will have certain sorts of freedom. In an appraisal account of emotion, my appraisal itself explains my emotion. In the account I adopt, in contrast, the actual evaluation is inconsequential. However, as I appraise the likelihood that I will have good health—or even the likelihood that

I will be able to enjoy my favorite foods—I imagine certain outcomes, and I experience certain memories. It is these concrete, particular imaginations and memories that lead to the emotion.[2]

In addition to generative principles, we have a set of "regulative principles." These are principles that modify our behavior as it has been motivated by the generative principles. The regulative principles operate primarily through the prefrontal cortex and may be, largely or even entirely, a matter of appraisal. Appraisal, in this context, is a process of systematic observation, imagination, and recollection. Ordinarily, I observe what is most salient in my environment; I imagine only a limited set of possible outcomes; and I access only a few memories. However, when the appraisal process begins, I scrutinize my conditions more carefully. I imagine a wider range of possible outcomes. I recruit a larger set of memories. Take a simple case. I am poised above a sumptuous triangle of cheesecake. My mouth is watering as I smell the cream cheese mingling with the warm scent of graham cracker crust. But I hesitate. I am evaluating my decision. I am determining whether or not I should eat the cheesecake. My perceptual scrutiny probably does not change much in this case. However, I may self-consciously recall my doctor's warnings about high cholesterol, and I may imagine the results of a heart attack (particularly if I have emotional memories associated with witnessing a heart attack).

Of course, the appraisal has consequences only through its emotional effects. In keeping with the preceding analysis, the effects of the appraisal are indirect. However, the appraisal process itself is central. Without this process, I would never imagine chest pains, debilitating bypass surgeries, grim afternoons in hospitals slowly walking up and down the yellowed corridors, bearing the wheeled pole with my intravenous fluid pouch, wearing nothing but crumpled socks and the faded, powder-blue, humiliating gown. Were it not for the appraisal, none of this would even occur to me, and I would have none of the associated feelings—and I could immerse myself, untroubled, in the bliss of cheesecake.

So, in this model, we have initial, motivational impulses (i.e., emotions) produced by perception, imagination, and emotional memories, and we have appraisal processes that qualify these emotions, through systematic perception, imagination, and recruitment of emotional memories. But how does it come about that the appraisal process occurs in the first place? Here, it seems that there must be some contradiction in the initial impulse. We often experience emotionally contradictory inputs (e.g., inputs conducive to anger and inputs conducive to fear). These may cycle through to a valenced output or may preserve ambivalence. Appraisal processes are triggered when there is an adequate degree of ambivalence. If I am going to evaluate whether or not to eat the cheesecake, I must have some sort of initial ambivalence about eating the cheesecake. This has the usual sources—perception, imagination, and memory. For example, the sight of the cheesecake may recall my visit to the doctor's office and the news about my cholesterol, along with his grimly vivid fables of coronary artery disease.

Though simple, this model does appear consistent with the neuropsychological evidence. The gist of the research is that a central function of the anterior cingulate cortex is to monitor contradictions in task performance.[3] When there is a contradiction, the anterior cingulate cortex activates the dorsolateral prefrontal cortex, which addresses the conflict through attentional orientation and "top-down" regulation (see MacDonald et al.). As Lieberman and Eisenberger put it, "when the tension level in [the dorsal anterior cingulate cortex] is high, it automatically triggers a signal to [the] lateral prefrontal cortex, which then exerts top-down control over the competing inputs" (179).[4] In keeping with this, Kondo and colleagues present evidence that attention shifting—a crucial component of regulative appraisal—is a function of interconnections between the anterior cingulate cortex and the dorsolateral prefrontal cortex.

This and related research suggests a picture along the following lines. There are two stages in any action.[5] The first stage involves an initial emotion or emotions. If there is a conflict in these emotions, that activates areas of the anterior cingulate cortex, which initiate the second stage of adjudication.[6] More exactly, emotional responses are generated in the ways discussed above. When these are not ambivalent, they simply produce actions. However, emotional responses that are ambivalent[7] activate parts of the anterior cingulate cortex—most obviously, the dorsal area bearing on pain-related distress (see Lieberman and Eisenberger 170). Once activation is passed on to the prefrontal cortex, the "top-down" processing begins,[8] thus the implementation of routines for adjudicating motivational conflicts. These routines are the "regulative principles" introduced above. (See Figures 10.1 and 10.2.)

There are many sorts of appraisal that may enter into our regulative processes. Often these regulative principles are simply a matter of elaborating on preferences. I go into a restaurant, read the menu, and find myself attracted to both the lamb saag and the fish curry. My appraisal

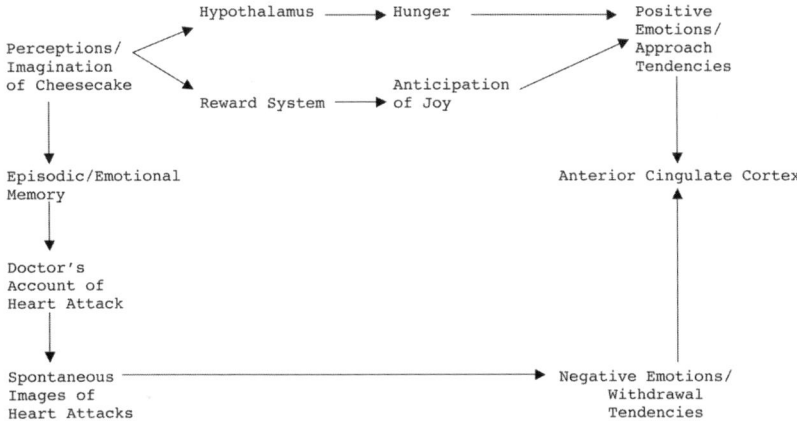

Figure 10.1. The development of ambivalence and activation of anterior cingulate cortex by emotion conflict. (Arrows mark excitatory connections.)

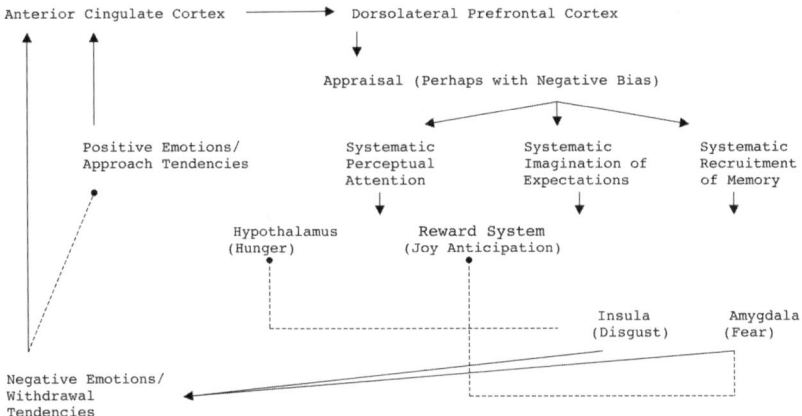

Figure 10.2. Processing of ambivalence by prefrontal cortex after activation by anterior cingulate cortex. (Dashed lines with blunt ends mark inhibitory connections. Arrows from systematic perception, imagination, and memory indicate excitatory connections to the hypothalamus, insula, etc.)

process may simply involve trying to remember if I have had either at the restaurant before and which I liked better.[9] We might refer to this as simple informational appraisal (with the understanding that the information at issue is emotion-relevant).

But it is clear that appraisal often involves somewhat different concerns. In the cases cited above, involving cholesterol, these concerns were prudential. But, of course, prudential concerns do not exhaust the sorts of regulative considerations that may affect my action. Consider, again, the servant's opposition to Cornwall. This is clearly not pruden- tial. Indeed, it is highly exemplary of a moral action, in part because it runs against the servant's own interests.

This suggests that we have (at least) three sorts of regulative prin- ciples—informational, prudential, and ethical. Note that this account leads to several straightforward predictions. First, emotional conflicts in each area should involve increased decision times, particularly when there is a conflict in the regulation principles themselves. Put simply, the process of cycling (depicted in Figures 10.1 and 10.2) takes time—time that is not required when the cycling does not take place. Moreover, this increase in decision times should be associated with increased activation in the anterior cingulate cortex and dorsolateral prefrontal cortex. In fact, this is just what we find. In cases of moral conflict, subjects take longer to make decisions (see Hauser 221). Moreover, this conflict is associated with greater anterior cingulate cortical activation and dorso- lateral prefrontal cortical activation (Hauser 222).

But the same points presumably apply to informational and pruden- tial decisions. Here, then, the question arises as to how we might under- stand and differentiate ethical principles in this context.

On the Nature and Operation of Ethical Regulative Principles: Guilt, Empathy, and the Self

The first thing to say about ethical principles here is that they are, in a sense, not ethical principles at all. I imagine that Kant expressed a common view when he wrote that "if any action is to be morally good, it is not enough that it should *conform* to the moral law—it must also be done *for the sake of the moral law*" (57–58). But, if the preceding account is correct, then nothing is ever truly done for the sake of the moral law. Our appraisal processes produce motivations, not by giving us rational reasons to act, freely chosen. Rather, they produce motivations by acti- vating emotional memories and emotion-eliciting imaginations with ad- equate intensity to overcome a motivational conflict that itself preceded and generated the appraisal. In this way, none of us is ever good in the Kantian sense.[10]

This does not mean, however, that there is no way of distinguishing ethical principles from informational or prudential principles. Most obviously, prudential appraisals concern oneself. They are egocentric. Logically, one would expect ethical appraisals, then, to concern what is not oneself. But clearly, ethical appraisal does not concern *everything* that is not oneself. Rather, it concerns what is, in relevant ways, parallel to oneself. This, I believe, is where consciousness enters.

I have argued elsewhere that the entire world is open to a pure physical reduction, to an account that makes reference only to material entities in strict causal relations. This includes me, you, every person everywhere. For example, there is no need to make any reference to a human mind, with freedom to make or not to make ethical decisions—or, for that matter, with freedom of any sort, or with subjectivity of any sort. There is no need for this—except in a single case, the case of the person making that causal account. We cannot give a reductionist account of the world that is complete, for that account cannot encompass itself. In this way, each one of us defines a separate world, with a narrow wedge of indeterminacy. From every observational standpoint, I am wholly fixed by my material nature. The words spilling out from me are produced by mere motor routines, caused by neural circuits, caused by the interaction of perception and other neural circuits, caused by external and genetic material events and conditions, stretching back endlessly. This is true from every observational standpoint—except my own. When I engage in prudential deliberations, I take up a particular perspective on the world, the perspective of myself. When I engage in moral deliberations, I take up a different perspective, the perspective of someone else. Of course, this does not have its effects through pure, rational deliberation. It has its effects through motivational feeling. But this feeling is, in a way, strange. It is based on shifting out of my egocentric perspective to the perspective of whomever or whatever I take to have subjectivity, whomever or whatever I take to define one of the countless worlds that allow for the overlapping, but never complete causality of the material world.

For me, then, moral sense is nothing other than a propensity to engage in regulative deliberations founded upon the imagination of worlds not defined by my subjectivity, but by the subjectivity of others.[11] Conversely, morality is a genuine issue and a possibility due to the multiplicity of worlds and the possibility of regulating my actions by reference to worlds different from my own.

This view receives some support from empirical research on guilt. Guilt, by this account, would be a response to a motivational conflict in which we act on impulses that contradict our ethical regulative principles. But just how could this be the case? If regulative principles do not produce emotions or actions through their inferential logic (e.g., by subsuming acts under general categories), how could guilt come about? By my account, guilt must come about by simpler mechanisms involving the activation of emotion circuits through perception, imagination, or emotional memories. Given the preceding identification of moral regulative principles with empathy, we would expect these mechanisms to involve empathy or related processes. In keeping with this, Tangney reports research linking guilt with "enhanced empathic responsiveness" and associating guilt with "reparative action" (see also Eisenberg). Neurocognitive research by Takahashi and colleagues may point in this direction as well. Specifically, Takahashi et al. explain that brain regions activated in guilt conditions "have been implicated in the neural substrate of social cognition or Theory of Mind." Their "results support the idea" that guilt is one of the "social emotions requiring the ability to represent the mental states of others."[12]

This account may also receive some support from its convergence with one part of Kantian ethical theory—undoubtedly one of the most important and highly regarded ethical theories in the history of philosophy.[13] Specifically, Kant defined the practical moral imperative in the following words: "Act in such a way that you always treat humanity, whether in your own person or in the person of any other, never simply as a means, but always at the same time as an end." He explains that things have "only a relative value." In contrast, "persons" cannot be reduced to "a value *for us*," for they have "*absolute* value" (96). Alternatively, a person "admits of no equivalent," thus has no "price," but instead has "dignity" (102). In terms of the preceding analysis, that absoluteness and dignity have their metaphysical groundwork in the absoluteness and non-equivalence of subjective worlds. That absoluteness and non-equivalence are, in turn, rendered psychologically functional through empathy.

Stories, Ethical Prototypes, and In-Group Empathy

Here the question arises as to just what process allows us to engage in ethical appraisals. It may seem from what I have just said that this is a simple matter of empathy. But that cannot be the case. Let us return again to the scene from *King Lear*. Clearly, the rebellious servant felt empathy with Gloucester. But he may have felt some empathy with Cornwall as well. Even if we decide that he did not (since he is a fiction, we can decide this as we like), it is still the case that someone might have empathy for both parties in such a situation. Cornwall may feel genuinely betrayed. He may feel genuine anger and fear. He may be experiencing a deep dread at the prospect of the French invasion. Empathy alone does not determine whether we favor Cornwall or Gloucester in this scene.

So what does determine that favor? One standard answer is *rules.* Rules give necessary and sufficient conditions for action. The crucial rule may be the utilitarian notion of producing the greatest happiness for the greatest number, or it may be Kant's categorical imperative, "Act only on that maxim through which you can at the same time will that it should become a universal law" (88). In any case, it is a general moral precept (or set of precepts) under which we subsume acts.

However, this seems unlikely. Ordinarily, we are more inclined to categorize objects and situations in terms of prototypes or roughly standard cases than in terms of necessary and sufficient conditions (see, for example, Holland et al., 182ff. and citations). Moreover, our emotions tend to be animated by concrete eliciting conditions. The more abstract the regulative principle, the less likely it is to excite our emotional response. In the context of evaluating human action, then, the most plausible candidates for regulative principles are narrative prototypes. In other words, it seems most likely that we do not evaluate situations by subsuming them under rules, but by comparing them to narrative prototypes. Our response to Gloucester and Cornwall, then, depends on precisely how we relate them to ethically relevant narrative prototypes.

Narrative prototypes can vary considerably. However, there are some that are fairly constant—not only within, but across societies and historical periods. In *The Mind and Its Stories*, I argued that to a great extent our moral evaluations are bound up with story structures. Specifically, there are three cross-culturally recurring narrative structures—sacrificial, romantic, and heroic tragi-comedy. Each of these narrative structures is connected with a moral prototype, which is to say, a standard case of moral behavior. I will be concerned with only one of these, the heroic prototype.

Heroic tragi-comedy has several components. Two are particularly important here. First, the main body of heroic tragi-comedy concerns a threat to the home society perpetrated by a foreign society. Commonly, there is a war in which the enemy society attacks the home society and achieves some preliminary success before our side finally defeats the invading army. This narrative section is associated with an ethical prototype that I have called "the ethics of defense." It is an ethics of protecting one's in-group against the aggression of an out-group.

A surprising aspect of heroic plots cross-culturally is that they often continue beyond the defeat of the enemy. In this continuation, members of the home society, including the main heroes, may suffer remorse or punishment for the suffering they inflicted during the preceding battles. This "epilogue of suffering" is also associated with an ethical prototype. I have referred to this prototype as "the ethics of compassion."

We all have all of these moral prototypes. They may be elicited in different contexts and inflected in different ways. In the case of *King Lear*, we might say that our response to Cornwall and Gloucester is at least in part the result of which prototype is triggered during the scene—the prototype associated with the cruelty of war and the ethics of compassion, or the prototype associated with the threat of enemy violence and the ethics of defense.

It is important to note that these prototypes operate not only at the level of regulative appraisal, thus in the second stage of cognition leading to action. They operate also in the first or initiating part of action. In other words, they contribute to the initial ambivalence (or lack thereof).

More exactly, I have been stressing that ethical principles are regulative, thus part of the second stage of cognition that leads to action. I have also been emphasizing that the motivational effects of such appraisal result from more basic and mechanical triggers than inferential logic. Finally, I have insisted on the centrality of empathy for ethical appraisal. However, spontaneous empathy is an important part of our initial response to situations also. It is well established that we have very swift and automatic emotional responses to the emotional expressions of other people. In part due to the "mirror neurons" that incline us to mimic the actions of others, we commonly experience emotional contagion—a pang of fear when faced with a companion who has become pale, wide-eyed, and trembles with terror; a surge of joy when we encounter a merry band of mirthful youths, laughing and yodeling together.[14] This empathy is, of course, at the basis of the ethics of compassion.

Our division of people into in-groups and out-groups is as fundamental to the heroic threat/defense prototype as empathy is to the epilogue of suffering. Moreover, this division is no less emotionally consequential and, in some cases, also almost immediate. As is well known, mere arbitrary assignment into groups is "sufficient to trigger in-group favoritism" (Hirschfeld 1), along with systematically biased evaluation of the characteristics, behaviors, and products of in-group and out-group members (see Duckitt 68–69). Moreover, in-group/out-group divisions produce hierarchizing behavior, such that "Group members . . . seek maximum relative advantage for the ingroup over the outgroup, even when this interferes with the achievement of maximum absolute outcomes for the subjects" (Duckitt 85). In the case of more or less directly perceptible differences, such as race, these effects may be very swift and automatic. Fiske and colleagues explain that feelings such as "disgust or contempt . . . appear rapidly, even in neuroimaging of brain activations to outgroups . . . . Categorization of people as interchangeable members of an outgroup promotes an amygdala response characteristic of vigilance and alarm and an insula response characteristic of disgust or arousal."[15] In some cases, mere exposure to the face of an out-group member may lead test subjects to behave in a hostile manner (Kunda 321–325). Such responses begin only a fraction of a second after exposure (see, for example, Ito et al., 193).

More exactly, there seem to be two particularly crucial parameters in our initial categorization of other people. One is in-group versus out-group. The second is familiar versus unfamiliar. These two parameters interact in complex ways. For example, we tend to distrust strangers. In keeping with this, exposure to unfamiliar faces tends to produce amygdala activity, implying fear or anger. Repeated exposure creates a sense of familiarity. One would expect this to reduce amygdala activity. It does—but only with regard to faces categorized as in-group members (at least in the case of race; see Oatley 73 and citations).

Our emotional responses to these differently defined groups may converge, reinforcing one another, and reinforced further by other emotion triggers—as when a stranger from an out-group glares at me in a threatening manner. However, they may contradict one another or further emotion triggers as well—as when a stranger from an out-group is in visible pain and appeals for my help. In other words, various factors of categorization and spontaneous empathy may give rise to conflicts that trigger painful activation of the anterior cingulate cortex, which, in turn, triggers regulative prefrontal activation. Such conflicts are more likely to arise in certain circumstances than in others. War is one of those circumstances. Indeed, the contradiction between narrow in-group loyalty and universal compassion (which necessarily includes compassion for suffering members of the out-group) is a recurring con-

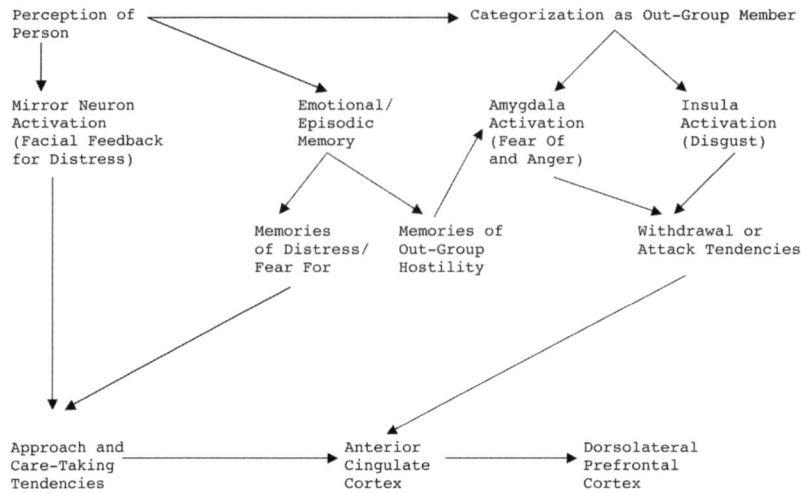

Figure 10.3. The development of ambivalence in relation to suffering "enemies."

cern in heroic tragi-comedy, as it is in real war. This is why that genre is associated with two mutually exclusive ethical prototypes—one bearing on in-group defense, the other bearing on unconstrained compassion. (See Figure 10.3.)

A situation of moral and empathic conflict is also likely to arise in cases of torture, to which we might now return.

Ethics, Prudence, and the Ambivalence of Cruelty

Based on the foregoing analysis, we might expect torture to be a two-stage process. Consider the torturer. We should expect an initial categorization of the victim as a member of an out-group (thus a trigger for amygdala and/or insula activation, an object of fear, hostility, and/or disgust). However, from the moment the torture begins, we should expect spontaneous empathic responses to be triggered as well. These contradictory emotions should activate the anterior cingulate cortex in a painful way that in turn activates the dorsolateral prefrontal cortex, leading to appraisal through regulative principles.

Again, specifically ethical appraisal is probably guided by moral prototypes.[16] In any given case, the precise prototypes that guide appraisal are the ones activated by the usual sorts of internal circuits and external events or conditions. Given this, we should expect that, across a range of cases, these prototypes will most often be consistent with socially domi-nant ideas and socially pervasive attitudes, imaginations, and memories. In short, they will, on the whole, cohere with dominant ideology. This dominant ideology will almost always be, not universalist, but strongly differentialist (in terms of in-group/out-group divisions). It will not be guided by an ethics of compassion, but rather by an ethics of defense. Put differently, in considering dominant ideologies—whether national, religious, racial, or other—we would expect to find, again and again, the sort of "ethics" that support social hierarchies. We would expect this in part for the simple reason that those who dominate social hierarchies are likely to have the strongest emotional commitment to hierarchy-preserving prototypes, and they are likely to have the greatest influence on social discourse. In circumstances where everyone has prototypes of both sorts, even a slight advantage can produce widespread and signifi-cant bias in favor of differentialist appraisal.[17] (See Figure 10.4.)

The preceding analysis appears to fit actual cases of torture very well. Consider, for example, Osiel's account of Argentinean torturers. These men experienced just the sort of motivational conflict we would expect. They felt "initial hesitations" (133) that triggered moral appraisal. Their moral appraisal relied on standard discourses of their time and place, discourses closely linked with paradigms of moral authority and differ-entialist ethical prototypes. The Catholic Church was particularly im-

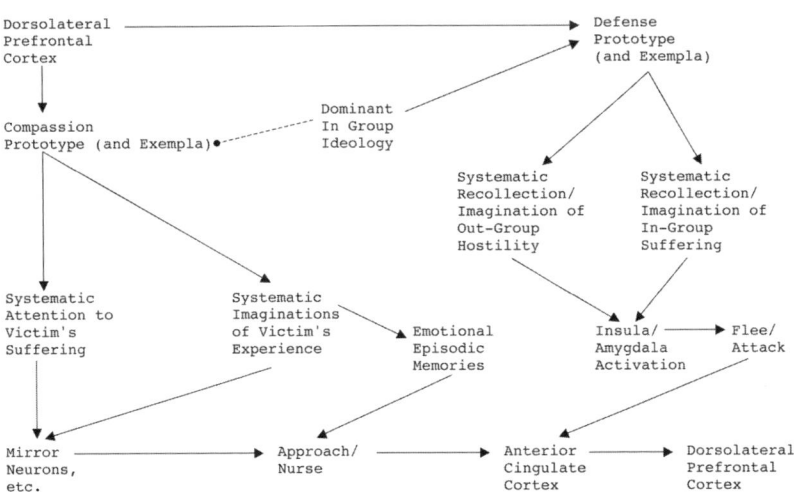

Figure 10.4. Ethical appraisal and the processing of ambivalence for suffering "enemies."

portant as a source of moral authority. As Osiel explains, "Church leadership viewed the military operations against the regime's opponents as a just war, warranting no qualms and requiring no apologies by those called to participate" (132). A just war is, of course, a war of defense against aggressors. Thus the right regulative principles guiding moral reflection, in this case, appeared to be those defined by the prototypical ethics of defense, with its ideal of protecting one's group against an enemy. This directly excluded the prototypical ethics of compassion, with its ideal of aiding all those who suffer.

Of course, our reasoning in moral and other issues is not confined to prototypes. It encompasses more fully specified instances as well. Moreover, these instances commonly enter into informational appraisal as well as ethical appraisal, since we tend to view them as accurate representations of real situations. This is true even when their source is fictional. Indeed, our appraisals are often strongly biased toward salient examples (see Nisbett and Ross 15). The point holds whether we are evaluating automobiles or torture. The ticking bomb scenario is perhaps the single most salient example of torture, at least for most Americans. It is routinely repeated, with small variations, not only in philosophical discussions of torture, but in television dramas, popular discourse, and elsewhere. The idea is this. The governing authority knows without a doubt that there is a ticking bomb that will kill many innocent people. He or she also knows without a doubt that Jones knows where the bomb is located. Finally, he or she knows that Jones will not give him any information unless tortured and that, if tortured, Jones will give accurate information. In short, the governing authority is omniscient. The only flaw in his or her omniscience is that he/she does not know where the bomb is located.[18]

The salience of this scenario makes it appear highly representative of cases of torture (or, rather, cases of torture perpetrated by the in-group). But, in fact, it is wildly unlikely that such a situation will occur. First, it is very unlikely that anyone will know with a high degree of certainty that Jones knows where the bomb is. There will be situations in which one might reasonably infer this. But the rhetorical force of the scenario relies on certainty, which is virtually never present. (Television programs give us a sense of certainty in such cases because they commonly show us Jones placing the bomb.) Moreover, in real life, it is usually not even entirely clear that there really is a clear and present danger. Again, in a television program, we know that there is a clear and present danger because the program cuts repeatedly to the timer on the bomb.

Furthermore, in real life, the "information" acquired by torture is not trustworthy. As a U.S. Army field manual explains, torture "yields unreliable results" (quoted in McCoy 102). Finally, there are other ways of acquiring information, ways that may in fact be foreclosed by the use of torture. As the field manual explains, torture "may damage subsequent collection efforts" (quoted in McCoy 102). In sum, this paradigmatic scenario—which almost certainly helps to guide our evaluation of torture—is wrong on every count.

Indeed, the operation of torture is almost precisely the reverse of what this scenario suggests. Take the case of Iraq. Authorities do not have certainty that a particular individual has information about a current, particular (thus known) threat. Rather, authorities infer that there are, at any given time, a number of threats that are being planned and perhaps executed. They infer that certain sorts of people (e.g., young Iraqi men) may be involved in or aware of such threats. In reality, then, the standard case of torture is highly uncertain and non-particular. It is very much a matter of loose statistical inference, not particular and certain information. For example, U.S. military personnel have regularly been the target of bombs. Thus it is likely that there will, in the future, be bombs targeting U.S. military personnel. In the past, the perpetrators of these bombings have largely been Iraqi men. Thus we can expect Iraqi men to be engaging in these bombings in the future. But, of course, we do not know which Iraqi men these are. Therefore, we need to round up a great number of Iraqi men. When we interrogate them, we find that they do not tell us anything about future bombings. There are two possible explanations for this. One is that they really do not know anything. The other is that they are rebels and are concealing information. But we do not know which are which. So, we torture them, on the assumption that, if they have information, they will give it under torture. Innocent people, who of course have not been trained to resist torture, say whatever they think the torturers want to hear. Those who have been trained may resist the torture and/or give misinformation. These are the ordinary conditions in which torture occurs, once it is allowed.

Note how, given all this, torture is likely to spread quickly and vastly. The paradigmatic ticking bomb scenario leads us to believe that torture is isolated and very narrowly constrained by certain knowledge. The actual situation in which torture operates has the opposite effect. It continually expands in scope. This may be one reason why the number of "mistaken" arrests in Iraq has been so high (between 70 percent and 90 percent, according to information gathered by the International Committee of the Red Cross [see "Report" 257]). As McCoy points out, torture "spread, in just months, from a few top Al Qaeda suspects in CIA custody to hundreds of ordinary Afghans and thousands of innocent Iraqis" (112).

Finally, prudential appraisal may enter here. Prudential evaluation is usually a non-issue in the public discourse surrounding torture. But the practice of torture can harm, not only the victim, but the torturer as well. Frantz Fanon reports a number of cases from French-occupied Algeria. For example, one European police officer began to hear the screams of his victims even when he was at home; they "prevented him from sleeping" and, even in summer, he "shut the shutters and stopped up all the windows" in the hope of keeping out the sound of his victims' pain (264–265).[19]

Of course, things are much easier for those who never witness the suffering of the victims—for example, us. But we may still not escape the deleterious effects of cruelty. The tendency of torture to expand means that it begins to affect larger and larger circles of people. Its use is extended to areas involving members of the in-group. General life in the home society may become increasingly violent. Ordinary people become habituated to the infliction of harm, and their sense of what constitutes cruelty is deformed. Moreover, this does not even touch on the retributive violence that torture is likely to foster in its victims, members of their families, and members of their in-groups.[20] As Patañjali wrote many centuries ago in his *Yoga-Sūtra*, "Hiṃsā (harm)—whether done, caused to be done, or approved; whether arising from greed, anger, or delusion; whether modest, medium, or beyond measure—bears fruit in sorrow and ignorance that are never ending" (83, altered).

Literature and Torture

The preceding section leads to somewhat pessimistic conclusions about the possibilities for opposing cruelty or for fostering ethical practices that are not mere rationalizations of social oppression, rendered apparently right and just by invocations of God or Nation. In practice, ethical appraisal is often a way of overcoming our natural inclination to empathize with those who suffer. As Bandura puts it, after "cognitively reconstruing injurious conduct so that it can be done free from the restraint of empathy," people are capable of doing "extraordinarily cruel things"

(25). That process of reconstrual is a process of regulation, prominently including ethical regulation.

Yet there are many cases where this does not occur, where our empathic inclinations are extended by appraisal, and cruelty is recognized as leading to endless suffering and ignorance. These cases prominently include literary instances or exempla. One case of this sort is *King Lear.* In conclusion, I would like to return to Shakespeare's remarkable depiction of torture, a depiction that subtly undermines ethical, prudential, and informational support for such a practice.

First, consider the prudential and informational appraisals. Shakespeare faces us with a version of the paradigmatic ticking bomb scenario. A foreign army has landed on the shore. Cornwall knows that Gloucester knows where the army is. Moreover, Cornwall's knowledge is certain. Thus it is that rare case when the government knows that a particular person has critical information. Indeed, Cornwall will be able to tell whether Gloucester is giving him accurate information. Indeed, he will be able to tell with absolute certainty. But Shakespeare takes this situation and shows its absurdity. Cornwall fits the role of the perfect interrogator only because he already has the crucial information. He already knows that the answer to his question about Lear's location is "Dover." Thus the very certainty that is necessary for the paradigmatic justification of torture is available—but it is available only in a way that makes the torture entirely pointless.[21]

Moreover, even if he did not have the information, Shakespeare makes it clear that torture is not necessary in this case. Gloucester answers Cornwall's question immediately, and truthfully. Yet Cornwall goes on to blind Gloucester and exile him anyway. All this serves to undermine another aspect of the paradigmatic justification. The pretense of the paradigm is that torture is tightly constrained and purely instrumental. But Shakespeare presents torture as wholly non-instrumental and unconstrained. Cruelty to Gloucester is something that Cornwall does because he takes pleasure in it, at least temporarily. Given recent neurological research, we can understand why. He has categorized Gloucester as a traitor, and we "derive satisfaction" from administering "effective punishment" (as opposed to mere "symbolic punishment") after a "defector's abuse of trust," as de Quervain and colleagues put it.

Finally, Shakespeare depicts all the effects of the act as contrary to Cornwall's interests. The cruelty inspires the opposition. As Regan explains, "Where [Gloucester] arrives he moves/All hearts against us" (IV.v.10–11). Moreover, it harms Cornwall himself, as he is mortally wounded by a servant. Interestingly, before he has completed the blinding of Gloucester and ordered that Gloucester be thrown out, there is no suggestion in the text that Cornwall has suffered a serious injury. It is as if the wound is a metaphor for the damage done by his performance of the torture. Finally, the entire cycle of violence leads to such devastation that no one can reasonably be viewed as triumphant. When I read the ending of the play, I cannot help but think of Iraq or Afghanistan. There too Albany's summation applies, "Our present business/Is general woe" (V.iii.320–321).

The ethical side of Shakespeare's portrayal here is no less important. Shakespeare fully develops our empathic response to Gloucester. We feel nothing but compassion for his suffering in humiliation, blinding, homelessness, and disorientation—in short, his suffering in torture. In other words, Shakespeare is careful to create a scene that will trigger our prototype for an ethics of compassion. This is remarkable because the larger situation depicted in the play should lead us very much toward the prototype for an ethics of defense. Specifically, in a play written for an English audience, Shakespeare has a foreign army invade England and he presents that audience with a prisoner who is in league with the invaders. This is precisely the sort of case that advocates of torture stress. The political situation is one of grave national danger—indeed, the prototypical case of such danger, the case which forms the basis of heroic tragi-comedy. Moreover, the prisoner is a traitor. Thus the prisoner is someone audience members would ordinarily cut off from all empathic identification, substituting delight in his suffering. Yet, despite all this, Shakespeare so thoroughly leads us to simulate Gloucester's world that we feel his suffering even in its incidentals. For instance, I cannot be the only one who finds it heartbreaking when Gloucester learns that Edmund has turned him in and, rather than reviling Edmund, says, "Then Edgar was abused./ Kind gods, forgive me" (III.vii.92–93). Gloucester's lament is poignant in part because, at that moment, he is simulating Edgar's feelings and thus operating on an ethics of compassion (for Edgar), rather than a vengeful ethics of defense (against Edmund). In this way, he provides, not only an object for our compassion, but a model for our ethical appraisal.

In short, Shakespeare presents us with the paradigmatic case that might have been used to support torture in defense of the nation. He shows us that, even in this extremity, there is no case to be made for torture. Thus a work such as *King Lear* suggests that there are opportunities for ethical appraisal that fall outside the bounds set by dominant political elites. Sometimes in literature other perspectives do emerge. Perhaps reading and discussing such literature in an age of torture may, in its own very small way, help to make it more possible for us to set aside cruelty, at least when our empathy and ambivalence are already strong, and to engage in appraisals that are not egocentric, but expand to all worlds equally. After all, our brains are strongly empathic and almost inexorably ambivalent in the infliction of pain. It is our societies—and our limited and limiting identifications with those societies—that lead us from ambivalence through moral rationalization to cruelty, including the cruelty of torture.[22]

Notes

1. See, for example, Preston and de Waal 12 and 20 on emotion, empathy, and imagination. There are, of course, differences between imagination and experience also—and, in relevant cases (such as emotion), between direct experience and observation (see Preston and de Waal 11, 20).

2. A good example of the difference between these accounts may be found in research on the feeling of social exclusion. Lieberman and Eisenberger explain that subjects feel socially excluded even when they know perfectly well that (despite certain superficial appearances) they are not excluded. In one study, subjects play catch on a computer screen with two other animated figures. At a certain point, the two other figures stop throwing the ball to the animated figure representing and controlled by the test subject. As Lieberman and Eisenberger explain, "Even when subjects are informed that the other two 'players' are really just computer players controlled by the program and that those players will exclude them part way through the game, subjects still report feeling social pain as a result of the experience" (173). In another study, Lieberman and Eisenberger report that the dorsal anterior cingulate cortex—associated with both social and physical pain—is activated when the test subject is told that there has been a technical glitch that prevents his or her participation in one round of cyber-catch. The activation is similar to that when the subject has every reason to believe that two other players are actually excluding him or her (175). These seem to be clear cases where the appraisal calculations themselves do not generate the emotion. Rather, associated perceptions, imaginations, and memories do.

3. Thus Carter and colleagues present research indicating that "the ACC monitors competition between processes that conflict during task performance" (748). MacDonald and colleagues concur, explaining that the anterior cingulate cortex is "more active when responding to incongruent stimuli, consistent with a role in performance monitoring" (1835).

4. Lieberman and Eisenberger are following Botvinick et al. Other research points to the orbitofrontal cortex (OFC). Thus Beer and her colleagues argue that this area is "particularly involved in the regulation of social behav-

ior" (595), perhaps because it is involved in motivating such regulation, perhaps because it is connected with "inferring others' mental states" (596). In connection with this, some writers have maintained that there is a "system reliant on the proper functioning of the OFC that is normally activated by perception or expectation of others' anger" (Moll et al.). This may not be a matter of regulation per se. Rather, it may be a source of the ambivalence that gives rise to regulative processes. There may also be involvement of the OFC in causal attribution (this at least appears to be suggested by some of the research reported by Beer et al., 596).

5. A number of authors have argued for some sort of duality along these lines. See, for example, Ito et al., whose "two-part system" (198) comprises "conflict monitoring" through the anterior cingulate cortex (199) and "implementation of cognitive control to reduce discrepancies" (198). See also Preston and de Waal on automatic activation of "autonomic and somatic responses" and the inhibition of those responses (4).

6. Of course, these two stages are not fully discrete in practice, since we are continually engaging in many activities with different emotional valences, different actional outcomes, and so forth, and these different activities are in continual interaction with one another. However, we may abstract the broad division of these two processes from the welter of ongoing actions.

7. Perhaps due in part to orbitofrontal activation.

8. Here, appraisal processes enter. These may be a matter of shifting attentional focus through anterior cingulate/dorsolateral coordination, as already noted. They may be a matter of initiating actions that alter the current situation so as to reduce the ambivalence. They may be a matter of suppressing negative feelings via the ventral prefrontal cortex. As Lieberman and Eisenberger explain, "right ventral prefrontal activity has been associated with explicit thought about negative affect and negatively evaluated attitude objects . . . and is also associated with inhibition of negative affective experience" (175–176). It may be a matter of altering our understanding of a situation (e.g., in one study, subjects "could reappraise an otherwise sad photograph of four women crying outside of a church as involving a wedding rather than a funeral," a reappraisal connected with decreased right amygdala activation and increased activation of "left dorsal and inferior PFC as well as anterior cingulate" (Ochsner 255).

9. In these cases, appraisal is, in effect, a more elaborated version of my initial response. It is parallel to the difference between the high road and the low road in LeDoux's model. The low road is information passed directly from the thalamus to the relevant emotion system, say the fear system, beginning with the amygdala. The high road is cortically mediated information—more detailed, more elaborate. The low road information may lead me to jump back in fright from something crouched and dark moving next to me in the bush. The high road information allows me to identify this as my own shadow.

10. Gazzaniga summarizes recent neuroimaging research on moral judgment, explaining that "when someone is willing to *act* on a moral belief, it is because the emotional part of his or her brain has become active when considering the moral question" ("The Ethical Brain," 167).

11. In this way, my view of ethical deliberation is related to the views of

writers who place the origin of morality in human empathic feeling (see, for example, Prinz). But I differ from writers such as Prinz, who ultimately expands moral feeling beyond empathy.

12. On the other hand, embarrassment involves the same sort of representation of others' mental states, as Takahashi et al. stress. Thus the mere activation of "theory of mind" states need not involve empathy.

13. Of course, Kantian ethics developed within the context of Western philosophy. There is also convergence with prominent ethical theories outside the European tradition. Consider, for example, the Confucian philosopher K'ang Yu-wei, who wrote that "The mind that cannot bear to see the suffering of others is humanity. . . . Everyone has it" (734). Moreover, if one "cuts off" the "love which is the mind that cannot bear to see the suffering of others, moral principles of mankind will be destroyed and terminated" (730).

14. It is important to stress that the mirror neuron system bears directly on emotion, and not simply on indifferent or neutral/instrumental actions (e.g., grasping). As Hauser explains, "Recent studies suggest that part of this system turns on when we directly experience a disgusting event or observe someone else experiencing disgust, with parallel findings for the experience of pain and empathy toward others in pain" (224–225). As Gallese puts it, "Sensations and emotions displayed by others can also be empathized with . . . through a mirror matching mechanism" (2005, 114). We engage in "affective mimicry," which is to say, we "mimic and synchronize expressions, vocalizations, postures, and movements with those of another person" (Hatfield, Cacioppo, and Rapson 48). This mimicry involves partially adopting the facial expressions, bodily postures, and other physical manifestations of the other person's emotion. Though these expressions, and so forth, are typically outcomes of emotions, they may also operate as triggers of emotions. In other words, when we adopt a certain facial expression, we tend to feel a version of the emotion that goes along with that expression (see Plantinga and citations). Moreover, the point is connected with ethical evaluation. Thus Hauser points out that "the mirror neuron system" has "a critical role in our moral judgments" (224).

15. Other evidence links amygdala response with fear and/or anger (see LeDoux and Phelps on the former and Panksepp 144, 146–147, on the latter).

16. More exactly, narrative prototypes. The connection of these narratives with the dorsolateral prefrontal cortex may not be merely incidental. Young and Saver point out that this region is important to our narrative abilities (76, 78). On the other hand, Young and Saver use the term *narrative* quite broadly. Their research may signal only that this region is important for planning and certain sorts of causal attribution.

17. Indeed, the general point holds not only for ethical appraisal. It holds for all forms of appraisal. Thus, in cases of motivational conflict, we would expect an adjudicative bias in favor of information beneficial to the in-group and in-group hierarchies. For example, information suggesting that the out-group has committed (fear- and anger-provoking) atrocities is likely to be more salient and more widely familiar than information that the in-group has committed (compassion-provoking) atrocities.

18. See Scarry 2004, 284, and McCoy 192.

19. Another European police officer tortured men at work, then came home and abused "his children, even the baby of twenty months, with . . . savagery" (267–268). He tied his wife to a chair, like someone he was interrogating. Before the torture, "he very rarely punished his children and at all events never fought with his wife" (268). After he realized what had happened to him, he asked Fanon, his psychiatrist, if there was any way that he could continue torturing Algerians "without any prickings of conscience" (269–270).

20. In addition, McCoy argues that those involved in torture may form a politically destabilizing element within the larger society, which may even "rise in revolt against the state it was supposed to defend" (86).

21. The situation is by no means confined to fiction. As Elaine Scarry notes, "torturing a person in order to get information about a ticking bomb is sometimes introduced in situations where the information is already available through means that require inflicting no cruelty" (285).

22. Readers wishing to engage in work to end torture will find information and opportunities at the website of Amnesty International (http://www.amnestyusa.org).

CHAPTER ELEVEN

Prophesying with Accents Terrible: Emotion and Appraisal in *Macbeth*

LALITA PANDIT HOGAN

Macbeth offers cognitive theory of emotion a unique opportunity for interdisciplinary crossover because within the experimental framework of emotion research, appraisal theory relies mostly on information about goals and plans of real people in real-life settings. In this play, the goals and the means adopted to attain these goals are, indeed, over the top, while an insistent attention is drawn to them as if they are not so over the top. In this way, while *defamiliarizing* the real, the play insistently depicts *familiar* patterns of reasoning that appraise situations, draw inferences, and elicit emotions and action outcomes. In doing so, the play foregrounds a cognitive analysis of evil, not a moral analysis.

Drawing on insights and models developed by Nico Frijda, Andrew Ortony, and others, the following discussion explores the role played by various categories of emotion, such as prospect-based emotions, attribution emotions, and fortune-of-others emotions in *Macbeth*—and what the play can show us about these categories of emotion. The prophecy with which *Macbeth* begins will be seen as an antecedent appraisal that elicits prospect-based emotions, activating a range of concerns and contingencies that explain why the Macbeths do what they do. The prefatory section presents an overview of the appraisal theory of emotions, highlighting features that are particularly relevant for a cognitive analysis of *Macbeth*.

Reading Strange Matters: Passion, Emotion, and Appraisal

As Nico Frijda (*Laws* 2007, 102, 110) and Thomas Wright (*Passions*, 31) indicate, early modern preoccupation with passions and today's cognitive

theory of emotion share a common goal: to give an account of emotion/
passion. In place of empirical data, early modern theorists of passion
frequently resort to exempla and analogies, such as comparisons with
hierarchical structures in society. For instance, Wright characterizes
passions and senses as "naughty servants" whose "love" for each other
creates an alliance against the master: the "Princesse in her throne,"
who is intent on considering the good of her "Kingdome." The "servants," that is, passions, for their part conspire to "hale her by force,"
unless she is willing to "condescend to what they demand" (Wright
7–8). Wright's choice of metaphor shows how *agents* are anthropomorphized in attributions of agency or causality. Cognitive theorists, such
as Ortony, Clore, and Collins, rely on empirical evidence to show how
common it is to "anthropomorphize the agent" in attributions of blame
and/or agency. Ortony talks about how frustration in trivial goal fulfillment can lead to blaming the "weather," or the "car" (54–55). Though
it is not the same as blaming fate, or the gods, the appraisal process involved is not dissimilar.

   In brief, appraisal theory assumes that emotions are always about
something—that is, there is always an object, agent, and event and a
pattern of reasoning involved. Events, objects, and agents do not in
themselves elicit emotions; our appraisal of events and our relation
to objects and agents as they impact our goals, or more generally, our
concerns, elicit emotion. In addition, appraisals do not need to be very
elaborate cognitive processes; they can be instantaneous, as in primary
or antecedent appraisal. In various kinds of "secondary" appraisal, appraisals that follow an emotion episode, evaluation can be more elaborate and may be characterized by thought trends that are likely to elicit
further emotion episodes which may or may not be conducive to goal-
fulfillment. Further, the notion that appraisal causes emotion does not
assume that appraisal is always correct, or that cognitive information
processing is error free. It only means that such mechanisms are part of
memory, of associative thinking and decision making. The automatic
nature of appraisal is due to activation of core relational themes as appraisal detectors. One emotion theorist, Richard Lazarus, defines a core
relational theme as "a terse synthesis of the separate appraisal components into a complex meaning centered whole" ("Relational Meaning
and Discrete Emotions," 64). For instance, the core relational theme for
envy is: "wanting what someone else has." Taking note of the narrative
format of many such cognitive processes, Lazarus argues strongly for a
narrative approach to emotions and says: "The core relational themes

are prototypes, which are especially useful in a narrative approach to the study of emotion" (65).

As one of the foremost theorists of emotion, Nico Frijda dedicates his most recent work to the formulation of nine laws of emotion based on various kinds of appraisal. One of the laws he discusses is the *Law of Concern* (7–8). According to this law, information about events in the environment that would trigger an antecedent appraisal would have to have some real or perceived impact on one's concerns about goal-fulfillment (or concerns about morality, group identification, values, and standards) in order for it to elicit emotion. In the absence of such concerns, the information that is processed would not be appraised in a way that is conducive to emotion elicitation. A related law is the *Law of Apparent Reality* (Frijda, 8–9). An event, or a possibility, that can lead to an emotion-eliciting appraisal has to, necessarily, *seem* real; one has to be able to imagine it as real. Thus, the role played by imagination in the "laws of emotion" suggests areas of vital overlap between real emotional experience and representational emotion.

However, it is important to keep in mind that appraisals of events and goals are not the only causal factors for emotion, because people are not always having goals and pursuing them. Emotional experience that results from attribution of praise and blame through the use of culturally derived *standards* and *values* covers significant areas of emotional experience (Ortony 46–47). In addition, a great deal of emotional experience relates to appraisals regarding other people's lives, or fates, such as the class of emotions described as *fortune-of-others* emotions (Ortony, 92–107). This class of emotions plays a significant role in cultural life and social cognition. In his essay "Emotion, Art, and the Humanities," in the *Handbook on Emotion*, Ed Tan identifies areas of overlap between emotion studies and imaginative literature (118–125). Some of these cover general premises, such as the link between cognitive processing and associative memory, primacy of imagination, mimesis as simulation, and so forth. More specifically, Tan lists emotion themes in popular literature, saying these "have the potential for eliciting emotion, because they can be hypothesized to touch on basic *concerns* and *contain core components of emotional meaning*, such as valence (including loss and gain) and difficulty" (Tan 121). For measuring of *valence* and *difficulty*, core relational themes are important as they highlight appraisal theory's particular suitability for literary study. Indexing of *valence* and *difficulty*, for instance, is an essential part of all narrative acts. In this context, Tan refers to Frijda's discussion of components of appraisal in his 1986 book on emotion (Frijda, 204–222), where valence is defined as "events, objects, and situations" that may "possess positive or negative valence, being evaluated for intrinsic attractiveness or aversiveness" (207). Difficulty of the "world of emotions," as explained by Frijda, is inherent in the emotion-situation, where there is no certainty about being able to find a solution, or a guarantee that a threat can be "countered or evaded, challenge met" (1986, 206). Defined in relation to goals and plans, story suspense, as we can recognize, is grounded in "difficulty" and/or uncertainty of this sort.

In Spinoza's discussion of emotion, as Frijda reminds us, valence plays a significant part. According to his construal, "amongst all the affects which are related to the mind in so far as it acts, there are none which are not related to joy or desire" (Spinoza 264 [Proposition LIX]). For Spinoza, desire is something like a disposition of the mind. Joy and sorrow are, for Spinoza, the addition and subtraction principle of the emotion system and are at the heart of appraisal that elicits emotion. He defines joy as "man's progression from a less to greater perfection," and sorrow as "man's passage from greater to less perfection" (267). *Confidence,* he thinks, is "joy arising from the idea of a past or future object from which cause for doubting is removed," while *despair* "is sorrow arising from the idea of a past or future object from which cause for doubting is removed" (271). Positive and negative valence would be determined by what the object is in terms of its desirability and aversiveness. For example, kingship is desirable to Macbeth, and his joy arises from *doubt* being removed from the prospect of achieving it, and his despair from not being able to *doubt* that his reign will come to an abrupt end. That is why the one function of the witches' final prophecies, or riddles, is to prolong *doubt* as long as possible.

As a parallel to the "humoral subjectivity" of the early modern era, the embodied nature of psychological experience is a key point in modern theories of emotion, though cultural constructions of bodily states also figure prominently. Jaak Panksepp, one of the exponents of the appraisal theory of emotions, reminds us that the "behavioral potentials of the nervous system are ingrained, although all must be environmentally molded to achieve any level of coordinated sophistication" ("Basics of Basic Emotion," 21). For this reason, appraisals are implicated in systems of social motivation. Systems of "social motivation" are context sensitive and work hand in hand with the more biologically based emotion systems. In a similar vein, Keith Oatley isolates four major social system motivation based themes that trigger appraisal mechanisms, elicit emotions, moods, action patterns, and (emotion) scripts. These are: assertion, aggression, affiliation, and attachment (*Emotions*, 81–90). For the purpose of understanding emotion in literature, the search for "life themes" anticipates a rethinking of intersections between biology and culture. For instance, Wilson Knight's isolation of what he considered typically Shakespearean "life themes" is an early attempt at detecting emotion themes in literature. According to Knight, in *Macbeth* the "life themes" are of war-honor, imperial splendor, sleep and feasting, ideas of creation and nature's innocence (130). From the cognitive point of view outlined above, war-honor and imperial majesty can be seen as emanating from the social system motivations of assertion through aggression; nature and sleep, in essence, are attachment themes; feasting is an affiliation theme. The search for life themes, however, is not the only way in which conventional Shakespeare criticism anticipates the study of emotion. With her focus on a dominant emotion in each major tragedy, Lily Campbell's 1930 study links emotion to genre. *Macbeth*, according to her, is "a study in the complementary pairs of passions of rash courage and fear" (238). Campbell draws on Aristotle, Plutarch, and Thomas Aquinas, as well as the many early modern treatises on melancholy, to suggest the extent to which Shakespearean drama takes part in the pervasive emotion discourses of the early modern, medieval, and classical ages (208–239).

Shakespearean drama is not, however, imprisoned in the modes of thought that were available in the early modern period; it also takes part in contemporary discourses on emotion. Gail Kern Paster's emphasis on "humoral subjectivity," in several of her seminal studies, uses cognitive theory in a historicist context. Paster's approach is grounded in her acute sense of the period's "persistent materialism of thought where conceptions of selfhood are involved," and the conviction that "selfhood," in "early modern terms" is "a self-experience intensely physical in kind and expression" (Paster 2002, 143). She considers "the intellectual dominance of Renaissance psychological materialism as a critical starting point" (2002, 21). Even though, at times, Paster overstates the case for early modern psychological materialism, her work has made it clear that the present moment of reading, dialogue, and interaction cannot be sealed off from the past. Similarly, in *Presentist Shakespeares*, a collection of essays intended as a presentist manifesto, Hawkes and Grady emphasize that the present is "not an obstacle to be avoided, or a prison to be escaped from," but it is "a factor to be sought out, grasped and perhaps, as a result, understood" (3). In an impassioned plea, they add that through a reversal of the "stratagems of new historicism, presentism will deliberately begin with the material present and allow that to set its interrogative agenda" (4). Long before Shakespearean critics felt it was necessary to launch an uncomfortably named movement, Hans-Georg Gadamer had stressed the continuity idea in his definition of *interpretation* as "the transformation of something alien and dead into total contemporaneity and familiarity." More specifically, he says interpretation "consists not in the restoration of the past, but in *thoughtful mediation with contemporary life*" (163, 168–169). In this regard, Shakespeare's own dramaturgy serves as a model. In *Macbeth*, he compressed Scottish and English history of several hundred years to situate the Macbeth story in relation to early modern discourses on, and anxieties about, kingship and succession (see William Carroll 1999, 185–191).

Assuming interpretation to be an act of mediation of the past with contemporary experience, section one of the following discussion interprets prophecy as *antecedent* appraisal that determines *situational meaning* of events (of civil war) and elicits prospect-based emotions and for-tune-of-others emotions, and invokes various laws governing appraisal and emotion elicitation: the law of apparent reality, the law of concern, and the laws of change and strangeness (Ortony and Frijda). Section two focuses on the idea of valence as it relates to the desirability value of the "lesser" and "greater" fortunes of Banquo and Macbeth (Spinoza). Section three gives a cognitive account of emotion intensity, with reference to local and global variables of emotion, such as the formation of a cognitive unit (Ortony, Clore, and Collins). Section four draws attention to regulative and procedural rules of emotion performance (Averill) in connection with differences in Macbeth's and Lady Macbeth's emotional experience. Core relational themes for emotion (Richard Lazarus) play a significant part in determining valence, and are referred to when needed. A concluding discussion on appraisal of emotions through conceptual metaphor (Kovecses) focuses on how Macbeth, toward the end, when his goals and plans are blocked, appraises emotion to give it a phenomenological tone, resolving the play's ethical mess and cognitive chaos in a sublime poetics of emotion.

There to Meet with Macbeth: Prophecy as Antecedent Appraisal

The play opens with defamiliarization of the familiar process of *antecedent* appraisal that elicits *prospect-based emotions*. The supernatural phenomenon separates appraisal from the agent (Macbeth) by making it strange and alien. It is surprising, therefore, that the opening scene of the play has given rise to much disagreement among critics. Cunningham considered the opening scene "spurious," and Granville-Barker agreed with him in thinking the scene is "pointless" and "un-Shakespearean," as many others down the line continued to think that this and other scenes involving the sisters were interpolations, attributing authorship of some of them to Middleton (Muir 3–5). Contradicting this view, other readers, notably Bradley and Coleridge, defended the witches and the scenes they appear in. Most important for a cognitive reading is Coleridge's notion that "the true reason for the first appearance of the Weird Sisters" is to strike "the keynote" of the "whole play" (Coleridge on *Macbeth* in Miola, 219). As a poet who theorized about the cognitive differences between *fancy* and *imagination*, Coleridge recognizes that imagination plays a role in eliciting emotion, and this process is thematized in Shakespeare's play. Continuity of insight from Shakespeare to Coleridge to modern theorists of emotion is evident in Frijda's explanation that non-real "information can be made implicational, schematic" through "imagination," and this can happen by the "influence of potent experiences such as visions" (*Laws*, 110). Characterized as "reactions (being pleased or displeased) to the prospect of an event," prospect-based emotions underscore the roles of imagination and speculation, because "a person can react to the prospect of an event counterfactually" by "thinking about, or imagining it" (Ortony 109).

This speculative nature of the appraisal agency in *Macbeth*, and its separation from the agent, functions like the defense mechanism of *splitting*, with its attendant tropes of *synecdoche* and/or *metonymy*. Even though cognitive accounts of emotion depart from psychoanalytic accounts, the idea of a defense mechanism is useful here. Everyone knows that denial through splitting is an instantaneous process, while the appraisal process sounds like conscious calculation, or something entirely external to the psyche. For this misunderstanding, Nico Frijda faults the "voluntaristic" language used by some appraisal theorists, which makes people think these are "conscious evaluations," though appraisals, as he reminds us, are "essentially thought to be non-conscious" (97). Insofar as the witches appropriate the *agency* of antecedent appraisal, while the *agent* is nonconscious about this activation and change, it is necessary that the witches establish their credentials before Macbeth comes onstage. Accordingly, their opening dialogue establishes the emotional *law of apparent reality* to create a storyworld. At the same time, the prophecy constitutes the witches, in partial diegetic terms, as embedded authors. Duncan, Macduff, Malcolm, and others are left out of this focalizing perspective, or diegesis. That is, they are not part of *this* storyworld. To Lady Macbeth the phenomenon is indirectly reported. Fully diegetic only for Macbeth (with Banquo as witness), the witches' vision, their initiative, their haunt in the forest, shape the perspective from which the hypothetical event (his becoming king) is brought into focus.

Through partial diegesis of this sort, the law of apparent reality works in conjunction with the other two laws: the *law of situational meaning* and the *law of concern*. The law of situational meaning refers to "patterns of information that represent the meaning of eliciting situations," with "different emotions arising in relation to different meanings." The law of concern is defined as emotion elicited "in response to events" that are construed as "important to the individual's concerns" (Frijda, *Laws*, 4–8). The opening scene is organized by various "patterns of information" that define the *situational meaning* for Macbeth and activate his concern. To be effective, the witches name the place "upon the heath" (that lies in the path of Macbeth's and Banquo's ride home), and situate themselves: in physical time, "ere set of the sun"; in social time, "when the battle is lost and won." Their choice of a junctural moment to meet with Macbeth, "when the hurlyburly is done," marks off a space that is joined to, and separated from, the battle just "lost" and "won" (1.1.4–8). Packing victory and defeat in one brief adverbial clause, the weird sisters implicate themselves in Macbeth's concerns, as much as they are implicated in the play's imperial theme. When associating the witches with anxieties about "female rule," from a historically bounded perspective, one must be sensitive to the dynamic view of the witches as embodiments of imagination and appraisal (Stallybrass, 202). In the early modern period, no doubt, the faculty of imagination was suspect. More specifically, through *lexia*, such as "specters and strange sights," imagination was linked to the biblical notion of evil (Kinney 2001, 56). Similarly, from a historical perspective, the *lexia* referring to blood may be seen to evoke the "primacy of the blood argument" for supporting hereditary monarchy (William Carroll, 189–191). If genealogical purity protects the divine right of kings, the function of demonic prophecy, in advising disbelief, is to pose questions, though Macbeth's weird sisters and the apparitions seem to have all the answers.

The weird sisters' speculation about who will be king, when no such speculation is expected or invited from them, conflates a *promise* to Macbeth with a *warning* to Duncan, though the latter is not their addressee. His sense of felicity when he approaches Macbeth's castle (1.4.1–9) relies on Duncan's not having access to "patterns of information" that would reactivate concerns about the loyalty of the Thanes. Yet the conflation of the general location of the witches with Duncan's locale implicates him in an antecedent appraisal that will lead to his murder. For instance, the witches show up immediately after Banquo has remarked on the remaining distance to Forres: "How far is't call'd to Forres? —What are these?" (1.3.39). Further, it is in a room in the palace in Forres that the king names Malcolm as Thane of Cumberland (his heir). Later, we will find out that the witches' haunt is, logically, in Forres. When uncertainty or the *difficulty* surrounding a hopeful situation elicits his desire to *know more*, Macbeth goes to meet the witches in Forres. In the beginning, the situation is simpler. Duncan is in a position to impact Macbeth's prospects by material rewards, the weird sisters through the immateriality of *antecedent* appraisal.

The way Shakespeare changes source information contributes to enhancing the role of imagination as it constitutes the law of apparent reality and establishes proximity variables for emotion intensity. In Buchanan's *History of Scotland* we are told that "Mackbeth" not only "conceived a secret hope of the kingdom" while he was fighting the Danes (not Norwegians), but that he was "further encouraged in his ambitious thoughts by a dream which he had." In the dream, "he saw three women, whose beauty was more august and surprising than . . . women's used to be" (quoted in Carroll, 130). Shakespeare follows Holinshed in presenting the weird sisters as "women in strange and wild apparel," but he attributes to their prophecies an immediate truth status. In Holinshed, the uncanny spectacle was first appraised as "vain and fantastical" "illusion" (Boswell-Stone, 23–24). Only "afterwards," the "common opinion," was that the "Weird Sisters" were the "goddesses of destiny" (24). The change from the dream (in Buchanan) to a real encounter (in Holinshed and Shakespeare), from beautiful to ugly, or just strange, constitutes a carefully orchestrated *antecedent* appraisal schema that can index valence with regard to desirable and undesirable prospect-based emotion. Hallucination can elicit higher cognitive activity than a dream. A real-seeming encounter, even though not real, is an event; it can be appraised in terms of valence much better than a dream.

Even as the supernatural agency that can be given ontological authority tempts Macbeth into believing in his prospects, various verbal processes undermine them. For instance, the way Macbeth is formally greeted a number of times draws attention to his good fortune, but the linguistic format of some of the greetings and other references to him not only mark this *event* as a joyful prospect, but foreshadow a reversal of fortune: the tragic peripeteia. To this effect, the joining and separating of the two Thanes' careers is suggested by Duncan's incidental rhyming of "death" with "Macbeth" when he commands: "Go pronounce his present death/And with his former title greet Macbeth" (1.2.66–67). Later in the play, Hecate rebukes the weird sisters for trading and trafficking with "Macbeth/In riddles and affairs of death" (3.5.4–5). As the name Macbeth becomes phonetically synonymous with death, temporal focalization is used to conflate events in the larger storyworld with appraisals and emotions of the inner sanctum of Macbeth and his weird sisters. Duncan's royal command (1.2.67–68) is issued after the witches' council (1.1.1–11), and it precedes their encounter with Macbeth (1.3.39–78), while the "real" news that Angus and Rosse bring follows it (100–115). In addition, the news that Rosse and Angus bring is coached in appraisal, not of events, but of people, through *attribution emotions*, where Cawdor's *blameworthiness*, through an invocation of *standards* and *values*, is measured for valence against Macbeth's *praiseworthiness*. The exigencies of Duncan's rule divide the Thanes into opposites of each other; the contraries will collapse when the exigencies are of Macbeth's rule.

In addition to prospect-based emotions and their foreshadowed reversal, in the opening scene Macbeth's destiny is already identified as an object of the weird sisters' own *fortune-of-others* emotions, as is shown by the excitement in their catcall of: "I come, Graymalkin," and the frolicksome frog motion of "Paddock calls,/Anon" (9–11). This branch of emotions involves social cognition and is defined, like others, around desirable and undesirable valence, where what is perceived as *desirable for the other* may lead to a "happy for" state of mind or "resentment," and what is perceived as *undesirable for the other* may lead to "gloating" or "pity" (Ortony 22–23). The witches are instigators of a play within the play. Thus, their interest and investment in the fortune-of-others emotions is more authorial, though an element of spectator psychology is not entirely absent. They do not exactly gloat, or commiserate, but remain interested. From their indifferent perspective, every battle is lost and won, and it does not matter who loses and who wins. For Cawdor, and later for Macbeth, it matters. The constitutive role that fortune-of-others emotions play is further suggested by how dominant the news motif is in the play. Topicality is, in other words, shown at its most universal in the link between fortune-of-others emotions and news.

To this purpose, Duncan's first words are spoken to the bloody man, who can "report" of the "revolt/The newest state" (1.2.1–3). The war news, clearly, touches on Duncan's and Malcolm's immediate concerns; yet we get only a brief reactive display of emotion from Duncan, and the focus is on news as narrative framing in terms of fortune-of-others and attribution emotions. The bloodied man's report, imbued with attribution-emotion–based moral commentary, distributes praise and blame, measures gain and loss, till he faints onstage (1.2.7–41). The weird sisters refer to themselves as "Posters of the sea and land" (1.3.33). Rosse and Angus's comparisons between the dead and the living Thanes are news-laden. Further, Rosse says the reports of Macbeth's brave deeds reached Duncan "as thick as hail" with the arrival of "post with post" (1.3.97–98). For indexing of valence in relation to innocence and piety, as opposed to the glamorous hurly-burly of news, "hail" is an important image, because it links later with Macbeth's image of "Pity" as "a naked new-born babe" "striding" the *blast* (1.4.21). Similarly, the embedded sound and visual metaphors of the horses' hooves, their bodies moving over hill and dale to bring fair and foul news, relate later to the fantastical idea that Duncan's horses went mad and ate each other. This uncanny occurrence is not only reported as "rumor" by the old man to Rosse (2.4.17–18); it reinforces the news motif. Contemplating Duncan's murder, Macbeth broods, "if th' assassination/Could trammel up the consequence, and catch/With his surcease success" (1.7.2–4. That is, what if the means (the assassination) should trammel up the ends (kingship and sovereignty)? *Trammel* can mean entangling a bird (partridge) in a net or "to fasten the legs of horses together so that they do not stray," and *surcease* is a legal term, meaning "halting of proceedings" (Muir, 36–37). In this figurative way, means and ends are conceptualized as birds caught in nets, or news-carrying horses subjected to *movement constraint*. Fulfilling the anticipation of Macbeth's worst fears, the old man attributes rebellious intent to Duncan's horses when he imagines they were all set "to contend against obedience" and make "war with mankind" (2.4.17–18). Contributing further to the retributive logic of narrative acts, Macduff's moment of grief also occurs in the backdrop of galloping horses and news about the massacre of his wife and children (5.3.204–230). In the news-crammed world of the play, the reader/viewer too is thickly implicated in the fortune-of-others emotion: his or her appetite for news and stories. Being implicated in this way makes the causality of evil less particular, more universal, more proximate, less distant—a part of the emotion system, not alien to it.

Why Hath It Given Me Earnest of Success: Promising More and Less

Locating the "proximity" variable in relation to emotion intensity, Ortony, Clore, and Collins maintain that appraisals and emotion episodes have to occur in close proximity to produce and maintain emotion intensity (Ortony, 62–64). Furthermore, the measuring of valence between satisfaction and dissatisfaction has to be a product of fast processing of information. In *Macbeth*, as we can see, emotion episodes and appraisals occur in unusually close proximity. As Macbeth's visual encounter stimulates imagination, at his bidding, "Speak, if you can: — what are you?" (1.3.47), the witches speak (48–50). The rhetorical trope of prosopopoeia (where inanimate objects speak) marks a change from the appraisal schema not being activated to its being activated. Though the witches are humanlike, not inanimate, Banquo does refer to them as "vanished bubbles of earth and water" (1.3.79–80). To Macbeth, the possibility of their speaking seems, at first, dubious. It is as if nature has become imbued with the historical event structure, its meanings, in order to speak to Macbeth. The first prophecy is, no doubt, a statement of the obvious, as Macbeth remarks, "By Sinel's death I know I am Thane of Glamis" (1.3.71). The second, about Macbeth winning Cawdor's title, is news. Only the third, "that shalt be King hereafter," promises what Lady Macbeth appraises, later, as the "golden round/which fate and metaphysical aid doth seem/To have crown'd thee withal" (1–5.28–29). It is important to note that the epistemological authority Macbeth gives to the witches is not only due to his assumption about their ontological privilege; it owes much to their appraisal and emotion eliciting words. For instance, the witches set up proximity variables and valence between *near* and *far*, what is *known* and what is *unknown* (Glamis, Cawdor, King); the *great* (Glamis), *greater* (Cawdor), and the *greatest* title (the King); superlative *happiness* measured against *relative* happiness. Following at the heels of the witches' "supernatural soliciting," when Rosse and Angus bring bits of conforming news, Macbeth appraises prophecy as truth. "Two truths are told," he thinks, appraising them as "happy prologues of the swelling act/Of the Imperial theme" (1.3.127–28). He thinks the "greatest," the promised truth, is "behind" (1.3.117). Using a sight meta-phor, he imagines other messengers right behind Rosse and Angus.

The visual linearity of happiness goals is, later, reversed when the ap-paritions show a pageantry of the Stuart kings and Macbeth says to the first: "Thy crown does sear my eyeballs" (4.1.113). At the passing of the fourth phantom king, Macbeth wonders: "What! Will the line stretch out to the crack of doom?" (4.1.117). A spectacle of royals separated from his own bloodline terrifies Macbeth, though he seeks kingship outside of the strict line of birth. Long before this moment of pain, an expedi-ent reversal from hope to denial of hope occurs when Malcolm is named "Prince of Cumberland," signifying *his* being the King hereafter (1.4.39). Macbeth's hope is grounded in a movement toward elective monarchy in Scotland, while Duncan takes a decisive step backward, toward he-reditary monarchy (Carroll, 120). Following social rules of affiliation in a society where services rendered by members of the aristocracy are rewarded by the monarch, Macbeth had just sworn services to Duncan as if they were "children and servants" of the king (1.4.25). In Macbeth's figurative articulation of *difficulty*, "The Prince of Cumberland!—That is a step/On which I must fall down, or else o'erleap/For in my way it lies" (1.4.49–50), the quick change is appraised as *movement constraint.* Remarking on research about animal emotion in response to "move-ment constraint," Frijda notes that a sophisticated, though automatic, appraisal is involved here, because stress results from "the clash between what the animal *can* do and what he *intends* or *wants* to do" (102, empha-sis added). Though he is not an animal, Macbeth experiences similar stress when he appraises the *difficulty* in terms of someone blocking his path on top of a precarious ladder from where he can either fall or over-leap. Inevitably, he concludes that he must *undo* what Duncan has *done.* Immediately after, Macbeth's apostrophe (formal, poetic address) to the "Stars," asking them to "hide [their] fires," and "Let that be,/Which the eye fears, when it is done, to see" (1.4.50–53), indicates the extent to which emotion has progressed to action readiness. If in Scottish histo-riography one of the arguments given for the shift from elective mon-archy to hereditary monarchy was so that "the ambition and strife of men might be stayed" (Thomas Craig, 228–229, quoted in Carroll, 186), Macbeth's reaction disproves that wisdom. Men's ambitions are not horses that can be chained to stay in one place.

Otherwise, why would it be that Malcolm's nomination does not prompt a re-appraisal of the prophecy as nothing more than a guess?

In part, the hedonic principle of appraisal and emotion is involved here. What the witches propose is attractive and addictive, while Duncan's nomination of Malcolm as heir apparent elicits aversion. The hedonic principle, however, works together with what Arthur Kirsch has described as the prophecies being "coached in inverse functions and equivocal contest" (84–85). To this effect, the prophecies pronounce what Macbeth will *lack*, what his *deficit* will be, as loudly as they proclaim what his *gain* will be. A full flowering of a more virulent rage, generated by envy, will take longer (3.1.55–70; 3.2.17–25); but the valence indexing potential of foreseen events is encrypted in the words spoken to Banquo: "Lesser than Macbeth, and greater/Not so happy, yet much happier/Thou shall get kings, though thou be none" (1.3.65–67). The first sentence starts with quantification of Banquo's happiness as lesser, but it ends with the emphasis on its being "greater." Similarly, the balancing contrast pattern of "not-so-happy," and yet, "much happier," with a hint of prolepsis (anticipatory arrangement of adjectives, or verbs) gives higher value to Banquo's share. At this moment, Macbeth's mood-congruent emotions are friendly to hope. He picks up the "two truths" of the "imperial theme," the metonyms of hope thrown in his direction, as if they are trophies to be given to the wife in order to procure her cooperation.

They Have More Than Human Knowledge: Appraising Truth and Consent

In a play that begins with appraisal of an event structure presented as prophecy, the formation of cognitive units involves treating it as "truth" and seeking others' confirmation of this "truth." For this he needs his wife's full consent and cooperation. Ortony defines cognitive unit formation as a local variable of emotion intensity and appraisal, where emotions elicited by the strength of the cognitive unit are often not related to one's own goals. He explains that the research purpose of the "*cognitive unit* variable is to accommodate cases in which the person experiencing the emotion is not the actual agent even though the emotion is characterized as involving the self" (77). What this means is that the formal self, as opposed to the actual self, is defined in "the context of the agents and actions" to be "*the self, or some other(s) with whom one perceives oneself to be in a unit*" (78). One of Ortony's examples is emotion elicited by the victory and defeat of athletic teams. These emotions, he says, are based on *cognitive unit* formation through allegiance and affiliation where the *formal* self and *actual* self converge. The convergence and divergence of cognitive units gives rise to intersecting appraisals, and "the situational structure of meaning" becomes *compound*. That is, it involves others who have their own goal trajectories, or who have adopted/appropriated others' goals. Nico Frijda calls the convergence, conflict, and appraisal associated with goals the "Multiplicity" component of appraisal (*Emotions* 1986, 208): the tangled web of intersubjectivity in which literary texts are always implicated.

For a consideration of the multiplicity factor, the distinction between the *formal* self and the *actual* self is valuable. In light of this perspective, the collaboration between Macbeth and Lady Macbeth goes beyond moral judgment about the wife tempting her husband to do evil, or the endorsement of a woman's pursuit of masculine power through subversion of gender roles. Whether we take it that "maternal power" in the play is "diffused" and distributed between Lady Macbeth and the witches (Adelman, 105), or agree that the play's emphasis on Banquo as the "father of a line of kings" (3.1.59), rather than Duncan, "has the effect" of "comfortably locating authority in the male body" (Kastan, 251), we are only speaking of how appraisal models are derived from cultural notions of gender, genealogy, and accession to royal authority. An appraisal model, no doubt, impacts appraisal, but it is not synonymous with the constitutive (and dramaturgical) role of the emotion that it elicits. Starting with a tight cognitive unit between Banquo and Macbeth, the play's action is predicated on the breaking up of this unit. Furthermore, the dissolution is a direct result of the prophecy, where the goal relevance for Banquo cannot elicit immediate concern, or hope. When they encounter them, both Thanes take the prophecies to be revealed truth. Banquo says: "I dreamt last night of the three Weird Sisters:/To you they have show'd some truth" (2.1.20–21), echoing what Macbeth asserts in his letter(s) to Lady Macbeth: "[I] learn'd by the perfect'st report" (1.5.2). Somewhat poignantly, he confesses: "When I burn'd in desire to question them further, they made themselves air" (1.5.2–5). While his "burning" desire to know more is an immediate reaction to Banquo's prospects, the disappearance enhances belief in their ontological access. Not involved in mystical ontology and the inner story of appraisal and emotions, Malcolm and Macduff form a cognitive unit in defense of hereditary monarchy, though all along they betray a sense of the rightness of elective monarchy. When they discuss resistance to Macbeth's rule, Malcolm and Macduff claim to be in support of "right rule" based on virtues of the ruler, not on birth. The strength of the cognitive unit allows them to appraise their own treason (in a play that begins with repelling an invasion, they seek help from the English to invade Scotland), and regicide as patriotic duty, as they take great care to highlight Macbeth's tyranny, and not his usurpation (4.3.1–110). The play insistently sets up and undermines a clear distinction between a "lawful king" and "a usurping tyrant" and Macduff is, indeed, "a royal version of Macbeth" (Kastan, 257).

The conflicting goal trajectories of the Scottish Thanes orient intersubjectivity within the Macbeth family in such a way that Lady Macbeth becomes a formal agent for the goal of usurpation through assassination. Macbeth's "soliciting" of his wife's participation in an endeavor he will alternately embrace and disavow initiates a political collaboration between husband and wife (on the Macbeths being "locked together," and/or being "locked away from each other," see De Quincy and Honigmann, in Honigmann, 129). On his day of success, Macbeth promptly shares the "news" with his wife, aligning himself with other news-persons in the play and extending much of the goal and concern relevance about the contemplated "deed" to her. To demonstrate this he addresses the wife as his "dearest partner of greatness," and says he did not want her to "lose the dues of rejoicing, by being ignorant of what greatness is promis'd [her]" (1.5.11–14).

In a sense, he elicits his wife's prospect-based emotions just as the weird sisters elicit his. Tempted by him, as he is tempted by them, she confesses that Macbeth's letters "have transported [her] beyond/The ignorant present, and [she feels] now/The future in the instant" (1.5.56–58). Emblematizing the titles in terms of birth, merit, and aspiration, also the past, the present, and the future, she repeats the greetings: "Great Glamis"; "Worthy Cawdor!" and adds: "Greater than both, by the all-hail hereafter!" (1.5.54–55). Lady Macbeth's readiness for regicide is evidenced by her involuntary startle response (similar to Macbeth's at the supernatural greeting) when the arrival of Duncan is announced. Visibly shaken, she turns to the messenger to say: "Thou'rt mad to say it" (1.5.31)

To achieve the ends that have become individualized in her, Lady Macbeth's notorious use of persuasive rhetoric relies on her having formal agency, anchored, appropriately, in *attribution emotions*, such as reproach. Reproach is an attribution emotion, and results from frustration, which is an event-based emotion (Ortony 56). Attribution emotions play a significant role in *cognitive unit* formation (Ortony 78). Anticipating her husband's hesitation, Lady Macbeth is prepared to pour her spirits in his "ear." She has already concluded that to chastise him with the "valour of [her] tongue" is the only way to have him be "crown'd" withal (1.5.26–27). As soon as he hesitates, as she had expected, she chastises: "Was the hope drunk,/Wherein you dressed yourself? Hath it slept since/And wakes it now to look so green and pale" (1.7.35–37). She envisions Macbeth himself as HOPE that is drunk and hung over. The vehemence with which "influence" is exerted draws attention to the constitutive function of two emotions: *frustration, an event-based emotion, leading to reproach, an attribution emotion.* Mirroring the early modern view that passion enters the mind through the eye and reason through the ear (Wright, 10), Lady Macbeth thinks of her exhortation as *reasoning* that must enter her husband's ear, while her reference to vision occurs in relation to *passion.* She imagines a vaguely seen "golden round" from afar, as promised by *metaphysical aid*, and sets it up against a close view of him being "crown'd" through *self effort* (1.5.28–30). Though historically grounded, Lady Macbeth's fictional being is an open system, and her reliance on attribution emotions of reproach and shame defines her as a *formal* agent, though at this point she acts like an *actual* agent.

Effective as Lady Macbeth is in seeking agency, her adoption of someone else's goals carries with it the danger of dislocating the self. Divergence of the self, as Ortony reminds us, occurs when actions are motivated by attribution emotions and cognitive unit formation, because one begins to view one's own self as the other (78). This kind of dislocation and displacement is evident in Lady Macbeth's rhetorical un-gendering of gendered selves. She begins to view Macbeth as female and herself as capable of being altered into a man with the help of the same "metaphysical aid" that promised him greatness. Separating Macbeth from his career of war killings, she swears that her Thane's *nature* is filled with the "milk of human kindness" (1.5.16). Concurrently, she appraises her own nature in terms that necessitate an apostrophe to the "spirits of the night" to "make thick the blood/To stop up the access and passage to remorse" (1.5.42–50). Inviting the "murthering ministers" to "come to [her] woman's breasts," to "take [her] milk for gall" (48), she repulses compassion and remorse, betraying a belief that remorse is intrinsic to human nature, or female nature. The dislocation of the self caused by the strength of the cognitive unit is reversed later in the haunting recurrence of the milk and blood *lexia* as well as a hallucinatory return to this moment through cognitive recall, a recall that does not omit the letter, repeating the actions of reading it, folding it, putting it away (5.1.3–8). In *Macbeth*, instances of tragic irony that elicit compassion are composed of just such poignant moments of cognitive-emotive peripeteia.

The strangeness associated with Lady Macbeth's plea to be unsexed, and later, her bravado about what she would do in order to attain kinship, marks a change from the habitual and, thus, elicits higher cognitive activity. One can see it simply as "the situational meaning components shaping the feature of interestingness" (Frijda, *Emotions*, 214), or, more sensationally, as "menacing heterogeneity of uncontrolled duplication that threatens the autonomy of power" (Goldberg, *Shakespeare's Hand*, 152–175). Either way, visual strangeness and/or the use of extreme rhetoric foregrounds physical sensation: the body and its emotions. For instance, Macbeth's startle response at the first sight of the witches and their fortune-telling is noted by Banquo and interpreted as "fear" and being "rapt withal" (1.3. 51–52; 57). In Macbeth's self-report to the wife, the fear is interpreted as wonder: "Whilst I stood rapt in the wonder of it, came missives from the King, who all-hailed me" (1.5.5). Appraisal theorists think that startle responses such as these cannot be explained away as biological reactions to stimuli, not involving appraisal. Startle responses can be triggered by "contents of long term memory," coming from "divergent sources," and "entered on a blackboard, continually supplanting, supplementing, and modifying what is already there" (Frijda, *Laws* 2007, 112). Frijda is quick to add that "blackboard is just a metaphor," and "appraisal experience is a fiction," qualifying further that "we do not experience appraisals, we experience events as appraised" (112), as Macbeth does in this case. Seen in this light, Lady Macbeth's extreme statement that she would be able to "pluck her nipple" from the "boneless gums" of a baby who is "smiling" in her face, and dash "its brains out" (1.7.55–57), a statement that has exercised critics to distraction, is hyperbole recruited to suggest to the imagination an event and a hypothetical emotion episode of extreme cruelty, with herself as its actual agent, to motivate action readiness through reproach. Lady Macbeth's use of shame and reproach as part of her persuasive rhetoric is so obvious that to interpret this utterance as evoking a sense of the uncanny, the "forbidden, the interdicted," that which "one must not see," would be an overstatement, though the play does deal with what one must not see (Garber, 122–123).

Cognitively more significant as they are, Coleridge's remarks on this passage are worth quoting in full. He says this passage though usually thought to prove a merciless and unwomanly nature, proves the direct opposite: she brings it as the most solemn enforcement to Macbeth as the solemnity of his promise to undertake the plot against Duncan. Had *she* so sworn, she would have done that which was most horrible to her feelings, rather than break the oath. . . . Had she regarded this [act of infanticide] with savage indifference, there would have been no force in the appeal; but her very allusion to it, and her purpose in this allusion, shows that she considered no tie so tender as that which connected her to her babe. (Coleridge quoted in Muir 42)

In other words, Lady Macbeth's conscious goal is not murder, but strengthening of the cognitive unit. Duncan is to her imagination, at this initial moment, no more than a storybook king. In contrast, her tie to the hypothetical "babe" invokes a strong attachment theme, as does her involuntary note to herself about the king looking like her father as he slept (2.2.13). An attachment theme based appraisal interferes with the cardboard-king schema by inscribing Duncan's sleeping face within the order of family, foreshadowing dissolution of the cognitive unit—though at this time she addresses Macbeth as "husband"—constituting herself in relation to various attachment themes: mother, daughter, and wife. For the rest of the play, she does not slip out of attachment themes. She revisits the scene of murder as a scene of trauma, remembers that the "Thane of Fife had a wife," wondering "where is she now?" (5.1.40–41). Her hermetic confession, "Hell is murky" (5.1. 34), contrasts with the wifely caution of an earlier utterance: "Get on your night-gown, lest occasion call us,/And show us to be watchers" (2.2. 69–71). This directive does not show Lady Macbeth's heartlessness; it is an early clue to the contrasting emotion regulation rules she and Macbeth follow.

To Beguile the Time Look Like the Time: Regulating Emotion

As a parallel to the early modern preoccupation with management of passions, cognitive theory of emotions identifies three "rules of emotion: constitutive, regulative, and procedural" (Averill, 266). Averill emphasizes that "like games and languages, emotions are both constituted and regulated by rules, and they require appropriate procedures for skilled performance" (266). Regulative and procedural rules are like "display rules" in relation to "feeling rules," that do not alter the nature of felt emotion, while they guide social expression of emotion (Frijda, *Emotions* 1986, 412–413). Averill's notion that emotion rules are like language rules is very important, because it assumes, rightly, that emotions are intrinsic to the brain, and socially adaptive, as language is. Just as the rules of grammar *constitute* a language, *regulative* rules help determine how to speak and write in social situations, and *procedural* rules determine the quality and effectiveness through rhetorical devices. Similarly, brain biology and consciousness *constitute* emotion; *regulative rules* guide how people "monitor behavior to conform to socially accepted standards," and *procedural rules* involve guiding and monitoring behavior for effectiveness (Averill, 266–267). Lady Macbeth's alarm, for instance, at the transparency of Macbeth's emotion-faces is elicited in response to her concern about the quality and effectiveness of his public image. An early modern preoccupation with bodily expression of passions, as evident in Thomas Wright's catalogue of how one can read inner passions in men's faces, gestures, clothes, words, and so forth (174–181), linked to his belief that passions can be *discovered* in "motions of the eies, pronunciation, managing of hands and body, manner of going" (131), no doubt finds an echo in Lady Macbeth's worry that Macbeth's face "is a book, where men/May read strange matters" (1.5.61–62). At the same time, it is continuous with the distinction between *display rules* and *feeling rules* in today's emotion theories. A book is a public document; a face is not. To present his face to the world as if it is an open book is not effective at a time when Macbeth is plotting to kill a king. The situation requires an unreadable face, not a transparent face.

Lady Macbeth's difference from Macbeth is marked by how consistently she follows procedural rules of emotional expression. It is clear that throughout Acts 1 and 2, her focus is on "quality and effectiveness" of the murder/assassination plot, informed by a concern to carry it off without public exposure. In contrast to the figurative overload of "look like the innocent flower,/But be the serpent under't" (1.5.65–66), an utterance that inevitably implicates Lady Macbeth in biblical accounts of evil, in a more sensible and practical way, she suggests that Macbeth should "wear welcome in [his] eye," in his "hand," in his "tongue," so as not to betray his intentions (1.5.63–64). The ironic reversal of her self-exposure induced by illness (5.1.30–60), for instance, is lexically linked to what she says, how she appraises, and what psychological resources she uses to manage a domestic contingency (2.2.15–71). Here too, the inclination is to be effective. Insofar as the quality of a psychotic outbreak resides in bringing back to memory, through hallucination, an emotion episode from which affect was forcibly withdrawn at the moment of its occurrence, the ritualism of Lady Macbeth's sleepwalking is hermetically effective. Accordingly, she creates graphic matches between references to "real" water and blood in the context of removing the evidence (2.2.63–71), initiating a belated effort to get rid of an imaginary smell of blood from her "little hand" (5.1.47–49).

In contrast, Macbeth, following regulative rules, monitors emotions, taking note of where they do or do not conform to social standards. For instance, Macbeth's famous soliloquy (2.1.33–61) documents regulative, ethical monitoring of complex emotion states, but it resolves in action readiness: "I go, and it is done: the bell invites me" (62). That is because the catalogue of ethical thoughts is only intended to shape emotional experience; it is not intended to change the plan of action. Similarly, part of Macbeth's concern in the preparatory scene is about the fear of failure and discovery, as expressed in: "If we should fail" (1.7.60). Toward the end of that sequence, he is "settled." Completing the crossbow figure she introduces, "screw your courage to the sticking place" (1.7.61), he declares he is now willing to "bend up" "each corporeal agent" to this "terrible feat" (1.7. 80–81). Calling it a "feat" likens it to battle, making it compatible with social standards. The priming of ruthless resolve in the earlier scene is counterbalanced when the imaginary dagger, "with its handle towards" Macbeth's hand, becomes sentient so that he can speak to it: "I see thee still:/And on thy blade, and dudgeon, gouts of blood/Which was not so before" (2.1.44–45). A conversation with the dagger "of the mind" (2.1.38) prepares him for action, since it helps unpack and reorganize emotional turmoil of this moment.

More specifically, Macbeth's dagger is no longer an object; it has transformed into an agent, and it sums up the situational meaning of the event structure. If the dagger, with "the handle towards [Macbeth's] hand," is aimed at Duncan, as it was formerly aimed at his enemies, what makes Duncan an enemy? The vision of blood on the "blade" and the "dudgeon" suggest not only blood, as in anticipated murder, but the blood of lineage that entitles Malcolm, and disentitles Macbeth. In connection with other kinds of blood that is shed, the word crops up in two scene-opening lines spoken by Duncan, "What bloody man is that?" (1.2.1) and "Is execution done on Cawdor?" (1.4.1). A monarch's *performative* speech acts, utterances that perform the actions of which they speak, quiet literally in this case, become *events*. Duncan's rule, like that of any other medieval king, requires rivers of enemy blood through war and expedient execution of traitors. When he becomes king, Macbeth will add tyranny to the bread and butter militarism of a medieval polity. In face of terrifying realities, the ritualized hallucinations associated with the two murders paradoxically arrange Macbeth's emotional experience into narrative scripts. Contrarily, his wife's ability to see only what is there to see infects her sight, later, to the extent that she can no longer see or sense the world around her. In the Banquet scene, Lady Macbeth's recall of the "air-drawn dagger" in conjunction with her appraisal of Macbeth's vision of Banquo's ghost as "the very painting of [his] fear," when she cautions—"You look but on a stool" (3.4.60–67)—marks the interiority of appraisal, its subjective particularity, even with regard to two people in a cognitive unit, tied inexorably to "A deed without a name" (4.1.49).

Lady Macbeth's concern in the banquet scene is with effective execution of the coronation feast: "You have displac'd the mirth, broke the good meeting/With the most admired disorder" (3.4.108), she says. Further, she personifies him as "Shame itself," and asks: "Why do you make such faces" (3.4.66). However, this "admired disorder" is what, from a regulative perspective, organizes emotion for Macbeth. Gazing at the ghost, he says he would prefer the "rugged Russian bear,/The arm'd rhinoceros, or th' Hyrcan tiger," than that "horrible shadow" (3.4.99–105). Imagining encounters with ferocious animals helps to compare *lesser* with *greater terror*, because Macbeth's main *concern* is to master his terror. By virtue of a disordered display, the scene rearranges emotion through reasoning and meta-emotion (emotion about emotion). Violating *display rules*, Macbeth stays true to *feeling rules*. In the context of Macbeth's developing tyranny, Malcolm suggests that Macduff should "dispute" the massacre of his wife and children, "like a man" (4.3.220). Clearly, Malcolm advocates a violation of the *feeling rule* for grief when he proposes that Macduff should appraise the massacre as an *offense against him:* the core relational theme for anger (Power et al., 90). On his part, Macduff affirms his natural right to grieve and, thus, defends the *feeling rule* for grief: "But I must also feel it as a man" (221). Once Macduff is allowed ten seconds of grief that involves a temporary violation of the display rule, Malcolm counsels: "Let grief/Convert to anger; blunt not the heart, enrage it" (228–229). Grief transformed into anger strengthens the converse cognitive unit between Macbeth's *antagonists* at a time when Macbeth, as monarch, stands alone in his tyranny, and, as an experiencing subject, he is alone in his despair.

The contrasting rules of emotion regulation the Macbeths follow are nowhere more evident than in the uncanny division of labor between them, with her focus on task-oriented efficiency and his on marking boundaries between bodily motions pertaining to the "deed" and verbal commentary that conforms to social values and expectations, dividing location of the body and the sites of moral reflection. Thus, while he refers to his hands as "hangman's hands" that "pluck out [his] eyes" (2.1.56–57), Lady Macbeth believes, "A little water clears us of this deed." From his perspective, as a part of him, this *deed* cannot be separated from the whole: "To know my deed, 'twere best not to know myself" (2.2.71), he muses. It is important to keep in mind, however, that Macbeth's conscience talk is palliative appraisal that facilitates murder and tyranny. Palliative appraisal is any type of construal that would make a painful and difficult emotion-situation bearable to the organism and, thus, reinforce adaptive ability. In this context, one striking example of palliative appraisal follows his vision of the dagger. Seeing himself as someone else, in this case an infamous legendary figure, Macbeth likens his own "stealthy pace" to "Tarquin's ravishing strides" moving toward his "design" "like a ghost" (2.1.55–56). The word *stride* is an important *lexia* here; it triggers a contrastive recall of the figure of "pity" as the "naked-new born babe" *striding* the blast. The Tarquin comparison, no doubt, feminizes Duncan, but this allusive appraisal has other cognitive implications as well. Tarquin's career is a "confirmed" part of Roman history. When something has once been done, even if it is not right, it becomes (perversely) acceptable. In this scene, subsequent objectification of appraisal is communicated through sound imagery, and we see the same complementary difference. Lady Macbeth does not hear the disembodied "voice" crying "Sleep no more"; she only hears ambient sound: "I heard the owl scream, and the crickets cry" (2.1.15–16; 34). The "voice" that Macbeth hears restores moral balance through a regulatory *expression* of socially appropriate emotion. As wife, housewife, and partner, Lady Macbeth contributes to the creation of this clinical-cathartic space for a compensatory moral discourse. It follows, therefore, that for herself she will require the hermetic clinical space of her illness.

If Shakespeare's tragedy is an analysis of evil, it is not a moral analysis in conventional terms, but a cognitive account of men's actions in an environment where prosperity and glory can be attained only through war killings. For Shakespeare's audiences the famous sleep murdering motif would have had further historical resonance. According to Holinshed, during one of the wars with the Danes, Duncan sent food to the enemy troops and his soldiers mixed "the juice of mikilwoort berries to it" that induces sleep and is poisonous. The hungry Danes ate the food gratefully. Subsequently, "Duncan sent unto Macbeth, commanding him with all diligence to come and set upon the enemies" (William Carroll, 140). The Danes were "so heavy with sleep, most part of them were slain and never stirred" (William Carroll, *Macbeth*, 140). For dramaturgical compression, Shakespeare deletes this part, but resurrects it in Macbeth's speech when he imagines that the voice said, "Glamis hath murdered Sleep and therefore, Cawdor/ Shall sleep no more, Macbeth shall sleep no more" (2.2.41–42). The historical Macbeth, on Duncan's orders, literally murdered sleep.

In light of this historical precedent, it is peculiar that Macbeth implicates two of his *titles* in the sleep-murdering act, naming the third as a title-less subjectivity: Macbeth. Sleep, as "sore labor's bath," "balm of hurt minds," and "Chief nourisher in life's feast" (2.2.37–3), is praised for its epicurean simplicity grounded in the stability of *attachment* and *affiliation* structures that bind individuals to community and regulate social and familial life, headed by the sovereign as god's deputy. He disburses favors and privileges in return for services rendered. And there is the rub. What is the nature of these services? On the day of Scottish victory, the Captain praised "brave Macbeth" for how he "unseam'd" the enemy "from the nave to the chops/and fixed his head upon our battlements" (1.2.22–23). In a converse parallel, aligned with the English, with Macbeth's head in his hand, Macduff will do the same honors at the end of the play (5.9.20–24). In contrast to acts that figuratively "unseam," take the seams out of tailored clothes, sleep "knits up the ravell'd sleeve of care" (2.2.36). Not consciously, but inadvertently, Macbeth inverts a value system and its blood argument so that the play, *Macbeth*, may draw attention to daily violations of an emotion-law of morality, a violation which is the very foundation of this culture and society. Such a decisive movement away from topicality and limiting historicity is facilitated by Shakespeare's "notorious ability to anticipate all possibilities" (Sinfield 184). More universally, then, the play suggests that the stability and comfort a group seeks through supposedly rule-governed violence sponsors actions that disrupt the same structures of stability and comfort in the lives of others, such as Duncan's sleepy enemies, whose need for food became an effective war strategy for the Scots. In other words, *fair* is always already *foul*.

There Is No Flying Hence, Nor Tarrying Here: Appraising the (Tragic) End

In a play that is predicated on prospect-based emotions, the moment of peripeteia is memorialized by Macbeth as his being stuck in a thick river of blood: "For mine own good, /All causes shall give way: I am in blood/Stepp'd in so far, that, should I wade no more,/Returning were as tedious as go o'er" (3.4.134–136). The tragic protagonist's inner experience of the moment of reversal of fortune is striking when contrasted with the modal self he projected when he "carv'd out a passage" with his "brandish'd steel" through another river of blood, leading to victory the "doubtful" outcome of a battle with "two spent swimmers, that do cling together/And choke their art" (1.2.9–10; 17–19). Now his art to resolve *doubt* is choked to the extent that if he decides to "wade no more" in this figurative river of blood, "returning" would be as "tedious" as going over. Tyrannical uses of power, here as elsewhere, are repetitive acts, as pathological as Lady Macbeth's repetitive actions, though with a different consequence for others. Even as Scottish history repeats its fouls, calling them fair, the foulness of the Macbeths' condition is ultimately physical. But first it is imagined, thought, as Frijda explains: "Emotion appraisal is shot through by cognition. Many events owe their emotional impact—their primary appraisal—by its meaning for future actions and future satisfaction of concerns" (*Laws* 102). Consistent with this account, Shakespeare's play begins with *emotion* appraisal by the weird sisters, as it was *shot through by cognition.* Subsequent events, as we have seen, *owe their emotional impact—their primary appraisal*—to how the meaning of the prophecies was determined *for future actions and future satisfaction of concerns.* The falling stage of dramatic action, very accurately, shows the protagonists experiencing *movement restraint,* because *what they can do clashes with they may intend or want to do,* as in the case of the hypothetical animal Frijda speaks of in this context. Like trapped animals, Lady Macbeth and Macbeth realize that in their attempt to extract the future from the present moment, through their "poor malice" (3.2.14), the promise of the future was strangled, even like that anonymous "birth strangled babe" the witches pour into their cauldron (4.1.130). The assassination "trammeled up" the consequence, as Macbeth had anticipated (1.7.2–3).

As a counterpoint to the emotion-eliciting appraisal schema of movement restraint, Frijda identifies "context components and action-relevant" components as appraisal components associated with the emotion-eliciting law of *change*. Summing up this idea, he says "events are changes." He reminds us of Schopenhauer's notion that all "happiness is a mere decrease of unhappiness" (Frijda, *Emotions* 1986, 209). Events in narrative are changes in states of feeling, thought, and emotion marking increase and decrease in satisfaction or dissatisfaction. That aspect of narrative emotion at this point in the play shifts to Malcolm and Macduff: Macbeth's *antagonists*. For Macbeth and his Lady there are no longer any events that can be appraised. Even her death cannot be appraised as *change* from a desirable state to an undesirable state; hence Macbeth's comment: "She should have died hereafter" (5.5.17). Within the realizational structure of the play, Macbeth appraises emotions as if they were events. For instance, when he is told Fleance has escaped, he compares a hypothetical *greater state of* satisfaction of concerns and goals to a significantly *lesser*. He says: "Then comes my fit again: I had else been perfect;/Whole as the marble, founded as the rock/As broad and general as the casing air:/But now, I am cabin'd, cribb'd, confin'd, bound in/To saucy doubts and fears" (3.4.20–23). While the *difficulty* of dwelling in "saucy," that is "insolent and importunate," doubts, are anthropomorphized as sentient agents, Macbeth appraises emotions not so much for goal fulfillment, but for their feeling tones to determine *valence* in relation to consciousness states.

Drawing on established theories of conceptual metaphor, Zoltan Kovecses explains that source domains (modes of expression) of metaphor are "concrete and physical" and more "clearly delineated concepts," while target domains (what is communicated) "tend to be fairly abstract and less delineated" (15). His large list of target domains identifies "emotion" as a "par excellence target domain" (Kovecses, 21). Source domains for emotion, he says, "Typically involve forces," such as "gravitational, magnetic, electric, mechanical," also natural, such as "waves, wind, storm, fire" (Kovecses, 19–21). In ordinary language as well as in literary language, the domains are mapped onto each other. Shakespearean drama presents a sizable archive of conceptual metaphors transformed into figurative metaphors through the processes of extension, where "a new conceptual element is added to the source domain"; through *elaboration*, where the poet may "elaborate on the source domain in an unusual way"; through *questioning*, where a poet may "call into question an everyday metaphor"; or through *combining*, where a series of everyday conceptual metaphors are combined (Kovecses, 47–50).

At a broad narrative level, the play is organized by a conceptual meta-

phor. The "weird sisters"—the word *weird* originating from *wyrd* (fate)—
constitute a source domain for the target domain of fate, or, in terms of
our analysis, they are a source domain for the target domain of appraisal.
The way an individual appraises events becomes his or her fate, because
the interpretation of the narrative of a life elicits emotions as fictional
narratives do, and emotions guide choices. Macbeth's situation may be
summed up as an individual's *will to power*, or "nature's revenge on the
ambition of man" (Watson, 177); either way, the weird sisters represent
a principle of causality. Once this causality is set in motion, and the protagonist comes to a point where he understands his situation in terms of
movement constraint, the figurative river of blood in which Macbeth is
stuck *combines* at least three conceptual metaphors: Life is a journey;
Water is life; Blood is death. Further, the metaphorical idea that life
is a journey is complicated, and *questioned*. Macbeth's wading in blood is
an *elaboration*, where a new element is added to the conventional source
domain of water. The prehistory of this metaphor is an *elaboration* of the
source domains of blood and water to envision a cosmic contamination
of all water everywhere when Macbeth asks: "Will all great Neptune's
ocean wash this blood/Clean from my hand? No, this my hand will
rather/The multitudinous seas incarnadine,/Making the green one red"
(2.2.60–63). If all water everywhere has been incarnadined, the particular river of blood allows Macbeth only to wade, not to swim. The elemental force of "casing air" is still a source domain for happiness characterized by a state of unobstructed security, certainty, and freedom,
while the similes "the whole as marble" and "founded as a rock" serve
as source domains for the target domains of "honour, love, obedience,
troops of friends," that "should accompany old age" (5.2.24–26), things
that Macbeth, by his own admission, lacks. Here again, the conceptual
metaphors are: Air is freedom; Stone is stability; Marble is fame;
Water is purification; Blood is contamination. The last metaphoric
concept questions the political apotheosis of blood as associated with
genealogy and royal lineage. Does that blood sanctify or contaminate
the institution of kingship?

In this later part of the play, Macbeth does not only appraise emotions, he revisits *antecedent* appraisal, as if to question it, challenge it,
and reproach it. The witches are approached at this stage for intrapsychic coping. Modes of intrapsychic coping and defensive appraisals
correspond to "constructing situational meaning structure in such a
fashion that the situation is appraised more favorably, less harmfully, or
more tolerably than the actual state of affairs warrants or imposes in the

first place" (Frijda, *Emotions* 1986, 420–421). In other words, the "supernatural soliciting" that elicited the prospect-based emotion of *hope* is now recruited to provide *palliative* appraisal to cure *despair.* Addressing the witches as "black, and midnight hags!" Macbeth challenges them: "I conjure you, by that which you profess" (4.1.48–49). As speaking subjects who have so far assumed epistemological privilege, the witches recede in order to project "the artificial sprites," who "by the strength of their illusion/Will draw him into his confusion" (3.5.28–30). The sadomasochism in the tone mirrors a thought process that will, in finding a coping strategy for something that one should not be able to cope with, turn against the self. Like Lady Macbeth's doctor, Macbeth's appraisal mechanism will not present effective ways of coping because the "dismal and fatal end" (3.5.21) is certain.

For palliative appraisal, at this stage, a seemingly plausible *doubt* has to be introduced to the *certainty* of a dismal end. Doubt can perpetuate hope, no matter how tenuous. Knowing this, the apparitions supply provisional *doubt* to defer the *certainty* of a dismal end. They invoke conditions and set up improbable possibilities, mixing assurance with warning. The first apparition warns: "Macbeth! Macbeth! Macbeth! beware Macduff" (4.1.71). The second tells him he cannot be killed by any man born of woman. The third apparition reassures that he cannot be harmed till "Great Birnam wood to high Dunsinane hill/Shall come against him" (4.1.93–94). From a cognitive perspective, the dizzying repetitions of three throughout the play, in combination with utterances using strategically placed rhymed couplets and epigrammatic syntax, contribute to the elicitation of belief because these verbal processes give the impression of a cause-effect tightness and closure. The susceptibility of the human brain to such constructions is in part due to the fact that the "analytical brain functions," as Jaak Panksepp maintains, "appear to have been designed, through evolutionary selection, to discern causal relations in the external world" ("The Basics of Emotion," 21). This capacity of the brain is, no doubt, inclined toward the development of science, but its hard-wired nature gives rise to an obsessive tendency to make causal connections in order to resolve doubt and uncertainty without delay, especially when one encounters threats from the environment. Rhymed couplets and rhetorical repetitions, also epigrams, because of their phonetic, syntactic, semantic closure and emphasis, may signal the "analytical brain" to construe cause-effect relations where none exist. It is very striking that Macbeth's witches and the apparitions use this format to induce belief in Macbeth because they suggest cause-effect connections through nothing more abstruse than end rhymes, syntactic balance and closure, indexing of valence through semantic features, cognitive marking of emphases through repetition: the "trading" and the "trafficking" through "riddles" (5.1.4–5).

Inducing belief by these means, the apparitions remove *doubt* about the agent (Macduff) and agency (the invading army), those who are about to execute the fatal end. However, it does not take long for the tenuous *doubt* to erode, and with it the hope, and Macbeth appraises the "Birnam wood's" approach to Dunsinane hill as his ability, finally "To doubt th' equivocation of the fiend,/That lies like truth" (5.5.43–44). His appraisal of movement restraint, "There is no flying hence, nor tarrying here" (5.5.48), makes the Dunsinane castle, which the Macbeth of history built at great cost, his prison and his stake. With an ambiguity in pronoun reference he says: "They have tied me to a stake: I cannot fly" (5.7.1). Locating himself in relation to witch-burning practices, Macbeth invokes an animal metaphor to emphasize his difference: "But, bear-like, I must fight the course" (5.7.2). This figure suggests struggle, but not of the will to seek freedom, only an automatic response elicited by the body's long-established patterns of physical combat. As he faces Macduff to know that the latter was "from his mother's womb/Untimely ripped" (5.8.15–16), Macbeth saves face by appraising the appraisers as "juggling fiends" who "keep the word of promise to our ear,/And break it to our hope" (5.8.18–21). Defiantly resigned, he says to Macduff: "I'll not fight with thee" (22). Macbeth's lingering *hope* was a provisional hope because the *doubt* about the certainty of the "dismal end" was provisional doubt.

The final evaluative comment, "this dead butcher, and his fiend like Queen" (5.9.34–39), is neither true nor false. It is apt, however, that this statement graces the royal lips of the new king, Malcolm. Had he possessed the "king-becoming graces" he would not *gloat* when the outcome is *desirable* to him and it is *undesirable* for the other. On the other hand, his comment is motivated by *attribution emotions* of everyone in the cognitive unit. The death of the "Thane" of Cawdor, the King that was Macbeth, is personal to the "Earls"; it brings them relief and joy. The dark and difficult emotion-knowledge the play taunts the reader with is that the "butcher and his fiend like Queen" were, after all, no better or worse than any of the others, or us. Yet the play does not confront us with this knowledge. It teases, because teasing with equivocation is the play's leitmotif. Every man and woman will not turn out to be a butcher and a fiend; yet everyone will be a *poor player* on the stage of his and her life; everyone's "yesterdays have lighted fools/the way to dusty death" (5.5.19–28). Cognitively speaking, this is what the play is about. The good man who does evil is everyone.

Glossary

The definitions below give the reader a thumbnail charaterization of certain specialized terms.

absorption: The mental state signifying *embodied transparency*, in which a person's body language reveals their true feelings. See also *embodied transparency*

achieved narrative: A narrative that is satisfying to its auditors because it helps them decide how to act by enabling prediction.

ambassadorial strategic empathy: An author's attempt to recruit "particular readers to a present cause through emotional confusion." See also *authorial strategic empathizing, bounded strategic empathy,* and *broadcast strategic empathy*

appraisal theory: The theory that emotions are "always about something" in the sense that they are caused by "an object, agent and event and a pattern of reasoning," and that our appraisal of how those objects, agents, and events affect our goals or concerns, which process is automatic and can be instantaneous, is what brings those emotions about.

authorial strategic empathizing: An author's attempt to evoke emotions in readers for the purpose of affecting the political views of that audience. See also *ambassadorial strategic empathy, bounded strategic empathy,* and *broadcast strategic empathy*

bounded strategic empathy: An author's attempt to evoke emotions within an in-group based on shared experiences, and leading ideally to closer bonds within that group. See also *ambassadorial strategic empathy, authorial strategic empathizing,* and *broadcast strategic empathy*

broadcast strategic empathy: An author's attempt to appeal to the widest possible audience, in the service of a specific group, by "*emphasizing our common human experiences, feelings, hopes, and vulnerabilities.*" See also *ambassadorial strategic empathy, authorial strategic empathizing,* and *bounded strategic empathy*

conceptual competence: "[T]he ability to form and combine thoughts." See also *Linguistic Competence* and *Spectatorial Competence*

embodied transparency: A state in which a person very briefly, spontaneously, and involuntarily reveals his or her true feelings through body language, often as a reaction to new information, a shock, or other outside stimulus. See also *absorption*

emotional contagion: "[T]he communication of one's mood to others." A form of empathy that causes us "a pang of fear when faced with a companion who has become pale, wide-eyed, and trembles with terror; a surge of joy when we encounter a merry band of mirthful youths."

emotional memories: "[U]nself-conscious episodic memories which, when activated, cause us to experience relevant emotions." See also *emotional contagion.*

empathic inaccuracy: "[A] potential effect of narrative empathy: *a strong conviction of empathy that incorrectly identifies the feeling of a literary persona.* Empathic inaccuracy occurs when a reader responds empathetically to a fictional character at cross-purposes with an author's intentions." See also *narrative empathy*

garden-path narrative: A narrative the end of which discloses a key piece of information that requires the reader to go back and reevaluate previous events in light of this new information, often producing a drastic change in the reader's assessment of a character or situation.

garden-path sentence: A sentence that at first invites us to apply the wrong syntactical template, the meaning of which can only be found by a recursive effort and the subsequent application of the correct template. Often effected by tricking readers into reading nouns as adjectives and vice versa, and leaving out definite and indefinite articles that would otherwise guide the reader to a correct interpretation.

generative principles of action: Patterns, originating in subcortical areas of the brain, that affect our motivational or emotional response to internal conditions, such as hunger; external conditions, such as a dangerous situation; or perceptual or imaginative conditions. See also *regulative principles*

grand syntagmatique: Christian Metz's typology of filmic shots "which could be used to map the syntagmatic structure of a film." See also *syntagma*

language of thought hypothesis: The hypothesis that "(l)anguage and thought cannot be separated because thoughts are encoded and expressed through language."

linguistic competence: "[T]he knowledge that we have about linguistic systems. We know, for instance, that some sequences of expressions do not belong to our language. Those sequences that do not belong to the set of well-formed expressions of a language are called ill-formed and can be detected by any speaker of a given language." See also *spectatorial competence* and *conceptual competence*

metarepresentational ability: The reader's ability to keep track of the source of the information s/he receives by way of tags, such as "she said," and thereby to evaluate the accuracy of the information based on his/her knowledge of the source; the ability to determine whether the author is representing or re-representing information.

mind-reading: See *Theory of Mind*

mirror neurons: Neurons that make up a "neural mirror system that demonstrates an internal correlation between the representations of perceptual and motor functionalities," indicating that while observing an action, certain parts of the brain are activated in the same way that they would be if the observer were performing the action

narrative empathy: A vicarious, spontaneous sharing of affect with a fictional character, which may be brought about or enhanced by one or more features of the narrative, including *character identification* and *narrative situation.* See also *empathic inaccuracy*

regulative principles: "[P]rinciples that modify our behavior as it has been motivated by the generative principles. The regulative principles operate primarily through the prefrontal cortex and may be, largely or even entirely, a matter of appraisal," and which may be informational, prudential, or ethical. See also *generative principles*

representational hunger: The need for narratives that offer solutions to culturally complicated problems that, no matter how many times they are re-represented, never seem to yield a fully satisfying solution

semantic memory: Shared cultural knowledge that we obtain from our cultural environment, and which we tend not to doubt.

source tags: Indicators of who is speaking, or who is the source of the represented information, that allow readers to differentiate the various voices in a text.

spectatorial competence: A viewer has spectatorial competence when he or she "understands and parses the 'grammatical' sequences of images and organizes them in constituents . . . [and is] able to understand plot/narrative arrangements, and to decode the association of musical cues to images and the resulting emotional effect." See also *linguistic competence* and *conceptual competence*

syntagma: A unit "of narrative autonomy within a text," of which there are eight types: the autonomous shot, which is a single shot; the parallel syntagma, which is "based on alternation"; the bracket syntagma, which is "organized according to a concept or an external way of arranging reality"; the descriptive syntagma, in which "shots are displayed in an order suggesting spatial coexistence" and describe "a fragment of reality"; the alternating syntagma, which "alternates shots from two (or more) events or spatial locations that are chronologically related"; the scene, which "is an arrangement that is chronological and linear, offering spatio-temporal continuity"; the episodic sequence, which "represents a succession of events that are chronologically related and glued together by other narrative mechanisms" such as music, or dissolves; and the ordinary sequence, which represents "a continuous action scenario, in which unimportant fragments have been elided." See also *grand syntagmatique*

Theory of Mind, or mind-reading: The theory that we naturally and unconsciously understand other people's behavior in terms of intentions that we attribute to them due to bodily cues, and the idea that we understand fictional characters in the same way; "our ability to explain behavior in terms of underlying thoughts, feelings, desires, and intentions."

Bibliography

Aarseth, Espen J. *Cybertext: Perspectives on Ergodic Literature.* Baltimore: Johns Hopkins University Press, 1997.

Abbott, H. Porter. "The Evolutionary Origins of the Storied Mind: Modeling the Prehistory of Narrative Consciousness and Its Discontents." *Narrative* 8, no. 3 (2000): 247–256.

———. "Closure." In *Routledge Encyclopedia of Narrative Theory.* Ed. David Herman, Manfred Jahn, and Marie-Laure Ryan. London: Routledge, 2005. 65–66.

———. "Immersions in the Cognitive Sublime: The Textual Experience of the Extratextual Unknown in García Márquez and Beckett." *Narrative* 17, no. 2 (2009): 131–142.

Adamson, Sylvia. "The Rise and Fall of Empathetic Narrative: A Historical Perspective on Perspective." In *New Perspectives on Narrative Perspective.* Ed. Willie van Peer and Seymour Chatman. Albany: SUNY Press, 2001. 83–99.

Adelman, Janet. "'Born of Woman': Fantasies of Maternal Power in *Macbeth*" (1991). In *Shakespearean Tragedy and Gender.* Ed. Shirley Nelson Garner and Madelon Sprengnether. Bloomington: Indiana University Press, 1996. 105–133.

Adorno, Theodor. *Philosophy of New Music.* Trans. Robert Hullot-Kentor. Minneapolis: University of Minnesota Press, 2006.

Ainslie, George, and John Monterosso. "Hyperbolic Discounting Lets Empathy Be a Motivated Process." *Behavioral and Brain Sciences* 25 (February 2002): 20–21.

Anderson, S. W., A. Bechara, H. Damasio, D. Tranel, and A. Damasio. "Impairment of social and moral behavior related to early damage in human prefrontal cortex." *Nature Neuroscience* 2 (1999): 1032–1037.

Andringa, Els, et al. "Point of View and Viewer Empathy in Film." In *New Perspectives on Narrative Perspective.* Ed. Willie van Peer and Seymour Chatman. Albany, N.Y.: SUNY Press, 2001. 133–157.

Anonymous. "Convention against Torture and Other Cruel, Inhuman or Degrading Treatment or Punishment." Adopted and opened for signature, ratification, and accession by General Assembly resolution 39/46 of 10 December 1984 (entry into force 26 June 1987). http://www.unhchr.ch/html/menu3/b/h_cat39.htm (accessed 13 June 2006).

Anonymous. "Documentary on Director Michael Haneke." *Caché*, DVD: Sony Pictures Classic Films, 2002.

Anonymous. "Report of the International Committee of the Red Cross (ICRC) on the Treatment by the Coalition Forces of Prisoners of War and Other Protected Persons by the Geneva Conventions in Iraq during Arrest, Internment and Interrogation (February 2004)." In *Torture and Truth: America, Abu Ghraib, and the War on Terror.* Ed. Mark Danner. New York: New York Review of Books, 2004. 251–275.

Arijon, Daniel. *Grammar of the Film Language.* Los Angeles: Silman-James Press, 1991.

Aristotle. *Poetics.* Trans. S. H. Butcher. *Criticism: The Major Texts.* Ed. Walter Jackson Bate. New York: Harcourt Brace Jovanovich, 1970. 19–39.

Auerbach, Erich. *Mimesis.* Princeton: Princeton University Press, 1991.

Averill, James R. "Emotions Unbecoming and Becoming." In *The Nature of Emotion: Fundamental Questions.* Ed. Paul Ekman and Richard J. Davidson. London: Oxford University Press, 1994. 265–269.

Ayotte, Julie, et al. "Congenital Amusia: A Group Study of Adults Afflicted with a Music-Specific Disorder." *Brain* 125 (2002): 238–251.

Azar, Beth. "How Mimicry Begat Culture." *Monitor on Psychology* 36, no. 9 (October 2005): 54–57.

Bakhtin, Mikhail. *Problems of Dostoevsky's Poetics.* Trans. Caryl Emerson. Minneapolis: University of Minnesota Press, 1984.

Bal, Mieke. *Narratology: Introduction to the Theory of Narrative.* Trans. C. van Boheemen. Toronto: University of Toronto Press, 1985.

Balter, Michael. "Study of Music and the Mind Hits a High Note in Montreal." *Science* 315 (9 February 2007): 758–759.

Bandura, Albert. "Reflexive Empathy: On Predicting More Than Has Ever Been Observed." *Behavioral and Brain Sciences* 25, no. 1 (2002): 24–25.

Banfield, Ann. "Beckett's Tattered Syntax." *Representations* 84 (November 2004): 6–29.

Baron-Cohen, Simon. *Mindblindness: An Essay on Autism and Theory of Mind.* Cambridge: MIT Press, 1995.

Barthes, Roland. *The Pleasure of the Text.* Trans. Richard Miller. New York: Hill and Wang, 1975.

———. "The Reality Effect." In Roland Barthes, *The Rustle of Language.* Trans. Richard Howard. Berkeley: University of California Press, 1989. 141–148.

Barwise, Jon, and John Perry. *Situations and Attitudes.* Cambridge, Mass.: MIT Press, 1983.

Bateson, Gregory. *Steps to an Ecology of Mind.* New York: Ballantine Books, 1972.

Batson, C. Daniel. *The Altruism Question: Toward a Social-Psychological Answer.* Hillsdale, N.J.: Erlbaum, 1991.

———. "Benefits and Liabilities of Empathy-induced Altruism." In *The Social Psychology of Good and Evil.* Ed. Arthur G. Miller. New York: Guilford Press, 2004. 359–385.

Batson, C. Daniel, and Adam A. Powell. "Altruism and Prosocial Behavior." In *Handbook of Psychology, Personality and Social Psychology, Vol. 5.* Ed. Theodore Millon and Melvin J. Lerner. New York: Wiley, 2003. 463–484.

Beckett, Samuel. "Dante . . . Bruno. Vico. Joyce." In Samuel Beckett, et al., *Our Examination Round His Factification for Incamination of Work in Progress.* London: Faber and Faber, 1929.

———. *Proust.* New York: Grove, 1931.

———. *Nohow On: Company, Ill Seen Ill Said, Worstward Ho.* New York: Grove, 1996.

Beer, Jennifer S., Erin A. Heerey, Dacher Keltner, Donatella Scabini, and Robert T. Knight. "The Regulatory Function of Self-Conscious Emotion: Insights from Patients with Orbitofrontal Damage." *Journal of Personality and Social Psychology* 85, no. 4 (2003): 594–604.

Behrmann, Marlene, and Galia Avidan. "Congenital Prosopagnosia: Face-blind from Birth." *Trends in Cognitive Science* 9, no. 4 (2005): 180–187.

Benjamin, Jessica. *The Bonds of Love: Psychoanalysis, Feminism and the Problem of Domination.* New York: Pantheon Books, 1988.

Bergman, Mats (n.d.). "Representationism and Presentationism." Commens Virtual Centre for Peirce Studies. http://www.helsinki.fi/science/commens/.

Bering, Jesse M. "The Existential Theory of Mind." *Review of General Psychology* 6, no. 1 (2002): 3–24.

Blakeslee, Sandra. "Cells That Read Minds." *New York Times* (10 January 2006): F1, F4.

Blood, Anne J., and Robert J. Zatorre. "Intensely Pleasurable Responses to Music Correlate with Activity in Brain Regions Implicated in Reward and Emotion." *Proceedings of the National Academy of Sciences* 98, no. 20 (2001): 11818–11823.

Blum, Lawrence A. *Friendship, Altruism and Morality.* London and New York: Routledge and Kegan Paul, 1980.

Booth, W. *The Rhetoric of Fiction*, 2nd ed. Chicago: University of Chicago Press, 1983.

Bordwell, David. *Making Meaning: Inference and Rhetoric in the Interpretation of Cinema.* Cambridge, Mass.: Harvard University Press, 1989a.

———. "A Case for Cognitivism." *Iris* 9 (Spring 1989b): 11–40.

———. "A Case for Cognitivism: Further Reflections." *Iris* 11 (Summer 1990): 107–112.

Bordwell, David, and Nöel Carroll. Eds. *Post-Theory: Reconstructing Film Studies.* Madison: University of Wisconsin Press, 1996.

Borenstein, Elhanan, and Eytan Ruppin. "The Evolution of Imitation and Mirror Neurons in Adaptive Agents." *Cognitive Systems Research* 6, no. 3 (2005): 229–242.

Bortolussi, Marisa, and Peter Dixon. *Psychonarratology: Foundations for the Empirical Study of Literary Response.* Cambridge and New York: Cambridge University Press, 2003.

Boswell-Stone, W. G. *Shakespeare's Holinshed: The Chronicle and the Historical Plays Compared.* London: Benjamin Bloom, 1966.

Botvinick, M. M., T. S. Braver, D. M. Barch, C. S. Carter, and J. D. Cohen.

"Conflict Monitoring and Cognitive Control." *Psychological Review* 108 (2001): 624–652.

Bourg, Tammy. "The Role of Emotion, Empathy, and Text Structure in Children's and Adults' Narrative Text Comprehension." In *Empirical Approaches to Literature and Aesthetics*. Ed. Roger J. Kreuz and Mary Sue MacNealy. Norwood, N.J.: Ablex, 1996. 241–260.

Boyd, Brian. "Jane, Meet Charles: Literature, Evolution, and Human Nature." *Philosophy and Literature* 22, no. 1 (1998): 1–30.

———. "The Origin of Stories: Horton Hears a Who." *Philosophy and Literature* 25, no. 2 (2001): 197–214.

Branigan, Edward. *Projecting a Camera: Language-Games in Film Theory*. New York: Routledge, 2006.

Brook, Andrew, and Kathleen Akins. Eds. *Cognition and the Brain: The Philosophy and Neuroscience Movement*. Cambridge: Cambridge University Press, 2005.

Brooks, Peter. *Troubling Confessions: Speaking Guilt in Law and Literature*. Chicago and London: University of Chicago Press, 2000.

Brothers, Leslie. "A Biological Perspective on Empathy." *American Journal of Psychiatry* 146 (1989): 10–19.

Bruner, Jerome. *Acts of Meaning*. Cambridge, Mass.: Harvard University Press, 1990.

———. "The Narrative Construction of Reality." *Critical Inquiry* 18, no. 1 (1991): 1–21.

Bryant, Jennings, and Dolf Zillmann. *Responding to the Screen: Reception and Reaction Processes*. Mahwah, N.J.: Erlbaum, 1991.

Buckland, Warren. *The Cognitive Semiotics of Film*. Cambridge: Cambridge University Press, 2000.

Bühler, Karl. *Theory of Language: The Representational Function of Language*. Philadelphia: I. Benjamins Pub. Co., 1990. Originally published in 1934 as *Sprachtheorie*.

Bunge, Mario. *Emergence and Convergence: Qualitative Novelty and the Unity of Knowledge*. Toronto: University of Toronto Press, 2003.

Burke, Edmund. *A Philosophical Enquiry into the Origin of Our Ideas of the Sublime and Beautiful*. Ed. J. T. Boulton. London: Routledge and K. Paul, 1958.

Butte, George. *I Know That You Know That I Know: Narrating Subjects from Moll Flanders to Marnie*. Columbus: Ohio State University Press, 2004.

Cacioppo, John T., and R. E. Petty. *Social Psychophysiology: A Sourcebook*. New York: Guilford Press, 1983.

———. "Just Because You're Imaging the Brain Doesn't Mean You Can Stop Using Your Head: A Primer and Set of First Principles." *Journal of Personality and Social Psychology* 85 (October 2003): 650–661.

Campbell, Lily B. "Macbeth: A Study of Fear." In *Shakespeare's Tragic Heroes: Slaves of Passion*. New York: Barnes and Noble, 1930. 208–239.

Cannon, Walter B. *The Wisdom of the Body*. New York: Norton, 1932.

Carroll, Joseph. *Evolution and Literary Theory*. Columbia and London: University of Missouri Press, 1995.

———. "Human Nature and Literary Meaning: A Theoretical Model Illustrated with a Critique of *Pride and Prejudice*." In *The Literary Animal: Evolu-*

tion and the Nature of Narrative. Ed. Jonathan Gottschall and David Sloane Wilson. Evanston, Ill.: Northwestern University Press, 2005. 76–106.

Carroll, Nöel. *Mystifying Movies: Fads and Fallacies in Contemporary Film Theory*. New York: Columbia University Press, 1988.

Carroll, William C. Ed. *Macbeth: Texts and Contexts*. New York: St. Martin, 1999.

Carter, Cameron S., Todd S. Braver, Deanna M. Barch, Matthew M. Botvinick, Douglas Noll, and Jonathan D. Cohen. "Anterior Cingulate Cortex, Error Detection, and the Online Monitoring of Performance." *Science* 280 (1 May 1998): 747–749.

Caughie, John. Ed. *Theories of Authorship*. New York: Routledge, 1981.

Changeux, Jean-Pierre. *The Physiology of Truth: Neuroscience and Human Knowledge*. Cambridge: Harvard University Press, 2004.

Chateaubriand, François-René. *The Genius of Christianity: or, The Spirit and Beauty of the Christian Religion*. 1856. Trans. Charles I. White. New York: H. Fertig, 1976.

Chatman, Seymour. *Coming to Terms: The Rhetoric of Narrative in Fiction and Film*. Ithaca: Cornell University Press, 1990.

Cheung, King-Kok. *Articulate Silences: Hisaye Yamamoto, Maxine Hong Kingston, Joy Kogawa*. Ithaca and London: Cornell University Press, 1993.

Chomsky, Noam. *Syntactic Structures*. The Hague: Mouton, 1957.

———. *Aspects of the Theory of Syntax*. Cambridge, Mass.: MIT Press, 1965.

———. *Language and Responsibility: Conversations with Mitsou Ronat*. New York: Pantheon Books, 1977.

———. *Lectures on Government and Binding*. Dordrecht: Foris, 1981.

———. *Knowledge of Language*. New York: Praeger, 1986.

———. *The Minimalist Program*. Cambridge, Mass.: MIT Press, 1995.

———. *New Horizons in the Study of Language and Mind*. Cambridge: Cambridge University Press, 2000.

Chomsky, Noam, Marc D. Hauser, et al. "The Faculty of Language: What Is It, Who Has It, and How Did It Evolve?" *Science* 298 (22 November 2002): 1569–1579.

Chupchik, G. C., and János Lázló. "The Landscape of Time in Literary Reception: Character Experience and Narrative Action." *Cognition and Emotion* 8 (1994): 297–312.

Clark, Andy. *Being There: Putting Brain, Body, and World Together Again*. Cambridge, Mass.: MIT Press, 1997.

———. *Mindware: An Introduction to the Philosophy of Cognitive Science*. Oxford: Oxford University Press, 2001.

Coetzee, J. M. *In the Heart of the Country*. London: Vintage Press, 1977.

Cohan, Steven. "Figures beyond the Text: A Theory of Readable Character in the Novel." *Novel: A Forum on Fiction* 17 (1983): 5–27.

Cohn, Dorrit. *Transparent Minds: Narrative Modes for Presenting Consciousness*. Princeton, N.J.: Princeton University Press, 1978.

———. *The Distinction of Fiction*. Baltimore: Johns Hopkins University Press, 1999.

Colin, Michel. "The *Grand Syntagmatique* Revisited." In *The Film Spectator:*

From Sign to Mind. Ed. Warren Buckland. Amsterdam: Amsterdam University Press, 1995. 45–85.

Commons, Michael Lamport, and Chester Arnold Wolfsont. "A Complete Theory of Empathy Must Consider Stage Changes." *Behavioral and Brain Sciences* 25, no. 1 (2002): 30–31.

Cook, Rufus. "Cross-Cultural Wordplay in Maxine Hong Kingston's *China Men* and *The Woman Warrior.*" *MELUS* 22 (1997): 133–146.

Cosmides, Leda, and John Tooby. "Cognitive Adaptations for Social Exchange." In *The Adapted Mind.* Ed. J. Barkow, Leda Cosmides, and John Tooby. Oxford and New York: Oxford University Press, 1992. 163–228.

———. "Evolutionary Psychology and the Emotions." In *Handbook of Emotions,* 2nd ed. Ed. Michael Lewis and Jeanette M. Haviland-Jones. New York: Guilford Press, 2000. 91–115.

———. "Consider the Source. The Evolution of Adaptations for Decoupling and Metarepresentation." In *Metarepresentations: A Multidisciplinary Perspective.* Ed. Dan Sperber. New York: Oxford University Press, 2000. 53–115.

———. "Does Beauty Build Adapted Minds? Toward an Evolutionary Theory of Aesthetics, Fiction and the Arts." *SubStance* 94/95 (2001): 6–27.

Craig, K. D. "Physiological Arousal as a Function of Imagined, Vicarious, and Direct Stress Experiences." *Journal of Abnormal Psychology* 73 (1968): 513–520.

Crane, Mary Thomas. *Shakespeare's Brain: Reading with Cognitive Theory.* Princeton: Princeton University Press, 2000.

Cranmer, Thomas. "Preface." *The Great Bible (the "Matthew" Bible).* 1549.

Culler, Jonathan. *Structuralist Poetics: Structuralism, Linguistics, and the Study of Literature.* Ithaca: Cornell University Press, 1975.

Culpeper, Jonathan. *Language and Characterisation: People in Plays and Other Texts.* London: Longman/Pearson Education, 2001.

Currie, Gregory. "The Long Goodbye: The Imaginary Language of Film." *British Journal of Aesthetics* 33, no. 3 (1993): 207–219.

Da Vinci, Leonardo. *Treatise on Painting.* Trans. A. Philip McMahon. 2 vols. Princeton: Princeton University Press, 1956.

Damasio, Antonio R., et al. "Somatic Markers and the Guidance of Behavior: Theory and Preliminary Testing." In *Frontal Lobe Function and Dysfunction.* Ed. H. S. Levin, H. M. Eisenberg, and A. L. Benton. Oxford and New York: Oxford, 1991. 217–229.

Damasio, Antonio R. *Descartes' Error: Emotion, Reason, and the Human Brain.* New York: Putnam, 1994.

———. *The Feeling of What Happens: Body and Emotion in the Making of Consciousness.* San Diego: Harcourt, 1999.

———. *Looking for Spinoza: Joy, Sorrow and the Feeling Brain.* New York: Harcourt, 2003.

Dapretto, Mirella, Mari S. Davies, Jennifer H. Pfeifer, Ashley A. Scott, Marian Sigma, Susan Y. Bookheimer, and Marco Iacobini. "Understanding Emotions in Others: Mirror Neuron Dysfunction in Children with Autism Spectrum Disorders." *Nature Neuroscience* 9, no. 1 (January 2006).

Darwin, Charles. *The Descent of Man, and Selection in Relation to Sex.* 2 vols. Princeton: Princeton University Press, 1981.

————. *The Expression of the Emotions in Man and Animals*, 3rd ed. Ed. Paul Ekman. Oxford and New York: Oxford University Press, 1998.

Davis, Mark H. "A Multidimensional Approach to Individual Differences in Empathy." *JSAS Catalog of Selected Documents in Psychology* 10 (1980): 85.

————. "Measuring Individual Differences in Empathy: Evidence for a Multidimensional Approach." *Journal of Personality and Social Psychology* 44 (1983): 113–26.

————. *Empathy: A Social Psychological Approach.* Boulder: Westview, 1994.

de Quervain, Dominique J.-F., Urs Fischbacher, Valerie Treyer, Melanie Schellhammer, Ulrich Schnyder, Alfred Buck, and Ernst Fehr. "The Neural Basis of Altruistic Punishment." *Science* 305, no. 5688 (27 August 2004): 1254–1258.

De Quincey, Thomas. *Confessions of an English Opium-Eater.* 1822. Otley, England: Woodstock Books, 2002.

Deacon, Terrence W. *The Symbolic Species: The Co-Evolution of Language and the Brain.* New York: W. W. Norton and Company, 1997.

Deigh, John. "Empathy and Universalizability." *Ethics* 105 (July 1995): 743–763.

Delgado, Richard. *The Coming Race War? and Other Apocalyptic Tales of America after Affirmative Action and Welfare.* New York: NYU Press, 1996.

Dennett, Daniel C. "Are Dreams Experiences?" In *Brainstorms: Philosophical Essays on Mind and Psychology.* Brighton: Harvester Press, 1981. 129–148.

————. *Consciousness Explained.* New York: Little, Brown, 1991.

————. *Darwin's Dangerous Idea: Evolution and the Meanings of Life.* New York: Simon and Schuster, 1995.

deSousa, Ronald. *The Rationality of the Emotions.* Cambridge, Mass.: MIT Press, 1987.

Diano, Sabrina, et al. "Ghrelin Controls Hippocampal Spine Synapse Density and Memory Performance." *Nature Neuroscience* 9, no. 3 (2006): 381–388.

Diderot, Denis. *Œuvres. Texte établi et annoté par André Billy.* Paris: Gallimard, 1951.

Diengott, Nilli. "Some Problems with the Concept of the Narrator in Bortolussi and Dixon's *Psychonarratology*." *Narrative* 12, no. 3 (October 2004): 306–316.

Dissanayake, Ellen. *Homo Aestheticus: Where Art Comes from and Why.* New York: Free Press, 1992.

————. "Becoming *Homo Aestheticus*: Sources of Aesthetic Imagination in Mother-Infant Interactions." *SubStance* 94/95 (2001): 85–103.

Dixon, Peter, et al. "Literary Processing and Interpretation: Towards Empirical Foundations." *Poetics* 22 (1993): 5–33.

————. "Literary Communication: Effects of Reader-Narrator Cooperation." *Poetics* 23 (1996): 405–430.

Doležel, Lubomír. *Heterocosmica: Fiction and Possible Worlds.* Baltimore: Johns Hopkins University Press, 1998.

Driscoll, D., S. W. Anderson, and H. Damasio. *Executive Function Test Performances following Prefrontal Cortex Damage in Childhood.* San Diego: Society for Neurosciences Abstracts, 2004.

Duckitt, John. *The Social Psychology of Prejudice.* New York: Praeger, 1992.

Eakin, Paul John. *How Our Lives Become Stories*. Ithaca and London: Cornell University Press, 1999.

———. "What Are We Reading When We Read Autobiography?" *Narrative* 12, no. 2 (2004): 121–132.

Edelman, Gerald M. *The Remembered Present: A Biological Theory of Consciousness*. New York: Basic Books, 1989.

Eibl, Karl. *Animal poeta. Bausteine der biologischen Kultur- und Literaturtheorie*. Paderborn: Mentis (Poetogenesis, 1: 2004). English abstract at http://www.lrz-muenchen.de/~eibl/animaeng_htm.html

———. "Epische Triaden. Über eine stammesgeschichtlich verwurzelte Gestalt des Erzählens." *Journal of Literary Theory* 2, no. 2 (2008): 197–208.

Eisenberg, Nancy, and Richard A. Fabes. "Children's Disclosure of Vicariously Induced Emotions." In *Disclosure Processes in Children and Adolescents*. Ed. Ken J. Rotenberg. Cambridge and New York: Cambridge University Press, 1995. 111–134.

Eisenberg, Nancy. "Emotion, Regulation, and Moral Development." *Annual Review of Psychology* 51 (2000): 665–697.

Eisenberg, Nancy, Richard A. Fabes, Denise Bustamante, and Robin M. Mathy. "Physiological Indices of Empathy." In *Empathy and Its Development*. Ed. Nancy Eisenberg and Janet Strayer. Cambridge and New York: Cambridge University Press, 1987. 380–385.

———. "The Development of Empathy-Related Responding." In *Moral Motivation through the Life Span*. Ed. Gustavo Carlo and Carolyn Pope Edwards. Lincoln: University of Nebraska Press, 2005. 73–117.

Ekman, Paul, and Erika L. Rosenberg. Eds. *What the Face Reveals: Basic and Applied Studies of Spontaneous Expression Using the Facial Action Coding System [FACS]*. Oxford: Oxford University Press, 2005.

Ekman, Paul, and Richard Davidson. Eds. *The Nature of Emotion: Fundamental Questions*. New York: Oxford University Press, 1994.

Ekman, Paul. *Emotions Revealed: Recognizing Faces and Feelings to Improve Communication and Emotional Life*. New York: Henry Holt, 2003.

Emmorey, Karen. *Language, Cognition, and the Brain: Insights from Sign Language Research*. Mahwah, N.J.: Lawrence Erlbaum Associates, 2002.

Emonds, Joseph. *Lexicon and Grammar: The English Syntacticon*. Berlin: De-Gruyter, 2000.

Evans, Malcolm. "Imperfect Speakers." In *Shakespeare's Tragedies*. Ed. Emma Smith. Oxford: Blackwell, 2004. 161–184.

Fanon, Frantz. *The Wretched of the Earth*. Trans. Constance Farrington. New York: Grove Press, 1963.

Feagin, Susan L. *Reading with Feeling: The Aesthetics of Appreciation*. Ithaca: Cornell University Press, 1996.

Feldman, Jerome A. *From Molecule to Metaphor: A Neural Theory of Language*. Cambridge, Mass.: MIT Press, 2006.

Feshbach, Norma D. "Studies of Empathic Behavior in Children." In *Progress in Experimental Personality Research*. Ed. Brendan A. Maher. New York: Academic Press. Vol. 8 (1978), 1–47.

Fisch, Harold. "Kafka's Debate with Job." In *New Stories for Old: Biblical Patterns in the Novel*. Houndmills: Macmillan, 1998. 81–99.

Fiske, Susan T., Lasana T. Harris, and Amy J. C. Cuddy. "Why Ordinary People Torture Enemy Prisoners." *Science* 306.5701 (26 November 2004): 1482–1483.

Fletcher, Pamela M. *Narrating Modernity: The British Problem Picture, 1895–1914.* Aldershot, England: Ashgate, 2003.

Fludernik, Monika. *The Fictions of Language and the Languages of Fiction.* London and New York: Routledge, 1993.

———. *Towards a "Natural" Narratology.* London: Routledge, 1996.

———. "Metanarrative and Metafictional Commentary: From Metadiscursivity to Metanarration and Metafiction." *Poetica* 35 (2003): 1–39.

———. "Natural Narratology and Cognitive Parameters." In *Narrative Theory and Cognitive Sciences.* Ed. David Herman. Stanford: CSLI Publications, 2003. 243–267.

Fludernik, Monika, and Uri Margolin. "Introduction." In *German Narratology.* Ed. Monika Fludernik and Uri Margolin. *Style* 38, nos. 2–3 (2004): 148–187.

Fodor, Jerry. *The Language of Thought.* Cambridge, Mass.: MIT Press, 1975.

———. *Representations.* Cambridge, Mass.: MIT Press, 1981.

———. *The Modularity of Mind.* Cambridge, Mass.: MIT Press, 1983.

von Foerster, Heinz. *Principles of Self-Organization—in a Socio-Managerial Context.* In *Self-Organization and Management of Social Systems: Insights, Promises, Doubts, and Questions.* Ed. Hans Ulrich and Gilbert J. B. Propst. Berlin: Springer, 1984. 2–24.

Forgas, Joseph P. Ed. *Handbook of Affect and Social Cognition.* Mahwah, N.J.: Erlbaum, 2001.

Forster, E. M. *Aspects of the Novel.* New York: Harcourt, 1927.

Foulkes, David. *Children's Dreaming and the Development of Consciousness.* Cambridge, Mass.: Harvard University Press, 1999.

Fox, Soledad. *Flaubert and Don Quixote: The Influence of Cervantes on Madame Bovary.* East Sussex: Sussex Academic Press. In press.

Foxton, Jessica M., et al. "Characterization of Deficits in Pitch Perception Underlying 'Tone Deafness.'" *Brain* 127, no. 4 (2004): 801–810.

Freeman, Walter. "A Neurobiological Role of Music in Social Bonding." In *The Origins of Music.* Ed. Nils L. Wallin, Björn Merker, and Steven Brown. Cambridge, Mass.: MIT Press, 2000. 411–424.

Freud, Sigmund. *The Interpretation of Dreams.* 1900. Trans. James Strachey. Harmondsworth: Penguin, 1976.

Fried, Michael. *Absorption and Theatricality: Painting and Beholder in the Age of Diderot.* Berkeley: University of California Press, 1980.

———. "Art and Objecthood." 1967. Reprinted in *Art and Objecthood: Essays and Reviews.* Chicago and London: University of Chicago Press, 1998.

Friedman, Susan Stanford. *Mappings: Feminism and the Cultural Geographies of Encounter.* Princeton, N.J.: Princeton University Press, 1998.

Frijda, Nico H. *The Emotions.* London: Cambridge University Press, 1986.

———. *The Laws of Emotion.* London: Lawrence Erlbaum, 2007.

Gadamer, Hans-Georg. *Truth and Method.* 2nd ed. Trans. Joel Weinssheimer and Donald Marshall. New York: Crossroad, 1989.

Gallese, Vittorio, and Alvin Goldman. "Mirror Neurons and the Simulation

Theory of Mind-Reading." *Trends in Cognitive Sciences* 2, no. 12 (December 1998): 493–502.

Gallese, Vittorio, et al. "The 'Shared Manifold' Hypothesis: From Mirror Neurons to Empathy." *Journal of Consciousness Studies* 8, nos. 5–7 (2001): 33–50.

———. "The Mirror Matching System: A Shared Manifold for Intersubjectivity." *Behavioral and Brain Sciences* 25 (February 2002): 35–36.

———. "A Unifying View of the Basis of Social Cognition." *Trends in Cognitive Science* 8, no. 9 (September 2004): 396–403.

———. "'Being Like Me': Self-Other Identity, Mirror Neurons, and Empathy." In *Perspectives on Imitation: From Neuroscience to Social Science. Volume 1: Mechanisms of Imitation and Imitation in Animals.* Ed. Susan Hurley and Nick Chater. Cambridge, Mass.: MIT Press, 2005. 101–118.

Garber, Marjorie. "Macbeth: The Male Medusa." In *Shakespeare's Ghost Writers: Literature as Uncanny Causality.* London: Methuen, 1987. 88–93.

Garfield, Jay L., Candida C. Peterson, and Tricia Perry. "Social Cognition, Language Acquisition and the Development of the Theory of Mind." *Mind and Language* 16, no. 5 (2001): 494–541.

Gazzaniga, Michael S. *Nature's Mind: The Biological Roots of Thinking, Emotions, Sexuality, Language, and Intelligence.* New York: Basic Books, 1992.

———. *The Ethical Brain: The Science of Our Moral Dilemmas.* New York: Harper, 2005.

———. *Human: The Science behind What Makes Us Unique.* New York: Ecco Press, 2008.

Geer, J. H., and L. Jarmecky. "The Effect of Being Responsible for Reducing Another's Pain on Subject's Response and Arousal." *Journal of Personality and Social Psychology* (1973): 232–237.

Genette, Gérard. *Narrative Discourse: An Essay in Method.* Trans. Jane E. Lewin. Ithaca, N.Y.: Cornell University Press, 1980.

———. *Narrative Discourse Revisited.* Trans. Jane E. Lewin. Ithaca, New York: Cornell University Press, 1988.

Gentner, Dedre, Keith J. Holyoak, and Boico N. Kokinov. *The Analogical Mind: Perspectives from Cognitive Science.* Cambridge, Mass.: MIT Press, 2001.

Gerrig, Richard J. "The Construction of Literary Character: A View from Cognitive Psychology." *Style* 24 (Fall 1990): 380–392.

———. *Experiencing Narrative Worlds: On the Psychological Activities of Reading.* New Haven: Yale University Press, 1993.

Goldberg, Jonathan. "Speculation: *Macbeth* and Source." In *Shakespeare's Hand.* London: University of Minnesota Press, 2003.152–175.

Goldstein, Avram. "Thrills in Response to Music and Other Stimuli." *Physiological Psychology* 8, no. 1 (1980): 126–129.

Gottschall, Jonathan, and David Sloan Wilson. *Literary Animal: Evolution and the Nature of Narrative.* Evanston: Nortwestern University Press, 2005.

Gould, Stephen, and R. C. Lewontin. "The Spandrels of San Marco and the Panglossian Paradigm: A Critique of the Adaptationist Programme." *Proceedings of the Royal Society of London.* B 205, 1979. 581–598.

Gourevitch, Philip. *We Wish to Inform You That Tomorrow We Will Be Killed with Our Families: Stories from Rwanda.* New York: Farrar, Straus, and Giroux, 1998.

Gratton, Lynn M., and Paul J. Elsinger. "High Cognition and Social Behavior: Changes in Cognitive Flexibility and Empathy after Cerebral Lesions." *Neuropsychology* 3 (1989): 175–185.

Greenblatt, Stephen. "Shakespeare Bewitched." In *New Historical Literary Study: Essays on Reproducing Texts, Representing History.* Ed. Jeffrey N. Cox and Larry J. Reynolds. Princeton, N.J.: Princeton University Press, 1993. 111–123.

Greene, Joshua. "From Neural 'Is' to Moral 'Ought': What Are the Moral Implications of Neuroscientific Moral Psychology?" *Nature Reviews* 4 (October 2003): 847–850.

Gross, Kenneth. *Shylock Is Shakespeare.* Chicago: University of Chicago Press, 2006.

Gutiérrez-Rexach, Javier. *Semantics: Critical Concepts.* New York: Routledge, 2003.

Hakemulder, Jèmeljan. *The Moral Laboratory: Experiments Examining the Effects of Reading Literature on Social Perception and Moral Self-Concept.* Utrecht Publications in General and Comparative Literature 34. Amsterdam and Philadelphia: John Benjamins, 2000.

———. "Foregrounding and Its Effect on Readers' Perception." *Discourse Processes* 38 (2004): 193–218.

Hamburger, Käte. *The Logic of Literature.* 2nd revised edition. Translated by Marilynn J. Rose. Bloomington: Indiana University Press, 1993. Originally published as *Die Logik der Dichtung,* 1957.

Haneke, Michael. *Caché.* Culver City: Sony Pictures, 2006.

Hart, Elizabeth F., and Bruce McConachie. Eds. *Performance and Cognition: Theatre Studies and the Cognitive Turn.* London: Routledge, 2006.

Hasson, Uri, et al. "Intersubject Synchronization of Cortical Activity during Natural Vision." *Science* 303 (12 March 2004): 1634–1640.

Hatfield, Elaine, John T. Cacioppo, and Richard L. Rapson. *Emotional Contagion.* Cambridge: Cambridge University Press, 1994.

Hauser, Marc D. *Moral Minds: How Nature Designed Our Universal Sense of Right and Wrong.* New York: HarperCollins, 2006.

Hauser, Marc D., Sanislas Dehaene, Jean-Rene Duhamel, and Giacomo Rizzolatti. Eds. *From Monkey Brain to Human Brain.* Cambridge: MIT Press, 2005.

Hawkes, Terence, and Hugh Grady. *Presentist Shakespeare.* London: Routledge, 2007.

Healy, Jane M. *Endangered Minds: Why Our Children Don't Think.* New York: Simon and Schuster, 1990.

———. *Failure to Connect: How Computers Affect Our Children's Minds—and What We Can Do about It.* New York: Simon and Schuster, 1998.

Herman, David. "Lateral Reflexivity: Levels, Versions, and the Logic of Paraphrase." *Style* 34 (2000): 293–306.

———. *Story Logic: Problems and Possibilities of Narrative.* Lincoln: University of Nebraska Press, 2002.

———. "Introduction." In *Narrative Theory and the Cognitive Sciences.* Ed. David Herman. Stanford: CSLI Publications, 2003. 1–30.

———. "Stories as a Tool for Thinking." In *Narrative Theory and the Cogni-*

tive Sciences. Ed. David Herman. Stanford, Calif.: CSLI Publications, 2003a. 163–192.

———. "Regrounding Narratology: The Study of Narratively Organized Systems for Thinking." In *What Is Narratology? Questions and Answers Regarding the Status of a Theory.* Ed. T. Kindt and H.-H. Müller. Berlin: de Gruyter, 2003b. 303–332.

———. "From Narrative Narcissism to Distributed Intelligence: Reflexivity as Cognitive Instrument in Joyce's *Finnegans Wake.*" *Frame* 17 (2004): 27–43.

———. "Narrative as Cognitive Instrument." *The Routledge Encyclopedia of Narrative Theory.* Ed. D. Herman, M. Jahn, and M.-L. Ryan. London: Routledge, 2005. 349–350.

———. "Narrative Theory and the Intentional Stance." *Partial Answers* 6, no. 2 (2008): 233–260.

Herman, David, Manfred Jahn, and Marie-Laure Ryan. Eds. *The Routledge Encyclopedia of Narrative Theory.* New York: Routledge, 2005.

Hernadi, Paul. "What Are Master-Pieces and Why Are There So Few of Them." In *Writings: 1932–1946.* Ed. Catharine R. Stimpson and Harriet Chessman. New York: Library of America, 1998. 355–363.

———. "Why Is Literature: A Coevolutionary Perspective on Imaginative Worldmaking." *Poetics Today* 23, no. 1 (2002): 21–42.

Hirschfeld, Lawrence A. *Race in the Making: Cognition, Culture, and the Child's Construction of Human Kinds.* Cambridge, Mass.: MIT Press, 1996.

Hobson, J. Allan. *Dreaming: An Introduction to the Science of Sleep.* Oxford: Oxford University Press, 2002.

Hoffman, Martin. "The Measurement of Empathy." In *Measuring Emotions in Infants and Children.* Ed. C. E. Izard. Cambridge and New York: Cambridge University Press, 1982. 279–296.

———. *Empathy and Moral Development: Implications for Caring and Justice.* Cambridge and New York: Cambridge University Press, 2000.

Hoffmann, E. T. A. *Musical Writings: Kreisleriana, The Poet and the Composer, Music Criticism.* Trans. Martyn Clarke. Ed. David Charlton. Cambridge: Cambridge University Press, 1989.

Hogan, Patrick Colm, and Lalita Pandit. Eds. *Cognitive Shakespeare: Criticism and Theory in the Age of Cognitive Science. College Literature* 33, no. 1 (2006).

Hogan, Patrick Colm. "The Epilogue of Suffering: Heroism, Empathy, Ethics." *SubStance* 30 (2001): 119–143.

———. *Cognitive Science, Literature, and the Arts: A Guide for Humanists.* London and New York: Routledge, 2003.

———. *The Mind and Its Stories: Narrative Universals and Human Emotion.* Cambridge: Cambridge University Press, 2003.

———. "Literature, God, and the Unbearable Solitude of Consciousness." *Journal of Consciousness Studies* 11, nos. 5–6 (May–June 2004): 116–142.

Holland, John, Keith Holyoak, Richard Nisbett, and Paul Thagard. *Induction: Processes of Inference, Learning, and Discovery.* Cambridge, Mass.: MIT Press, 1986.

Honigmann, E. A. J. "Macbeth: Murderer as Victim." In *Shakespeare: Seven Tragedies Revisited: The Dramatist's Manipulation of Response.* London: Palgrave, 1976, 2002. 126–149.

Howe, Stephen. *Empire: A Very Short Introduction.* Oxford and New York: Oxford, 2002.

Hume, David. *A Treatise of Human Nature.* Ed. L. A. Selby-Bigge. Oxford: Clarendon Press, 1978.

Hurley, Susan, and Nick Chater. Eds. *Perspectives on Imitation: From Neuroscience to Social Science.* Vol. 2: *Imitation, Human Development, and Culture.* Cambridge: MIT Press, 2005.

Huron, David. *Sweet Anticipation: Music and the Psychology of Expectation.* Cambridge, Mass.: MIT Press, 2006.

Hutchinson, Steven. *Cervantine Journeys.* Madison: University of Wisconsin Press, 1992.

Iacoboni, Marco, et al. "Cortical Mechanisms of Human Imitation." *Science* 286 (1999): 2526–2528.

Ickes, William. Ed. *Empathic Accuracy.* New York: Guilford, 1997.

Iser, Wolfgang. *The Act of Reading: A Theory of Aesthetic Response.* Baltimore: Johns Hopkins University Press, 1978.

Isherwood, Charles. "Review of *The Merchant of Venice* by the Theater for a New Audience." *New York Times.* 5 February 2007.

Islas, Arturo, and Marilyn Yalom. "Interview with Maxine Hong Kingston." In *Conversations with Maxine Hong Kingston.* Ed. Paul Skenazy and Tera Martin. Jackson: University Press of Mississippi, 1998. 21–32.

Ito, Tiffany A., Geoffrey R. Urland, Eve Willadsen-Jensen, and Joshua Correll. "The Social Neuroscience of Stereotyping and Prejudice: Using Event-Related Brain Potentials to Study Social Perception." In *Social Neuroscience: People Thinking about Thinking People.* Ed. John T. Cacioppo, Penny S. Visser, and Cynthia L. Pickett. Cambridge, Mass.: MIT Press, 2006. 189–208.

Ito, Tiffany, and John Cacioppo. "Affect and Attitudes: A Social Neuroscience Approach." In *Handbook of Affect and Social Cognition.* Ed. Joseph Forgas. Mahway, N.J.: Lawrence Erlbaum, 2001. 50–74.

Jahn, Manfred. "Speak Friend and Enter: Garden Paths, Artificial Intelligence, and Cognitive Narratology." In *Narratologies: New Perspectives on Narrative Analysis.* Ed. David Herman. Columbus: Ohio State University Press, 1999. 167–194.

Jameson, Fredric. *The Political Unconscious: Narrative as a Socially Symbolic Act.* Ithaca, N.Y.: Cornell University Press, 1981.

Jauss, Hans Robert. *Aesthetic Experience and Literary Hermeneutics.* Trans. Michael Shaw. Minneapolis, Minn.: University of Minnesota Press, 1982.

Jeannerod, Marc. *Motor Cognition: What Actions Tell the Self.* Oxford: Oxford University Press, 2006.

Jenkins, Lyle. *Biolinguistics: Exploring the Biology of Language.* Cambridge: Cambridge University Press, 2000.

Jose, P. E., and W. F. Brewer. "Development of Story Liking: Character Identification, Suspense, and Outcome Resolution." *Developmental Psychology* 20 (1984): 911–924.

Jost, François. *L'œil-caméra. Entre film et roman.* Lyon: Presses Universitaires de Lyon, 1987.

K'ang, Yu-wei. "K'ang Yu-wei's Philosophy of Great Unity." In *A Source Book in*

Chinese Philosophy. Trans. and ed. Wing-tsit Chan. Princeton, N.J.: Princeton University Press, 1963.

Kanizsa, G. *Organization in Vision: Essays on Gestalt Perception.* New York: Praeger, 1979.

Kant, Immanuel. *Critique of Judgment.* Trans. Werner S. Pluhar. Indianapolis: Hackett, 1987.

———. *Groundwork of the Metaphysic of Morals.* Trans. H. J. Paton. New York: Harper and Row, 1964.

Kastan, David Scott. "*Macbeth* and the 'Name of King.'" In *Shakespeare Tragedies.* Ed. Emma Smith. London: Blackwell, 1988. 248–257.

Kaufmann, Walter. *Tragedy and Philosophy.* New York: Doubleday, 1968.

Keen, Suzanne. "A Theory of Narrative Empathy." *Narrative* 14, no. 3 (October 2006): 207–236.

———. *Empathy and the Novel.* New York: Oxford, 2007.

Kenner, Hugh. "Who Else but Sam?" *New York Times Book Review.* 4 December 1977.

———. "Beckett at Eighty." In *Historical Fictions.* San Francisco: Northpoint Press, 1990. 289–295.

Kern, Stephen. *Eyes of Love: The Gaze in English and French Paintings and Novels, 1840–1900.* London: Reaktion Books, 1996.

Keysers, Christian, et al. "Demystifying Social Cognition: a Hebbian Perspective." *Trends in Cognitive Science* 8 (November 2004): 501–507.

Kingston, Maxine Hong. *The Woman Warrior.* New York: Vintage Books, 1975.

———. *China Men.* New York: Knopf, 1980.

Kinney, Arthur. *Lies Like Truth.* Detroit: Wayne State University Press, 2001.

Kirsch, Arthur. "Macbeth." In *Passions of Shakespeare's Tragic Heroes.* London: University of Virginia Press, 1990. 84–85.

Klemenz-Belgardt, Edith. "American Research of Response to Literature." *Poetics* 10 (1981): 357–380.

Knight, Wilson G. *The Wheel of Fire: Interpretations of Shakespearean Tragedy with Three New Essays.* London: Methuen, 1960. 140–159.

———. *The Imperial Theme: Further Interpretations of Shakespeare's Tragedies Including the Roman Plays.* London: Methuen, 1961.

Knights L. C. *Explorations: Essays in Criticism Mainly on the Literature of the Seventeenth Century.* New York: George W. Stewart Inc., 1964.

Koenigs, Michael, et al. "Damage to the Prefrontal Cortex Increases Utilitarian Moral Judgements." *Nature* 446 (2007): 908–911.

Kolbert, Elizabeth. "Goodnight Mush." *New Yorker.* 4 December 2006: 91.

Kondo, Hirohito, Naoyuki Osaka, and Mariko Osaka. "Cooperation of the Anterior Cingulate Cortex and Dorsolateral Prefrontal Cortex for Attention Shifting." *NeuroImage* 23 (2004): 670–679.

Konijn, Elly A., and Johan F. Hoorn. "Reality-based Genre Preferences Do Not Direct Personal Involvement." *Discourse Processes* 38 (2004): 219–246.

Kovecses, Zoltan. *Metaphor: A Practical Guide.* London: Oxford University Press, 2002.

Krebs, Dennis. "Empathy and Altruism." *Journal of Personality and Social Psychology* 32 (1975): 1134–1146.

Kreiswirth, Martin. "Narrative Turn in the Humanities." *Routledge Encyclopedia of Narrative Theory*. Ed. David Herman, Manfred Jahn, and Marie-Laure Ryan. Abingdon: Routledge, 2005. 377–382.

Kruger, Daniel J. "Evolution and Altruism: Combining Psychological Mediators with Naturally Selected Tendencies." *Evolution and Human Behavior* 24 (2003): 118–125.

Kruger, Daniel J., Maryanne Fisher, and Ian Jobling. "Proper Hero Dads and Dark Hero Cads: Alternate Mating Strategies Exemplified in British Romantic Literature." In *The Literary Animal: Evolution and the Nature of Narrative*. Ed. Jonathan Gottschall and David Sloane Wilson. Evanston, Ill.: Northwestern University Press, 2005. 225–243.

Kuiken, Don, et al. "Locating Self-Modifying Feelings within Literary Reading." *Discourse Processes* 38 (2004): 267–286.

Kunda, Ziva. *Social Cognition: Making Sense of People*. Cambridge, Mass.: MIT Press, 1999.

LaBerge, S., and D. J. DeGracia. "Varieties of Lucid Dreaming Experience." In *Individual Differences in Conscious Experience*. Ed. R. G. Kunzendorf and B. Wallace. Amsterdam: John Benjamins, 2000. 269–307.

Labov, W. *Language in the Inner City—Studies in the Black English Vernacular*. Philadelphia: University of Pennsylvania Press, 1972.

Lakoff, George, and Mark Johnson. *Metaphors We Live By*. Chicago: University of Chicago Press, 1980.

Lanthony, Philippe. "Seurat's Pointillism: Optical Mixture and Color Texture." In *The Eye of the Artist*. Ed. Michael F. Marmor and James G. Ravin. St. Louis: Mosby, 1997. 118–129.

Lawson, Mark. "Guilt, Lies and Videotape." *The Guardian*, 20 January 2006. http://arts.guardian.co.uk/features/story/0,,1690428,00.html

Lazarus, Richard S. "Relational Meaning and Discrete Emotions." In Klaus R. Scherer, Angela Schorr, and Tom Johnstone, eds. *Appraisal Processes in Emotion: Theory, Methods, Research*. London: Oxford, 2001. 37–67.

Lazarus, R. S., et al. "A Laboratory Study of Psychological Stress Produced by a Motion Picture Film." *Psychological Monographs* 76 (1962): 1–35.

Leavis, F. R. *Revaluation: Tradition and Development in English Poetry*. 1947. New York: Norton, 1963.

LeDoux, Joseph E., and Elizabeth A. Phelps. "Emotional Networks in the Brain." In *Handbook of Emotions*. Ed. Michael Lewis and Jeannette M. Haviland-Jones. 2nd ed. New York: Guilford Press, 2000. 157–172.

LeDoux, Joseph E. *The Emotional Brain: The Mysterious Underpinnings of Emotional Life*. New York: Simon and Schuster, 1996.

Lee, Vernon, and C. Anstruther-Thomson. "Beauty and Ugliness." *Contemporary Review* 72 (October 1897): 544–569 (November 1897): 669–688.

Lee, Vernon [Violet Paget]. *The Beautiful: An Introduction to Psychological Aesthetics*. Cambridge: Cambridge University Press, 1913.

Lehrer, Jonah. "Built to Be a Fan." *Seed Magazine*, 19 February 2006. http://www.seedmagazine.com; originally posted Summer 2004.

Lejeune, P. *On Autobiography*. Trans. K. M. Leary. Minneapolis: University of Minnesota Press, 1988.

Lennon, Randy, and Nancy Eisenberg. "Gender/Age Differences in Empathy/

Sympathy." In *Empathy and Its Development*. Ed. Nancy Eisenberg and Janet Strayer. Cambridge and New York: Cambridge University Press, 1987. 195–217.

Levenson, Robert, and Anna Ruef. "Physiological Aspects of Emotional Knowledge and Rapport." In *Empathic Accuracy*. Ed. William Ickes. New York: Guilford Press, 1997. 44–72.

Levitin, Daniel J., and Vinod Menon. "Musical Structure Is Processed in 'Language' Areas of the Brain: A Possible Role for Brodmann Area 47 in Temporal Coherence." *Neuroimage* 20 (2003): 2142–2152.

Levitin, Daniel J. *This Is Your Brain on Music: The Science of a Human Obsession*. New York: Dutton, 2006.

Lewis, Michael, and Jeannette M. Haviland-Jones. Eds. *Handbook of Emotions*. Second Edition. New York: Guilford Press, 2000.

Li, M., and P. Vitányi. *An Introduction to Kolmogorov Complexity and Its Applications*. New York: Springer Verlag, 1993/1997.

Lieberman, Matthew D., and Naomi I. Eisenberger. "A Pain by Any Other Name (Rejection, Exclusion, Ostracism) Still Hurts the Same: The Role of Dorsal Anterior Cingulate Cortex in Social and Physical Pain." In *Social Neuroscience: People Thinking about Thinking People*. Ed. John T. Cacioppo, Penny S. Visser, and Cynthia L. Pickett. Cambridge, Mass.: MIT Press, 2006. 167–187.

Lindenberger, Herbert. *Opera in History: From Monteverdi to Cage*. Stanford: Stanford University Press, 1998.

Lipps, Theodor. "Das Wissen von Fremden Ichen." *Psychologische Untersuchungen* 1 (1906): 694–722.

———. *Zur Einfühlung*. Leipzig: Engleman, 1913.

Livingstone, Margaret. *Vision and Art: The Biology of Seeing*. New York: Abrams, 2002.

Lodge, David. *The Modes of Modern Writing: Metaphor, Metonymy, and the Typology of Modern Literature*. Chicago: University of Chicago Press, 1977.

———. *Consciousness and the Novel: Connected Essays*. Cambridge, Mass.: Harvard University Press, 2002.

Longinus. *On the Sublime*. Trans. W. Rhys Roberts. In *Criticism: The Major Texts*. Ed. Walter Jackson Bate. New York: Harcourt Brace Jovanovich, 1970. 62–75.

Lothe, Jakob. *Narrative in Fiction and Film: An Introduction*. Oxford: Oxford University Press, 2000.

Louwerse, Max, and Don Kuiken. *Thematics: Interdisciplinary Studies*. Philadelphia: John Benjamins Publishing Co., 2002.

———. "The Effects of Personal Involvement in Narrative Discourse." *Discourse Processes* 38 (2004): 169–172.

MacDonald, Angus W., Jonathan D. Cohen, V. Andrew Stenger, and Cameron S. Carter. "Dissociating the Role of the Dorsolateral Prefrontal and Anterior Cingulate Cortex in Cognitive Control." *Science* 288, no. 5472 (9 June 2000): 1835–1838.

Mallen, Enrique. *The Visual Grammar of Pablo Picasso*. New York: Peter Lang, 2003.

Mankiewicz, Frank, and Joel Swerdlow. *Remote Control: Television and the Manipulation of American Life.* New York: Times Books, 1978.

Marcus, Mitchell P. *Theory of Syntactic Recognition for Natural Languages.* Cambridge, Mass.: MIT Press, 1980.

Marcus, Robert F. "Somatic Indices of Empathy." In *Empathy and Its Development.* Ed. Nancy Eisenberg and Janet Strayer. Cambridge and New York: Cambridge University Press, 1987. 374–379.

Margolin, Uri. "Of What Is Past, Is Passing, or to Come: Temporality, Aspectuality, Modality, and the Nature of Literary Narrative." In *Narratologies: New Perspectives on Narrative Analysis.* Ed. David Herman. Columbus: Ohio State University Press, 1999. 142–166.

———. "Cognitive Science, the Thinking Mind, and Literary Narrative." In *Narrative Theory and the Cognitive Sciences.* Ed. David Herman. Stanford: CSLI Publications, 2003. 271–294.

Marling, Karal Ann. *As Seen on TV: The Visual Culture of Everyday Life in the 1950s.* Cambridge, Mass.: Harvard University Press, 1994.

Marmor, Michael F. "Illusion and Optical Art." *The Eye of the Artist.* Ed. Michael F. Marmor and James G. Ravin. St. Louis: Mosby, 1997. 147–165.

———. "The Eye and Art: Visual Function and Eye Disease in the Context of Art." In *The Eye of the Artist.* Ed. Michael F. Marmor and James G. Ravin. St. Louis: Mosby, 1997. 2–25.

———. "Ophthalmology and Art: Simulation of Monet's Cataracts and Degas' Retinal Disease." *Archives of Ophthalmology* 124 (December 2006): 1764–1769.

Marr, David. *Vision: A Computational Investigation into the Human Representation and Processing of Visual Information.* San Francisco: W. H. Freeman, 1982.

Martinez, Matias, and Michael Scheffel. *Einführung in die Erzähltheorie.* München: Verlag C. H. Beck, 1999.

Massai, Sonia. Ed. *World-wide Shakespeares: Local Appropriations in Film and Performance.* New York: Routledge, 2005.

Mathews, Paul, and Jeffrey McQuain. *The Bard on the Brain: Understanding the Mind through the Art of Shakespeare and the Science of Brain Imaging.* New York: Dana Press, 2003.

McCoy, Alfred W. *A Question of Torture: CIA Interrogation, from the Cold War to the War on Terror.* New York: Henry Holt and Company, 2006.

McHale, Brian. *Postmodernist Fiction.* London: Routledge, 1987.

Mehrabian, Albert. "Relations among Personality Scales of Aggression, Violence, and Empathy: Validational Evidence Bearing on the Risk of Violence Scale." *Aggressive Behavior* 23 (1997): 433–445.

Mellmann, Katja. *Emotionalisierung—Von der Nebenstundenpoesie zum Buch als Freund. Eine emotionspsychologische Studie zur Literatur der Aufklärungsepoche.* Paderborn: Mentis-Verlag, 2006.

———. "Objects of Empathy. Characters (and Other Such Things) as Psychopoetic Effects." In *Characters in Fictional Worlds.* Ed. Jens Eder, Fotis Jannidis, and Ralf Schneider. Forthcoming.

Menon, Vinod, and Daniel J. Levitin. "The Rewards of Music Listening: Response and Physiological Connectivity of the Mesolimbic System." *Neuro-image* 28 (2005): 175–184.

Merleau-Ponty, Maurice. *Phenomenology of Perception*. Trans. Colin Smith. London: Routledge and Kegan Paul, 1962.

Metz, Christian. *Film Language: A Semiotics of the Cinema*. Cambridge: Cambridge University Press, 1974a.

———. *Language and Cinema*. The Hague: Mouton, 1974b.

Meyer-Lindenberg, Andreas. "Neural Mechanisms in Williams Syndrome: A Unique Window to Genetic Influences on Cognition and Behavior." *Nature Reviews Neuroscience* 7 (May 2006): 380–393.

Miall, David S. "Affect and Narrative: A Model of Responses to Stories." *Poetics* 17 (1988): 259–272.

———. "Beyond the Schema Given: Affective Comprehension of Literary Narratives." *Cognition and Emotion* 3 (1989): 55–78.

———. *Literary Reading: Empirical and Theoretical Studies*. New York: Peter Lang, 2006.

Miall, David S., and Don Kuiken. "What Is Literariness? Three Components of Literary Reading." *Discourse Processes* 28 (1999): 121–138.

Michaels, Walter Benn. *The Shape of the Signifier: 1967 to the End of History*. Princeton: Princeton University Press, 2004.

Miller, Joseph D. "The 'Novel' Novel: A Sociobiological Analysis of the Novelty Drive as Expressed in Science Fiction." In *Biopoetics: Evolutionary Explorations in the Arts*. Ed. Brett Cooke and Frederick Turner. Lexington, Ky.: ICUS Books, 1999. 315–334.

Millett, Kate. *The Politics of Cruelty: An Essay on the Literature of Political Imprisonment*. New York: W. W. Norton, 1994.

Milton, John. *Complete Prose Works*. 8 vols. New Haven: Yale University Press, 1935–1982.

Minami, Masahiko, and Alyssa McCabe. "Rice Balls and Bear Hunts: Japanese and North American Family Narrative Patterns." *Journal of Child Language* 22 (1995): 423–445.

Miola, Robert S. Ed. *Macbeth*. New York: Norton, 2004 edition. London: Guilford Press, 2000.

Mithen, Steven. "The Evolution of Imagination: An Archeological Perspective." *SubStance* 94/95 (2001): 28–53.

———. *The Singing Neanderthals: The Origins of Music, Language, Mind and Body*. London: Weidenfeld and Nicolson, 2005.

Moll, Jorge, Roland Zahn, Ricardo de Oliveira-Souza, Frank Krueger, and Jordan Grafman. "The Neural Basis of Human Moral Cognition." *Nature Reviews Neuroscience* 6 (2005): 799–809.

Muir, Kenneth. Ed. *Macbeth*. London: Arden, 2001.

Mullan, John. *Sentiment and Sociability: The Language of Feeling in the Eighteenth Century*. Oxford: Clarendon Press, 1988.

Neidich, Warren. *Blow-Up: Photography, Cinema and the Brain*. New York: Distributed Art Publishers, Inc., 2003.

Neisser, Ulric. "The Self Perceived." In *Ecological and Interpersonal Sources of Self-Knowledge*. Ed. Ulric Neisser. Cambridge: Cambridge University Press, 1993. 3–21.

Nell, Onora. *Acting on Principle: An Essay on Kantian Ethics.* New York: Columbia University Press, 1975.

Ngai, Sianne. *Ugly Feelings.* Cambridge, Mass.: Harvard University Press, 2005.

Nichols, John, A. Robert Martin, Bruce G. Wallace, and Paul A. Fuchs. *From Neuron to Brain.* Sunderland: Sinauer Associates, 2001.

Nisbett, Richard E., and Lee Ross. *Human Inference: Strategies and Shortcomings of Social Judgment.* Englewood Cliffs, N.J.: Prentice-Hall, 1980.

Nowak, Martin. *Evolutionary Dynamics: Exploring the Equations of Life.* Cambridge, Mass.: Harvard University Press, 2006.

Nowak, Martin, Nathalia Komarova, and Partha Nigoyi. "Evolution of Universal Grammar." *Science* 291 (2001): 114–118.

———. "Computational and Evolutionary Aspects of Language." *Nature* 417 (2002): 611–617.

Nussbaum, Martha C. *Love's Knowledge: Essays on Philosophy and Literature.* Oxford and New York: Oxford University Press, 1990.

———. *Upheavals of Thought: The Intelligence of Emotions.* Cambridge: Cambridge University Press, 2001.

Oatley, Keith. *Best Laid Schemes: The Psychology of Emotions.* Cambridge: Cambridge University Press, 1992.

———. "A Taxonomy of the Emotions of Literary Response and a Theory of Identification in Fictional Narrative." *Poetics* 23 (1994): 53–74.

———. *Emotions: A Brief History.* Malden, Mass.: Blackwell, 2004.

Ochsner, Kevin N. "Characterizing the Functional Architecture of Affect Regulation: Emerging Answers and Outstanding Questions." In *Social Neuroscience: People Thinking about Thinking People.* Ed. John T. Cacioppo, Penny S. Visser, and Cynthia L. Pickett. Cambridge, Mass.: MIT Press, 2006. 245–268.

O'Neill, Patrick. "Points of Origin: On Focalization in Narrative." *Canadian Review of Comparative Literature* 19 (1992): 331–350.

Opdahl, Keith M. *Emotion as Meaning: The Literary Case for How We Imagine.* Lewisburg, Pa.: Bucknell University Press, 2002.

Ortony, Andrew, Gerald L. Clore, and Allan Collins. *The Cognitive Structure of Emotions.* London: Cambridge University Press, 1988; 1992; 1994.

Osiel, Mark. "The Mental State of Torturers: Argentina's Dirty War." In *Torture: A Collection.* Ed. Sanford Levinson. Oxford: Oxford University Press, 2004.

Oster, Judith. *Crossing Cultures: Creating Identity in Chinese and Jewish American Literature.* Columbia and London: University of Missouri Press, 2003.

Palmer, Alan. "The Mind beyond the Skin." In *Narrative Theory and the Cognitive Sciences.* Ed. David Herman. Stanford: CSLI Publications, 2003. 322–348.

———. *Fictional Minds.* Lincoln and London: University of Nebraska Press, 2004.

Palmer, Stephen. *Vision Science.* Cambridge, Mass.: MIT Press, 1999.

Panksepp, Jaak. "The Basics of Basic Emotion." In *The Nature of Emotion: Fundamental Questions.* Ed. Paul Ekman and Richard J. Davidson. New York: Oxford University Press, 1994. 21–24.

————. "The Emotional Sources of 'Chills' Induced by Music." *Music Perception* 13, no. 2 (1995): 171–207.

————. "Emotions as Natural Kinds within the Mammalian Brain." In *Handbook of Emotions*. Ed. Michael Lewis and Jeannette M. Haviland-Jones. 2nd ed. New York: Guilford Press, 2000. 137–156.

Paster, Gail Kern. *Humoring the Body: Emotions and the Shakespearean Stage*. Chicago and London: University of Chicago Press, 2004.

Patañjali. *The Yoga-Sutra of Patañjali: A New Translation and Commentry*. Ed. and trans. Georg Feuerstein. Rochester, Vt.: Inner Traditions International, 1989.

Peretz, Isabelle, et al. "Dissociations entre musique et langage après atteinte cérébral: un nouveau cas d'amusie sans aphasie." *Canadian Journal of Experimental Psychology* 51, no. 4 (1997): 354–368.

————. "Music and Emotion: Perceptual Determinants, Immediacy, and Isolation after Brain Damage." *Cognition* 68 (1998): 111–141.

————. "Cortical Deafness to Dissonance." *Brain* 124, no. 5 (2001): 928–940.

————. "Brain Specialization for Music: New Evidence from Congenital Amusia." In *The Cognitive Neuroscience of Music*. Oxford: Oxford University Press, 2003. 192–203.

Perloff, Marjorie. *The Poetics of Indeterminacy: Rimbaud to Cage*. Evanston, Ill.: Northwestern University Press, 1983.

Perrson, Per. *Understanding Cinema: A Psychological Theory of Moving Imagery*. Cambridge: Cambridge University Press, 2003.

Phelan, James. "Who's Here? Thoughts on Narrative Identity and Narrative Imperialism." *Narrative* 13, no. 3 (2005): 205–210.

————. *Living to Tell about It: A Rhetoric and Ethics of Character Narration*. Ithaca: Cornell University Press, 2005.

Phelan, Peggy. "Reciting the Citation of Others; or, A Second Introduction." In *Acting Out: Feminist Performances*. Ed. Lynda Hart and Peggy Phelan. Ann Arbor: University of Michigan Press, 1993. 13–31.

Phelps, Elizabeth A., and Joseph E. LeDoux. "Contributions of the Amygdala to Emotion Processing: From Animal Models to Human Behavior." *Neuron* 48 (2005): 175–187.

Pinker, Steven. Ed. *Visual Cognition*. Cambridge, Mass.: MIT Press, 1985.

————. *How the Mind Works*. New York: Norton, 1997.

————. *The Blank Slate*. New York: Viking, 2002.

Plantinga, Carl, and Greg Smith. Eds. *Passionate Views: Film, Cognition, and Emotion*. Baltimore: Johns Hopkins University Press, 1999.

Plantinga, Carl. "The Scene of Empathy and the Human Face on Film." In *Passionate Views: Film, Cognition, and Emotion*. Ed. Carl Plantinga and Greg M. Smith. Baltimore: Johns Hopkins University Press, 1999. 239–255.

Pope, Alexander. "An Essay on Criticism." In *The Best of Pope*. Ed. George Sherburn. Revised edition. New York: Ronald Press, 1929. 53–74.

Popper, Karl. "Evolutionary Epistemology." In *Evolutionary Theory: Paths into the Future*. Ed. J. W. Pollard. Chichester: Wiley, 1984. 239–254.

Porton, Richard. "Collective Guilt and Individual Responsibility: An Interview with Michael Haneke." *Cinéaste* 31 (2005): 50–51.

Posner, Richard. "Against Ethical Criticism." *Philosophy and Literature* 21 (1997): 1–27.

Postman, Neil. *Amusing Ourselves to Death: Public Discourse in the Age of Show Business.* New York: Viking Penguin, 1985.

Pound, Ezra. *Selected Poems.* New York: New Directions, 1957.

Power, Mick, and Tim Dalgleish. *Cognition and Emotion: From Order to Disorder.* Essex, UK: Taylor and Francis, 1997.

Pratt, Mary Louise. *Toward a Speech Act Theory of Literary Discourse.* Bloomington, Ind.: Indiana University Press, 1977.

Premack, D. G., and G. Woodruff. "Does the Chimpanzee Have a Theory of Mind?" *Behavioral and Brain Sciences* 1 (1978): 515–526.

Preston, Stephanie D., and Frans B. M. de Waal. "Empathy: Its Ultimate and Proximate Bases." *Behavioral and Brain Sciences* 25 (February 2002): 1–20, 49–71.

Prince, Gerald. *Narratology: The Form and Functioning of Narrative.* The Hague: Mouton, 1982.

Prinz, Jesse J. "Imitation and Moral Development." In *Perspectives on Imitation: From Neuroscience to Social Science.* Vol. 2: *Imitation, Human Development, and Culture.* Ed. Susan Hurley and Nick Chater. Cambridge, Mass.: MIT Press, 2005. 267–282.

Rabinowitz, Peter J. *Before Reading: Narrative Conventions and the Politics of Interpretation.* Ithaca, N.Y.: Cornell University Press, 1987.

———. "Truth in Fiction: A Reexamination of Audiences." In *Narrative/Theory.* White Plains, N.Y.: Longman, 1996. 209–226.

Radford, Michael. *The Merchant of Venice.* Culver City: Sony Pictures, 2004.

Radway, Janice A. *Reading the Romance: Women, Patriarchy, and Popular Literature.* With a new introduction by the author. Chapel Hill: University of North Carolina Press, 1991.

Raggatt, Peter T. F. "Multiplicity and Conflict in the Dialogical Self: A Life Narrative Approach." In *Identity and Story: Creating Self in Narrative.* Ed. Dan P. McAdams, Ruthellen Josselson, and Amia Lieblich. Washington, D.C.: American Psychological Association, 2006.

Ramachandran, V. S. *A Brief Tour of Human Consciousness: From Impostor Poodles to Purple Numbers.* New York: Pi Press, 2004.

Ransdell, Joseph. "On Peirce's Conception of the Iconic Sign." In *Iconicity: Essays on the Nature of Culture.* Ed. P. Boouissac et al. Tübingen: Stauffenburg-Verlag, 1986. 51–74.

Ravin, James G., and Philippe Lanthony. "An Artist with a Color Vision Defect: Charles Meryon." In *The Eye of the Artist.* Ed. Michael F. Marmor and James G. Ravin. St. Louis: Mosby, 1997. 101–107.

Reed, Stephen K. *Cognition: Theory and Applications,* 6th ed. Belmont, Calif.: Wadsworth, 2004.

Restak, Richard. *The New Brain: How the Modern Age Is Rewiring Your Mind.* Emmaus, Pa.: Rodale, 2003.

Richardson, Alan, and Francis Steen. "Literature and the Cognitive Revolution: An Introduction." *Poetics Today* 21, no. 1 (2002): 1–8.

Richardson, Alan, and Ellen Spolsky. Eds. *The Work of Fiction: Cognition, Culture, and Complexity.* Aldershot, Hampshire, UK: Ashgate, 2004.

Richardson, Alan. *British Romanticism and the Science of Mind.* Cambridge: Cambridge University Press, 2001.
———. "Studies in Literature and Cognition: A Field Map." In *The Work of Fiction: Cognition, Culture, and Complexity.* Ed. Alan Richardson and Ellen Spolsky. Aldershot: Ashgate, 2004. 1–30.
———. "Facial Expression Theory from Romanticism to the Present." In *Introduction to Cognitive Cultural Studies.* Ed. Lisa Zunshine. Baltimore: Johns Hopkins University Press, forthcoming.
Richardson, Brian. "The Other Reader's Response: On Multiple, Divided, and Oppositional Audiences." *Criticism* 39, no. 1 (Winter 1997): 31–53.
———. *Unnatural Voices: Extreme Narration in Modern and Contemporary Fiction.* Columbus: Ohio State University Press, 2006.
———. "Singular Text, Multiple Implied Readers." *Style* 41 (2007): 259–274.
Richardson, Samuel. *Clarissa or The History of a Young Lady.* Ed. Angus Ross. New York: Penguin, 1985.
Ricoeur, Paul. *Time and Narrative.* 3 vols. 1983–1985. Trans. Kathleen McLaughlin and David Pellauer. Chicago: University of Chicago Press, 1984–1988.
Rimmon-Kenan, S. *Narrative Fiction: Contemporary Poetics.* London: Methuen, 1983.
Rizzolatti, Giacomo, Leonardo Fogassi, Vittorio Gallese, and Luciano Fagiga. "Action Recognition in the Premotor Cortex." *Brain* 119 (1996): 593–609.
———. "Resonance Behaviors and Mirror Neurons." *Archives Italiennes de Biologie* 137 (1999): 85–100.
———. "Neuropsychological Mechanisms Underlying the Understanding and Imitation of Action." *Nature Reviews Neuroscience* 2, no. 9 (2001): 661–670.
Roach, Joseph. "Culture and Performance in the Circum-Atlantic World." In *Performativity and Performance.* Ed. Andrew Parker and Eve Kosofsky Sedgwick. New York: Routledge, 1995. 45–63.
Robbe-Grillet, Alain. *Dans le labyrinthe.* Paris: Les Editions de Minuit, 1959.
Rock, I. *The Logic of Perception.* Cambridge, Mass.: MIT Press, 1983.
Rolls, Edmund T. "A Theory of Emotion, Its Functions, and Its Adaptive Value." In *Emotions in Humans and Artifacts.* Ed. Robert Trappl, Paolo Petta, and Sabine Payr. Cambridge, Mass.: MIT Press, 2002. 11–34.
Rosen, Howard J., et al. "Emotion Comprehension in the Temporal Variant of Frontotemporal Dementia." *Brain* 125 (October 2002): 2286–2295.
Roston, Murray. *Tradition and Subversion in Renaissance Literature: Studies in Shakespeare, Spenser, Johnson, and Donne.* Pittsburgh: Duquesne University Press, 2007.
Rubin, David C., and Daniel Greenberg. "The Role of Narrative in Recollection: A View from Cognitive Psychology and Neuropsychology." In *Narrative and Consciousness: Literature, Psychology, and the Brain.* Ed. Gary D. Fireman, Ted E. McVay, and Owen J. Flanagan. Oxford: Oxford University Press, 2003.
Ryan, Marie-Laure. *Possible Worlds, Artificial Intelligence, and Narrative Theory.* Bloomington: Indiana University Press, 1991.

———. *Narrative as Virtual Reality: Immersion and Interactivity in Literature and Electronic Media*. Baltimore: Johns Hopkins University Press, 2001.

Sanders, Barry. *A Is for Ox: Violence, Electronic Media, and the Silencing of the Written Word*. New York: Pantheon Books, 1994.

Sankey, Derek. "The Neuronal, Synaptic Self: Having Values and Making Choices." *Journal of Moral Education* 35, no. 2 (2006): 163–178.

Sapolsky, Barry, and Dolf Zillmann. "Experience and Empathy: Affective Reactions to Witnessing Childbirth." *Journal of Social Psychology* 105 (1978): 133–144.

Sarris, Andrew. *The American Cinema: Directors and Directions, 1929–1968*. New York: Dutton, 1968.

Saussure, Ferdinand de. *Cours de linguistique général*. Paris: Gallimard, 2005.

Scarry, Elaine. *Dreaming by the Book*. New York: Farrar, Strauss, Giroux, 1999.

———. "Five Errors in the Reasoning of Alan Dershowitz." In *Torture: A Collection*. Ed. Sanford Levinson. Oxford: Oxford University Press, 2004.

Schacter, Daniel L., and Donna Rose Addis. "The Ghosts of Past and Future." *Nature* 445, no. 4 (January 2007): 7123–7127.

Schacter, Daniel L. *Searching for Memory: The Brain, the Mind, and the Past*. New York: Basic Books, 1996.

———. *The Seven Sins of Memory: How the Mind Forgets and Remembers*. Boston: Houghton Mifflin, 2001.

Schatz, Sara, and Javier Gutiérrez-Rexach. *Conceptual Structure and Social Change*. New York: Praeger, 2003.

Scherer, Klaus R., Angela Schorr, and Tom Johnstone. Eds. *Appraisal Processes in Emotion: Theory, Methods, Research*. London: Oxford, 2001.

Schjeldahl, Peter. "Words and Pictures." *New Yorker* 82, no. 32 (2005): 162–168.

Schleiermacher, Friedrich. *On Religion: Speeches to Its Cultured Despisers*. Trans. Richard Crouter. New York: Cambridge University Press, 1988.

Schneider, Ralf. "Toward a Cognitive Theory of Literary Character: The Dynamics of Mental-Model Construction." *Style* 35 (2001): 607–642.

Schoenberg, Arnold. *Structural Functions of Harmony*. New York: Norton, 1969.

Schroeder, Timothy, and Carl Matheson. "Imagination and Emotion." In *The Architecture of the Imagination: New Essays on Pretence, Possibility, and Fiction*. Ed. Shaun Nichols. New York: Oxford University Press, 2006. 19–40.

Schroeder, Timothy. *Three Faces of Desire*. Oxford: Oxford University Press, 2004.

Shakespeare, William. "The Tragedy of King Lear." Ed. Russell Fraser. New York: Signet, 1963.

———. "Julius Caesar." Ed. G. Blakemore, Evans. In *The Riverside Shakespeare*. Boston: Houghton Mifflin, 1974.

———. "Macbeth." Ed. Kenneth Muir. London: Arden, 1984.

———. "Merchant of Venice." Ed. Leah Marcus. New York: Norton, 2005.

Sheehan, E. P., et al. "Reactions to AIDS and Other Illnesses: Reported Interactions in the Workplace." *Journal of Psychology* 123 (1989): 525–536.

Sheehan, Richard Johnson, and Scott Rode. "On Scientific Narrative: Stories of

Light by Newton and Einstein." *Journal of Business and Technical Communication* 13, no. 3 (July 1999): 336–358.

Shelley, Percy Bysshe. *Shelley's Poetry and Prose.* Ed. Donald H. Reiman and Neil Fraistat. New York: Norton, 2002.

Shen, Dan. "Why Contextual and Formal Narratologies Need Each Other." *JNT: Journal of Narrative Theory* 35, no. 2 (2005): 141–171.

Shenk, David. *Data Smog: Surviving the Information Glut.* New York: Harper-Edge, 1997.

———. *The End of Patience: Cautionary Notes on the Information Revolution.* Bloomington: Indiana University Press, 1999.

Sherman, Nancy. "Empathy and Imagination." In *Philosophy of Emotions. Midwest Studies in Philosophy, XXII.* Ed. Peter French and Howard K. Wettstein. South Bend, Ind.: University of Notre Dame Press, 1998. 82–119.

Shields, Stephanie A. *Speaking from the Heart: Gender and the Social Meaning of Emotion.* Cambridge and New York: Cambridge University Press, 2002.

Shore, Bradd. *Culture in Mind: Cognition, Culture, and the Problem of Meaning.* New York: Oxford University Press, 1996.

Shyamalan, M. Night. *The Sixth Sense.* Burbank: Buenavista Pictures, 1999.

Sinfield, Alan. "History, Pedagogy, and Intellectuals." In *Macbeth.* Ed. Sylvan Barnet. New York: Signet, 1988. 180–185.

Singer, Tania, Daniel Wolpert, and Chris Frith. "Introduction: The Study of Social Interactions." In *The Neuroscience of Social Interaction: Decoding, Imitating, and Influencing the Actions of Others.* Ed. Christopher D. Frith and Daniel Wolpert. Oxford: Oxford University Press, 2004. xiii–xxvii.

Singer, Tania, et al. "Empathy for Pain Involves the Affective but Not Sensory Components of Pain." *Science* 303 (20 February 2004): 1157–1162.

Slote, Michael. *Morals from Motives.* New York: Oxford University Press, 2001.

Smith, Adam. *The Theory of Moral Sentiments.* In *Glasgow Edition of the Works and Correspondence of Adam Smith, Vol. 1.* Ed. D. D. Raphael and A. L. Macfie. Oxford: Clarendon Press, 1976.

Smith, Frank. *To Think.* New York: Teachers College Press, 1990.

———. *Understanding Reading: A Psycholinguistic Analysis of Reading and Learning to Read.* 5th ed. Hillsdale, N.J.: Lawrence Erlbaum Associates, 1994.

Smith, Greg. *Film Structure and the Emotion System.* Cambridge: Cambridge University Press, 2007.

Smith, Murray. "Aesthetics and the Rhetorical Power of Narrative." In *Moving Images, Culture and the Mind.* Ed. Ib Bondebjerg. Bedfordshire: University of Luton Press, 2000. 157–166.

Solso, Robert. *Cognition and the Visual Arts.* Cambridge, Mass.: MIT Press, 1994.

Sperber, Dan. *Explaining Culture: A Naturalistic Approach.* Oxford: Blackwell, 1997.

Sperber, Dan, and Deirdre Wilson. *La pertinence.* Paris: Les Editions de Minuit, 1989.

———. *Relevance: Communication and Cognition.* 2nd ed. Oxford: Blackwell, 1995.

Spinoza, Baruch. Ethics. In *Spinoza Selections*. Ed. John Wild. New York: Scribner, 1958.

Spivak, Gayatri Chakravorty. "Subaltern Studies: Deconstructing Historiography." In *The Spivak Reader: Selected Works of Gayatri Chakravorty Spivak*. Ed. Donna Landry and Gerald MacLean. New York and London: Routledge, 1996. 203–235.

Spolsky, Ellen. *Gaps in Nature: Literary Interpretation and the Modular Mind*. Albany: SUNY Press, 1993.

———. "Doubting Thomas and the Senses of Knowing." *Common Knowledge* 3, no. 2 (1994): 111–129.

———. "Elaborated Knowledge: Reading Kinesis in Pictures." *Poetics Today* 17, no. 2 (1996): 157–180.

———. *Satisfying Skepticism: Embodied Knowledge in the Early Modern World*. Aldershot: Ashgate, 2001.

———. "Darwin and Derrida: Cognitive Literary Theory As a Species of Post-structuralism." *Poetics Today* 23, no. 1 (2002): 43–62.

———. "Cognitive Literary Historicism: A Response to Adler and Gross." *Poetics Today* 24, no. 2 (2003): 161–183.

———. "Women's Work Is Chastity: Lucretia, *Cymbeline*, and Cognitive Impenetrability." In *The Work of Fiction: Cognition, Culture, and Complexity*. Ed. Alan Richardson and Ellen Spolsky. Aldershot, Hampshire, UK: Ashgate, 2004. 51–83.

———. "Purposes Mistook: Failures Are More Tellable." Talk delivered at the panel on "Cognitive Approaches to Narrative" at the annual meeting of the Society for the Study of Narrative, Burlington, Vermont, 2004.

———. *Word vs. Image: Cognitive Hunger in Shakespeare's England*. Houndmills: Palgrave Macmillan, 2007.

———. "Frozen in Time." *Poetics Today* 28, no. 4 (2007): 807–816.

Stallybrass, Peter. "Macbeth and Witchcraft." In *Focus on Macbeth*. Ed. John Russell Brown. London, Boston, and Henley: Routledge, 1982. 189–209.

Stam, Robert, Robert Burgoyne, and Sandy Flitterman. *New Vocabularies in Film Semiotics*. London: Routledge, 1992.

Stanzel, F. K. *Narrative Situation in the Novel: Tom Jones, Moby-Dick, The Ambassadors, Ulysses*. Trans. James P. Pusack. Bloomington: Indiana University Press, 1971.

———. *A Theory of Narrative*. Trans. Charlotte Goedsche. Cambridge and New York: University of Cambridge Press, 1984.

Stein, Gertrude. *Look at Me Now and Here I Am: Writings and Lectures 1909–45*. Ed. Patricia Meyerowitz. Harmondsworth: Penguin, 1971.

———. "Tender Buttons." In *Writings: 1903–1932*. Ed. Catharine R. Stimpson and Harriet Chessman. New York: Library of America, 1998. 313–355.

———. "Composition as Explanation." In *Writings: 1903–1932*. Ed. Catharine R. Stimpson and Harriet Chessman. New York: Library of America, 1998. 520–529.

———. "What Are Master-Pieces and Why Are There So Few of Them." In *Writings: 1932–1946*. Ed. Catharine R. Stimpson and Harriet Chessman. New York: Library of America, 1998. 355–363.

Steiner, Wendy. *Exact Resemblance to Exact Resemblance: The Literary Portraiture of Gertrude Stein.* New Haven: Yale University Press, 1978.

Stenning, Keith. *Seeing Reason: Image and Language in Learning to Think.* Oxford: Oxford University Press, 2002.

Stern, Barbara. "Classical and Vignette Television Advertising Dramas: Structural Models, Formal Analysis, and Consumer Effects." *Journal of Consumer Research* 20 (1994): 601–615.

Stockwell, Peter. *Cognitive Poetics.* London: Routledge, 2002.

Stoll, Clifford. *High-Tech Heretic: Reflections of a Computer Contrarian.* New York: Anchor Books, 1999.

Stotland, Ezra, et al. *Empathy, Fantasy and Helping.* Beverly Hills: Sage, 1978.

Sugiyama, Michelle Scalise. "On the Origins of Narrative: Storyteller Bias as a Fitness-Enhancing Strategy." In *Human Nature* 7 (1996): 403–425.

———. "Food, Foragers, and Folklore: The Role of Narrative in Human Subsistence." *Evolution and Human Behavior* 22, no. 4 (2001): 221–240.

———. "Lions and Tigers and Bears: Predators as a Folklore Universal." In *Heuristiken der Literaturwissenschaft. Disziplinexterne Perspektiven auf Literatur.* Ed. Uta Klein, Katja Mellmann, and Steffanie Metzger. Paderborn: Mentis, 2006. 319–331.

Sussman, Henry, and Carol Jacobs. Eds. *Acts of Narrative.* Stanford: Stanford University Press, 2003.

Takahashi, Hidehiko, Noriaki Yahata, Michihiko Koeda, Tetsuya Matsuda, Kunihiko Asai, and Yoshiro Okubo. "Brain Activation Associated with Evaluative Processes of Guilt and Embarrassment: An fMRI Study." *NeuroImage* 23 (2004): 967–974.

Tan, Ed S. "Emotion, Art, and the Humanities." In *Handbook of Emotions.* Second Edition. Ed. Michael Lewis and Jeannette M. Haviland-Jones. London: Guilford Press, 2000. 116–134.

Tancredi, Laurence. *Hardwired Behavior: What Neuroscience Reveals about Morality.* New York: Cambridge University Press, 2005.

Tangney, June Price. "Moral Affect: The Good, the Bad, and the Ugly." *Journal of Personality and Social Psychology* 61, no. 4 (1991).

Taylor, Charles. "The Politics of Recognition." In *Multiculturalism: A Critical Reader.* Ed. David Theo Goldberg. Cambridge: Blackwell, 1994.

Taylor, Marjorie, et al. "The Illusion of Independent Agency: Do Adult Fiction Writers Experience Their Characters as Having Minds of Their Own?" *Imagination, Cognition and Personality* 22 (2002/2003): 361–380.

Tettamanti, Marco, et al. "Listening to Action-related Sentences Activates Fronto-parietal Motor Circuits." *Journal of Cognitive Neuroscience* 17 (February 2005): 273–281.

Thelen, Esther, and Linda B. Smith. *A Dynamic Systems Approach to the Development of Cognition and Action.* Cambridge, Mass.: MIT Press, 1994.

Titchener, E. B. *Experimental Psychology of the Thought Processes.* London: Macmillan, 1909.

Trevarthen, Colwyn. "The Self Born in Intersubjectivity: The Psychology of an Infant Communicating." *The Perceived Self: Ecological and Interpersonal Sources of Self-Knowledge.* Ed. Ulric Neisser. Cambridge: Cambridge University Press, 1993. 121–173.

Trimble, Michael R. *Soul in the Brain: The Cerebral Basis of Language, Art, and Belief.* Baltimore: Johns Hopkins University Press, 2007.

Tsur, Reuven. "Two Critical Attitudes: Quest for Certitude and Negative Capability." *College English* 36/37 (March 1975): 776–788.

———. *Toward a Theory of Cognitive Poetics.* Amsterdam: North Holland, 1992.

Tulving, Endel. "Episodic and Semantic Memory." In *Organization of Memory.* Ed. E. Tulving and W. Donaldson. New York: Academic Press, 1972.

Turner, Mark. *The Literary Mind.* Oxford: Oxford University Press, 1996.

Ulmer, W. A. "Wordsworth, the One Life, and *The Ruined Cottage.*" *Studies in Philology* 93 (1996): 304–333.

Van Peer, Willie, and H. Pander Maat. "Perspectivation and Sympathy: Effects of Narrative Point of View." In *Empirical Approaches to Literature and Aesthetics.* Ed. Roger J. Kreuz and Mary Sue MacNealy. Norwood, N.J.: Ablex, 1996. 143–154.

Van Peer, Willie, and Seymour Chatman. Eds. *New Perspectives on Narrative Perspective.* Albany: SUNY Press, 2001.

Van Peer, Willie. "Justice in Perspective." In *New Perspectives on Narrative Perspective.* Ed. Willie van Peer and Seymour Chatman. Albany, N.Y.: SUNY Press, 2001. 325–338.

Vermeule, Blakey. "God Novels." In *The Work of Fiction: Cognition, Culture, and Complexity.* Ed. Alan Richardson and Ellen Spolsky. Aldershot: Ashgate, 2004. 147–166.

Wallace, Michelle. "The Search for the 'Good Enough' Mammy: Multiculturalism, Popular Culture, and Psychoanalysis." In *Multiculturalism: A Critical Reader.* Ed. David Theo Goldberg. Cambridge: Blackwell, 1994. 259–268.

Walsh, Richard. "Why We Wept for Little Nell: Character and Emotional Involvement." *Narrative* 5 (October 1997): 306–321.

Warhol, Robyn. *Having a Good Cry: Effeminate Feelings and Pop-Culture Forms.* Columbus, Ohio: Ohio State University Press, 2003.

Watson, Robert N. "Tragedies of Revenge and Ambition." In *Shakespearean Tragedy.* Ed. Claire McEachern. London: Cambridge University Press, 2002. 160–181.

Wexler, Bruce E. *Brain and Culture: Neurobiology, Ideology, and Social Change.* Cambridge: MIT Press, 2006.

Wheatley, Catherine. "Secrets, Lies, and Videotapes." *Sight and Sound* 16, no. 2 (2006): 32–36.

Wheeler, Michael. *Reconstructing the Cognitive World.* Cambridge, Mass.: MIT Press, 2005.

Williams, Raymond. *Culture and Society, 1780–1950.* Reprint with a new introduction. New York: Columbia University Press, 1983.

Wilson, E. O. *Consilience: The Unity of Knowledge.* New York: Knopf, 1998.

Wimsatt, W. K., and M. C. Beardsley. "The Intentional Fallacy." In *The Verbal Icon.* Lexington: University of Kentucky Press, 1954.

Winerman, Lea. "The Mind's Mirror." *Monitor on Psychology* 36, no. 9 (October 2005): 48–53.

Winn, Marie. *The Plug-In Drug.* New York: Viking Press, 1977.

Winnicott, Donald Woods. *Playing and Reality*. Harmondsworth, Middlesex: Penguin Books, 1971.

Wispé, Lauren. "History of the Concept of Empathy." In *Empathy and Its Development*. Ed. Nancy Eisenberg and Janet Strayer. Cambridge and New York: Cambridge University Press, 1987. 17–37.

Wollen, Peter. *Signs and Meanings in the Cinema*. Bloomington: Indiana University Press, 1972.

Wood, Marcus. *Slavery, Empathy and Pornography*. Oxford and New York: Oxford, 2002.

Woolf, Virginia. *Flush. A Biography*. London: Hogarth Press, 1983.

Wordsworth, William. "Preface 1800 Version (with 1802 Variants)." In *Lyrical Ballads*. Ed. R. L. Brett and A. R. Jones. London: Methuen, 1963. 235–266.

Wright, Thomas. *The Passions of the Minde in Generall*. (1604) Ed. Thomas O. Sloan. Urbana: University of Illinois Press, 1971.

Young, Kay, and Jeffrey L. Saver. "The Neurology of Narrative." *SubStance* 94/95 (2001): 72–84.

Zahn-Waxler, Carolyn, et al. "Empathy and Prosocial Patterns in Young MZ and DZ Twins: Development and Genetic and Environmental Influences." In *Infancy to Early Childhood: Genetic and Environmental Influences on Developmental Change*. Ed. Robert M. Emde and John K. Hewitt. Oxford and New York: Oxford University Press, 2001. 141–162.

Zeki, Semir. *Inner Vision: An Exploration of Art and the Brain*. Oxford: Oxford University Press, 1999.

Zillmann, Dolf. "Empathy: Affect from Bearing Witness to the Emotions of Others." In *Responding to the Screen: Reception and Reaction Processes*. Ed. Dolf Zillmann and Jennings Bryant. Mahwah, N.J.: Erlbaum, 1991. 135–167.

———. "Mechanisms of Emotional Involvement with Drama." *Poetics* 23 (1994): 33–51.

Žižek, Slavoj. *The Art of the Ridiculous Sublime: On David Lynch's* Lost Highway. Seattle: University of Washington Occasional Papers, 2000.

Zunshine, Lisa. "Theory of Mind and Experimental Representations of Fictional Consciousness." *Narrative* 11 (October 2003): 270–291.

———. "Richardson's Clarissa and a Theory of Mind." In *The Work of Fiction: Cognition, Culture, and Complexity*. Ed. Alan Richardson and Ellen Spolsky. Aldershot, Hampshire, and Burlington, Vt.: Ashgate, 2004. 127–146.

———. *Why We Read Fiction: Theory of Mind and the Novel*. Columbus: Ohio State University Press, 2006.

———. "Theory of Mind and Fictions of Embodied Transparency." *Narrative* 11, no. 3 (October 2008): 270–291.

———. "Lying Bodies of the Enlightenment: Theory of Mind and Cultural Historicism." Introduction to *Cognitive Cultural Studies*. Ed. Lisa Zunshine. Baltimore: Johns Hopkins University Press, forthcoming.

Zwaan, Rolf A. "Effect of Genre Expectations on Text Comprehension." *Journal of Experimental Psychology: Learning, Memory and Cognition* 20 (1994): 920–933.

Contributors

H. PORTER ABBOTT is Research Professor of English at the University of California, Santa Barbara. He teaches and publishes in the areas of narrative, cognitive, and evolutionary approaches to literature and the arts, modernist and postmodernist literature, and autobiography. His most recent books are *Beckett Writing Beckett: the Author in the Autograph* (1996), *On the Origin of Fictions: Interdisciplinary Perspectives* (edited, 2001), and *The Cambridge Introduction to Narrative* (2002, second edition 2008). Many of his recent essays have focused on audience-resistant strategies in literature and film.

JAVIER GUTIÉRREZ-REXACH is Professor of Hispanic Linguistics and Linguistics at the Ohio State University. He received his Ph.D. in Linguistics from UCLA, and in Philosophy from the University of Madrid. He also holds a degree in Film/TV direction (IORTV, Spain). He is interested in formal linguistics, cognitive science, and film studies. He is the author or editor of several books: *Semantics: Critical Concepts* (2003), *Conceptual Structure and Social Change* (2002), and *From Words to Discourse* (2001), among others. He has also published more than one hundred articles, book chapters, reviews, and papers in conference proceedings.

LALITA PANDIT HOGAN is a poet and professor of English at the University of Wisconsin–La Crosse. She was guest co-editor and contributing author to two special issues of *College Literature*, including the Winter 2006 Special Issue: *Cognitive Shakespeare: Criticism and Theory in the Age of Neuroscience*. She has also co-edited and co-authored *Rabindranath Tagore: Universality and Tradition* (2003).

PATRICK COLM HOGAN is a Professor in the Department of English, the Program in Comparative Literature and Cultural Studies, and the Program in Cognitive Science at the University of Connecticut. He has published twelve books, including *The Politics of Interpretation* (1990), *Colonialism and Cultural Identity* (2000), *The Culture of Conformism* (2001), *The Mind and Its Stories: Narrative Universals and Human Emotion* (2003), *Cognitive Science, Literature, and the Arts* (2003), *Empire and Poetic Voice* (2004), and *Understanding Indian Movies* (2008). He is currently editing *The Cambridge Encyclopedia of the Language Sciences.*

SUZANNE KEEN, Thomas H. Broadus Professor of English at Washington and Lee University, teaches the novel, narrative, and contemporary and postcolonial literature. She is the author of *Empathy and the Novel* (2007), *Narrative Form* (2003), *Romances of the Archive in Contemporary British Fiction* (2001), and *Victorian Renovations of the Novel: Narrative Annexes and the Boundaries of Representation* (1998).

HERBERT LINDENBERGER, Avalon Foundation Professor of Humanities Emeritus at Stanford University, has written books on Wordsworth, Büchner, Trakl, historical drama, critical theory, and opera, and is currently working toward a collection of essays on the interrelation of the arts, of which the chapter in this volume will be a part.

KATJA MELLMANN is a research fellow in the German Literature department at the University of Göttingen. Her scholarship has focused on integrating theory and research from cognitive and evolutionary psychology to analyze the emotional potential of text structures that emerged in eighteenth-century literature. Her publications include *Emotionalisierung* (2006). Other research interests include poetic meter, narratology, and popular women's literature in the nineteenth century.

KLARINA PRIBORKIN completed her Ph.D. at Bar-Ilan University. She writes about the problems of cross-cultural communication between ethnic mothers and their American-born daughters in contemporary ethnic American literature.

ELLEN SPOLSKY is Professor of English at Bar-Ilan University, Israel. She is a literary theorist with an appetite for biological theories such as cognitive cultural theory, iconotropism, and performance theory. Her books and essays have worked toward understanding both the universal and the historically local aspects of Renaissance art, poetry, and drama.

RICHARD WALSH is a senior lecturer in English and Related Literature at the University of York, where he teaches primarily narrative theory, early film, and American literature. His first book, *Novel Arguments: Reading Innovative American Fiction*, argued for the positive rhetorical force of non-realist narrative modes, and opened up the line of inquiry that has defined his subsequent research in the field of narrative theory. This work has culminated in *The Rhetoric of Fictionality: Narrative Theory and the Idea of Fiction* (2007). This book proposes a fundamental reconceptualization of the role of fictionality in narrative, and in doing so challenges many of the core assumptions of narrative theory.

LISA ZUNSHINE is the Bush-Holbrook Professor of English at the University of Kentucky, Lexington. She is a former Guggenheim fellow and the author of *Bastards and Foundlings: Illegitimacy in Eighteenth-Century England*, *Why We Read Fiction: Theory of Mind and the Novel*, and *Strange Concepts and the Stories They Make Possible: Cognition, Culture, Narrative*, and editor or co-editor of six collections of essays, including *Approaches to Teaching the Novels of Samuel Richardson* (with Jocelyn Harris) and *Introduction to Cognitive Cultural Studies*.

Index

Abbott, H. Porter, 4–5, 24, 134, 137–138, 205–226, 313
aboutness, 212, 216–217
Abraham, F. Murray, 55–56
Absorption and Theatricality (Fried), 179–183, 191–202
Adamson, Sylvia, 77
Addis, Donna Rose, 43
Adonais (Shelley), 24–25
Adorno, Theodor, 28
aesthetics, 69, 89n30
affective responses. *See* emotions
alienation effect, 25–26, 65
Althusserian Marxism, 98
altruism, 62, 67–68, 86–87nn5–6
ambassadorial strategic empathy, 71, 84
ambivalence, 230–234, 240–244
American Psycho, 196–197
Amnesty International, 250n22
amusia, 31
amygdala, 21, 34n2, 88n23; and fear/anger, 13, 15, 67, 249n15
analogy, 40, 44. *See also* metaphors
Anders, Günther, 216
Anderson, S. W., 117n12
antecedent appraisal, 256–262, 277–278
anterior cingulate cortex, 232, 247n3, 248n5
anthropological model reader, 138n2
anthropomorphic fallacy, 121

anthropomorphic focalization, 124
apparent reality, law of, 257–259
appetite stimulant hormone, 41–42
appraisal theory, 251–256, 281
Aquinas, Thomas, 255
Areopagitica, 39
Argento, Dario, 117n7
Arijon, Daniel, 117n11
Aristotle, 2, 38; and catharsis theory of tragedy, 13–16, 22, 24, 25, 34n3; on emotions, 255; on instinct of imitation, 15–16; *Poetics* by, 13–16; on tragic hero, 16, 31
arts, 24, 33, 64; and brain disorders, 31, 35n13; evolutionary role of, 23–24, 35n9; and Modernism, 9, 24–30; as natural, 22–23; neuroscience and cognitive science on, 13–34. *See also* drama; humanities; music; painting
Atonement (McEwan), 217
attribution emotions, 260, 266–267
Austen, Jane, 76–77, 187
auteur theory, 97
authorial strategic empathizing, 83
autism, 31
autoactivation, 47
autonomic nervous system, 88n23
Averill, James R., 269
Ayotte, Julie, 31

Baillie, Johanna, 188–189, 191, 194
Bakhtin, Mikhail, 155, 169

Balanced Emotional Empathy Scale (BEES), 88n20
Bandura, Albert, 244–245
Banfield, Ann, 213–214
Barthes, Karl, 87n11, 125
Bazin, André, 100, 116n5
Beckett, Samuel, 24, 211–217
Beer, Jennifer S., 247–248n4
BEES. *See* Balanced Emotional Empathy Scale (BEES)
Beethoven, Ludwig van, 17–18, 21, 22
Belisaire (David), 182, 193–194
Belisaire (Peyron), 182, 193
Belisaire (Vincent), 182, 193
Belisarius Receiving Alms (Borzone), 193–194
Benjamin, Jessica, 168
Bering, Jesse M., 184–185
Bible, 39, 46, 130–131
biolinguistic approach to visual computation, 105–107
Birth of a Nation, The, 117
Blood, Anne J., 20–21
Blum, Lawrence, 89n32
body language, 47–49, 58n12, 80, 87n8, 162–163, 185–186. *See also* transparency, embodied
Booth, Wayne, 75, 76–77, 86, 91n62
Bordwell, David, 95–96, 98, 116n4
Bortolussi, Marisa, 73, 75, 86, 90n39, 90n43, 91n48, 93n79, 119, 121, 138n2
Borzone, Luciana, 193
brain, 29; disorders of, 31, 35n13; imaging of, 6, 20–22, 61, 65–66, 88n21. *See also* neuroscience; *and specific areas of brain*
Branigan, Edward, 218
Brecht, Bertolt, 25–26, 64–65
Broadway Boogie Woogie (Mondrian), 26
Broca region (language center), 7
Bruner, Jerome, 37, 38, 39–40, 43
Buchanan, George, 259
Bühler, Karl, 116n6, 132
Bunge, Mario, 9
Burke, Edmund, 19

Caché, 218–223, 226nn8–10
Cacioppo, John T., 63, 249n14
Cage, John, 9, 29–30
Campbell, Lily, 255
Cannon, Walter B., 57n7
Carroll, Joseph, 6, 121
Carroll, Nöel, 95–96, 98, 116n4
Carroll, William, 256, 258, 259, 263, 274
Carter, Cameron S., 247n3
catharsis theory, 13–16, 22, 25
Changeux, Jean-Pierre, 47, 59n21
character identification, 70, 72–76, 90n39, 91n48, 91n50, 206
Chardin, Jean-Baptiste-Simeon, 179–181, 191, 192, 194, 198–199, 200, 202n1
Chateaubriand, François-René, 19
Cheung, King-Kok, 177nn2–3
children, 37, 41, 45, 47, 49, 58n14, 117n12, 164
China Men (Kingston), 165
Chinese-American culture. See *Woman Warrior, The* (Kingston)
Chomsky, Noam, 2, 102–107, 111, 113–114
Christie, Agatha, 217
cinema. *See* film studies
Citizen Kane, 117n9
Clarissa (Richardson), 200
Clark, Andy, 45, 50, 53, 57n5
classics, 45–46
Clore, Gerald L., 252, 262
Coetzee, J. M., 224
cogmotions, 89n29
Cognitive Approaches to Literary Studies, 68–69
cognitive epistemology, 44–47
cognitive science: and appraisal theory of emotions, 251–256; and arts, 13–34; and dreaming, 144–145; emotions and cognition, 68–69, 89n29; and metacognition, 148–150; and narrative, 39–44. *See also* Theory of Mind
cognitive unit formation, 264–266
Cohn, Dorrit, 77

Coleridge, Samuel Taylor, 257, 268–269
Colin, Michel, 103, 104–105, 109, 111
Collins, Allan, 252, 262
color-blindness, 31–32
colors, in painting, 16–18, 22, 26, 27–28, 31–32
Commons, Michael, 177–178n5
competence/performance distinction, 109–111
concepts, 108–109
conceptual competence, 109–110
concern, law of, 253, 258
Confessions of an English Opium Eater (De Quincey), 19
Constructivist art, 30
Cook, Rufus, 165
Corot, Jean-Baptiste-Camille, 28
Cosmides, Leda, 68, 129–130, 132–133
Cranmer, Thomas, 39
Critique of Judgment (Kant), 19
Culler, Jonathan, 226n11
culturalism, 96
Cunningham, Henry, 257
Currie, Gregory, 137

Damasio, Antonio, 7, 15, 41, 45, 58n14, 68
Dans le labyrinth (Robbe-Grillet), 223–224
Darwin, Charles, 23, 35n9, 87n8, 108
David, Jacques-Louis, 182, 193–194, 197
David Copperfield (Dickens), 90n42
Deacon, Terrence, 145
de Certeau, Michel, 186
deconstruction, 116n1
Defoe, Daniel, 53
Degas, Edgar, 32
DeGracia, D. J., 148–150
Deigh, John, 87n8
Dennett, Daniel, 58n11, 58n18, 141–142
De Quincey, Thomas, 19, 266
Derrida, Jacques, 116n1

Descent of Man (Darwin), 23
DeSousa, Ronald, 68
De Stijl movement, 26
De Waal, Frans B. M., 66–67, 247n1, 248n5
Dickens, Charles, 90n42
Diderot, Denis, 125–126, 182–183, 196, 203n10
digestive system, 42–43, 44
discourse, free indirect, 77, 167–168
Dissanayake, Ellen, 205
Dixon, Peter, 73, 75, 86, 90n39, 90n43, 91n48, 93n79, 119, 121, 138n2
dorsomedial midbrain, and sublimity, 21
drama: by Brecht, 25–26; and catharsis, 13–16, 22, 25, 34n3; Diderot on, 203n10; and Modernism, 24, 25–26; and prosopagnosia, 33. *See also* Shakespeare, William; tragedy; *and specific plays*
dreaming, 144, 146, 156; affective response to, 156–157; cognitive dimension of, 144–145; Dennett on, 141–142; fictionality of, 151–152; Freud on, 142–143, 145, 146; lucid dreaming, 148–150, 154; medium of, 155–156; and metacognition, 148–150; and narrative theory, 141–157; narrativity of, 152–153; narrator of, 153–154; physiological perspective on, 143–144; psychological perspective on, 143–146; and REM versus non-REM sleep, 143–144, 149; self in, 147–148; story and discourse of, 153; voice of, 154–155
Driscoll, D., 117n12
drugs, and drug abuse, 19, 21

Eakin, Paul, 174
Easterlin, Nancy, 6
Eibl, Karl, 131, 132, 133, 136–137
Eine kleine Nachtmusik (Mozart), 21
Eisenberg, Nancy, 87n9, 88n20, 236, 248n8

Eisenberger, Naomi I., 232, 247n2, 247n4
Ekman, Paul, 87n8
elaboration, 276–277
Eliot, George, 53, 56
Emonds, Joseph, 213
emotions, 249n14, 253, 254, 260–262, 267, 276; affective response to dreams, 156–157; appraisal of, 231–234, 248nn8–9, 251–256; and attribution, 260, 266–267; cognition, 68–69, 89n29; conflicting, 231–232; constitutive rules of, 269–274; and contagion, 63–64; despair as, 254, 278; and generative principles of action, 230–231; in *Macbeth*, 251–280; and memories, 230–231, 234, 236; procedural rules of, 269–274; prospect-based emotions, 256–262; rules of, 269–276; and storytelling, 63–64; valence and difficulty, 253–254, 262–264
empathy, 15, 62, 80, 87n10, 87n14, 91n60; accuracy of, 80–82, 89; and aesthetics, 69, 89n30; and altruism, 62, 67–68, 86–87nn5–6; of authors, 71, 79–80, 82–84; bounded strategic, 71, 83–84; and the brain, 13, 14–15, 22, 88n18, 88n23; broadcast strategic, 71–72, 84–85; and character identification, 70, 72, 73–76, 90n39, 91n48, 91n50, 206; definition of, 62–65; development of, 177–178n5; and ethics, 236–240, 248–249n11; and fiction, responses to, 70–72; and guilt, 235–236; in Kingston's *Woman Warrior*, 89–90n38; and mirror neurons, 15–16, 34–35nn4–5, 61–62, 86n3, 89n25, 238; and narrative, 73, 70–78, 89–90n38; and neuroscience, 13, 14–15, 34–35nn4–5, 61–62, 65–67, 86n3, 87n8, 88n21, 88–89nn23–25, 89n29; and readers, 79–85; physiological measures of, 65, 88n19;

psychologists' study of, 65–70; scales, 65, 66, 79, 88n20, 88n22; and similarity/in-groups, 69–70, 89nn33-35, 238–240, 249n17; sympathy distinguished from, 62–63, 87n9; and tragic heroes, 13–16, 25, 26, 34n3
Engel, Johann Jacob, 189, 194
Enigma (Leviant), 27
Eroica (Beethoven), 19
ethics, 234–236, 238, 240–244, 248–249n11, 249nn13–14; and ambivalence, 230–234, 240–244; Kant on, 234, 236, 237, 249n13; and regulative principles, 234–236; and stories, 236–240, 249n17; and torture, 227–228, 240–244
Europeras 1 & 2 (Cage), 29
evolution: and language development, 108; and narrative, 40–41, 49–50; role of arts in, 23–24, 35n9

face blindness. *See* prosopagnosia
Fanon, Frantz, 244, 250n19
Fauvism, 27–28
fear, amygdala associated with, 13, 15, 67, 249n15
feelings. *See* emotions
Feldman, Jerome A., 57–58n9
feminist criticism, 65, 81, 86n5, 90n44
Feshbach, Norma D., 90n41
Fictional Minds (Palmer), 165
Fielding, Henry, 53
Fifth Symphony (Beethoven), 18–19, 21, 22
filmic sign, 100–101
film studies, 33, 97–98, 116n5, 135; biolinguistic approach to visual computation, 105–107; current state of, 95–96; and Film Theory, 95–96; garden-path effects in films, 217–223; and generative enterprise, 103–105; and *grand syntagmatique*, 98, 101–105, 111–113, 117n14; and language, 99–109; and linguistic compe-

tence, 109–111, 115; and linguistic dialects, 114–115, 117n15; Marxist approach to, 96, 98; and neuroscientific research, 15, 30; post-theory stage of, 95–96, 98, 116n4; psychoanalytic approach to, 96, 98; semiotics of film, 95–116; and social challenge, 114–115, 117n15; and spectatorial system, 111–113; strong/weak approaches to, 95, 116n1; and structuralism, 98–101; and universal grammar, 107–109; and voice and perception, 135–138. *See also* voice and perception; *and specific films and directors*
Fils ingrat, Le (Greuze), 181, 199
Fils puni, Le (Greuze), 181, 199
Finnegans Wake (Joyce), 212
first-person narration, 72, 76, 77–78, 85, 89–90nn38-39
Fisch, Harold, 46
Fiske, Susan T., 229
Flaubert, Gustave, 125
Fludernik, Monika, 140n9
Flush (Woolf), 120–129
fMRI (Functional Magnetic Resonance Imaging), 21, 61, 65–66, 88n21
focalization, 119–120, 122–125, 133–134, 138n1, 138–139nn4-5
Fodor, Jerry, 103, 107
Forster, E. M., 75
Foulkes, David, 143, 144–146
Freeman, Walter, 23
Frege, Gottlob, 112
Freud, Sigmund, 89n30, 142–143, 145, 146
Fried, Michael, 179–183, 191–202
Frijda, Nico, 251–254, 257–258, 265, 268, 275–278
fusiform gyrus, 30, 32

Gadamer, Hans-Georg, 256
Gallese, Vittorio, 34n4, 86n3, 162, 249n14
Gattaca, 187
gaze, 49, 140n11, 189, 202–203n8

Gazzaniga, Michael, 6–7, 248n10
Genette, Gérard, 119–120, 122–123, 125, 133, 138, 138nn1-2, 138–139nn4-7, 155
generative principles of action, 230–231
Gerrig, Richard J., 75, 206
ghrelin, 41–42
Gilgamesh epic, 130–131
Goldstein, Avram, 19–20
Goodman, Nelson, 225n5
Gottschall, Jonathan, 6
Gould, Stephen, 41–42, 58n13
Grady, Hugh, 255–256
grammar, universal, 107–109
grand syntagmatique, 98, 101–105, 111–113, 117n14
Greene, Joshua, 7
Greuze, Jean-Baptiste, 181, 196
Griffith, D. W., 117n8
Gross, Kenneth, 60n38
Gutiérrez-Rexach, Javier, 3–4, 95–117, 313

Hakemulder, Jèmeljan, 73, 74, 91n57
Handel, George Frideric, 19
Haneke, Michael, 218–223, 226n10
Hart, F. Elizabeth, 6
Hatfield, Elaine, 63, 249n14
Hauser, Marc D., 7, 109, 110, 234, 249n14
Hawkes, Terence, 255–256
Haydn, Joseph, 19
Heidegger, Martin, 116n1
Herman, David, 37–38, 39, 43, 142, 153, 155
hermeneutic theory, 44–47
Hernadi, Paul, 41, 49–50
heroic tragi-comedy, 237–238
hippocampus, 21, 41
Hobson, J. Allan, 143–145
Hoffman, Martin, 67
Hoffmann, E. T. A., 18–19, 21, 22, 24
Hogan, Lalita Pandit, 5, 35n5, 251–280, 313
Hogan, Patrick Colm, 5, 75–76, 92n65, 225n2, 227–250, 314

Holinshed, Raphael, 259, 273–274
homeostasis, 39, 57n7
Homer, 19, 119, 130–131
Homere endormi (David), 182
Homere recitant (David), 182
Horace, 38
Hugh Selwin Mauberley (Pound), 25
human information consumption
theory, 43–44
humanities, 1–9. *See also* arts; drama;
music; narrative; painting
human rights, 92n74
Hume, David, 66, 69
Huron, David, 6, 28, 35n12, 35n14

Ill Seen Ill Said (Beckett), 211–215
imitation, 15–16, 212
individualism, 173–176, 178n7
Ingres, Jean Auguste Dominique, 33
in-groups/out-groups, 69–70,
89nn33-35, 238–240, 249n17
Interpersonal Reactivity Index (IRI),
79, 88n20, 88n22
Intersubjectivity, Theory of, 168–
169
In the Heart of the Country (Coetzee),
224
IRI. *See* Interpersonal Reactivity
Index (IRI)
Iser, Wolfgang, 91n55
Ito, Tiffany, 239, 248n5

Jackendoff, Ray, 218
Jacques le Fataliste (Diderot), 125–126
Jahn, Manfred, 142, 153, 217, 218,
219, 225n6
James, Henry, 124
Jeannerod, Marc, 40
Job, Book of, 46
Johnson, Mark, 58n9, 121
Joseph Andrews (Fielding), 53
Jost, François, 135
Joyce, James, 212–214, 216
Jurassic Park, 101

K'ang Yu-wei, 249n13
Kanizsa, G., 117n13

Kant, Immanuel, 19, 234, 236, 237,
249n13
Kastan, David Scott, 265, 266
Keats, John, 24–25, 216
Keen, Suzanne, 3, 35n5, 61–93, 314,
343
Keiller, Patrick, 136
Kenner, Hugh, 212–213
Kern, Stephen, 187–188
King Lear (Shakespeare), 46, 227–230,
234, 236–238, 245–247
Kingston, Maxine Hong, 161–177
Kirsch, Arthur, 264
Knight, Wilson, 255
knowledge: acquisition and devel-
opment of, 1–2, 8; as embodied,
44–45, 49
Koenigs, Michael, 14–15
Kolbert, Elizabeth, 37
Kondo, Hirohito, 232
Kovecses, Zoltan, 276
Kuiken, Don, 73–74, 91n52

LaBerge, S., 148–150
Labov, William, 114
Lacan, Jacques, 98, 116n4
Lakoff, George, 58n9, 121
language, 99, 132, 162; biolinguistic
approach to visual computation,
105–107; Chomsky on, 2, 102–107,
111, 113–114; faculty, 103–104;
and film, 99–109; and linguistic
competence, 109–111, 115; and linguistic dialects, 114–115, 117n15;
music and language processing,
23, 35n7; organ, 105–106; and
pointing gesture, 101, 116n6;
and universal grammar, 107–109
language of thought hypothesis, 107–
108
Lawson, Mark, 221–222
Lazarus, Richard, 252–253
learning, 40, 41, 42, 45, 49–52. *See also*
cognitive science; neuroscience
Leavis, F. R., 25
LeDoux, Joseph, 6, 67, 68, 248n9,
249n15

Lee, Vernon, 64, 89n30
Le Guin, Ursula, 217
Leonardo da Vinci, 16–17, 22, 24
Leviant, Isia, 26–27
Levitin, Daniel J., 21–22, 23, 29, 31, 35n16
Lewontin, R. C., 41–42, 58n13
Lieberman, Matthew D., 232, 247n2, 247n4, 248n8
Lindenberger, Herbert, 3, 5, 7, 13–35, 314
linguistic competence, 109–111, 115
linguistics. *See* language
linguistic sign, 100
Lipps, Theodor, 64, 87n12, 89n30
literary narratives. *See* narrative
Livingstone, Margaret, 17, 26, 27
Lodge, David, 78, 209, 210
Long Day's Journey into Night (O'Neill), 14
Longinus, 19
Lorig, Tyler S., 89n29
Lost Highway, 116n4
Louwerese, Max, 74
Luhmann, Niklas, 139–140n8
Lynch, David, 116n4

Macbeth (Shakespeare), 14, 51, 257, 272; antecedent appraisal in, 256–262, 277–278; appraisal of truth and consent in, 264–269; attribution emotions in, 260, 266–267; conceptual metaphors in, 276–277; dagger in, 271–272; ending of, 275–280; fortunes-of-others emotions in, 253, 260–262; and Holinshed, 259, 273–274; Lady Macbeth in, 266–273, 275, 276, 279; life themes in, 255, 256; movement restraint in, 261, 263, 275, 279; prospect-based emotions in, 256–262; proximity variable in relation to emotion intensity in, 262–264; rules of emotion in, 269–276; witches' prophecy in, 251, 256–265, 268, 275, 277–279

MacDonald, Angus W., 232, 247n3
Mallen, Enrique, 110
Marcus, Leah, 60n38
Marcus, Mitchell, 207
Marmor, Michael F., 17, 27, 32
Marr, David, 106, 110
Marx, Karl, 2
Marxism, 89, 96
Massai, Sonia, 60n38
"Master-Pieces" (Stein), 216
"Mazes" (Le Guin), 217
McConachie, Bruce, 6
McCoy, Alfred W., 243, 244, 250n20
McEwan, Ian, 217
McGill University, 20–21, 30
McHale, Brian, 209, 224
Mehrabian's Balanced Emotional Empathy Scale, 88n22
Mellmann, Katja, 4, 119–140, 138nn2–3, 314
memory, 41, 68, 146, 164; emotional memories, 230–231, 234, 236; Schachter on, 42, 43
Menon, Vinod, 21–22, 29
Merchant of Venice, The (film), 54–55
Merchant of Venice, The (Shakespeare), 54–56, 60n36, 60n38
Merchant of Venice, The (theater production), 55–56
Meryon, Charles, 32
metacognition, 148–150
metaphors: physiological ground of, 57–58n9; in Shakespeare's plays, 276–277; source domains of, 276. *See also* analogy
metarepresentational ability, 6, 172
Metz, Christian, 98, 99–105, 112
Meyer-Lindenberg, Andreas, 31, 35n16
Miall, David S., 73, 75, 77, 91n46, 91n52
Michaels, Walter Benn, 196–197
Middleton, Thomas, 257
Milton, John, 19, 39
mimesis, 152
mind-reading. *See* Theory of Mind
Minimalist Program, 111–113

mirror neurons, 7, 89n25, 249n14; and empathy, 15–16, 34–35nn4–5, 61–62, 86n3, 89n25, 238; and narrative as nourishment, 47–48, 51; and Theory of Mind, 47–48, 51, 59n26, 184

Mithen, Steven, 6, 23

modern art. *See* painting

Modernism: and drama, 25–26; and music, 9, 28–30, 35n14; and poetry, 24–25; and visual arts, 26–28, 30

Moll, Jorge, 7, 15, 248n4

Mondrian, Piet, 26, 27, 28

Monet, Claude, 18, 32

monologues, 77, 169–170

moral sense, 234–235. *See also* ethics

movement restraint, 261, 263, 275, 279

Mozart, Wolfgang Amadeus, 19, 21

Murder of Roger Ackroyd, The (Christie), 217

music, 6, 21, 24, 29; and amusia, 31, 35n13; atonal and contratonal music, 28–30; of Beethoven, 17–18, 21, 22; consonant and dissonant pitch combinations in, 28–29; evolutionary basis for, 23, 35n9; and language processing, 23, 35n7; and Modernism, 9, 28–30, 35n14; of Mozart, 19, 21; research on, 6, 19–22, 23, 28–31; sublimity of, 18–22, 29; Williams syndrome and musicality, 31, 35n16

naloxone, 20

narrative: achieved, 40–41; activity theory of, 37–38, 39, 43; Bruner on, 37, 38, 39–40, 43; cognitive and neurosciences regarding, 39–44; cognitive epistemology and hermeneutic theory on, 44–47; and concept of "narrator," 121–122; and dreaming, 141–157; evolutionary role of, 40–41, 49–50; garden-path sentence and garden-path narrative, 206–225; Genette's comparison of literature

and film, 119–120, 122–123, 125, 138, 138nn1–2, 138–139nn4–7; Herman's activity theory of, 37–38, 39, 43; as nourishment, 37–57; relationship of literature to fundamental human questions, 52–53; tellability of, 46–47, 51; and themes expressing culturally complicated problems, 49–57; and Theory of Mind, 47–49, 51; usefulness of, for individuals and communities, 37–38; voice and perception in, 119–138. *See also* narrative empathy; narrator; *and specific authors and titles of works*

narrative empathy, 72, 73, 79, 80–82, 83, 86, 90–91n46; and altruism, 67–68, 86–87n6; ambassadorial strategic empathy, 71, 84; and author's empathy, 71, 79–80, 82–84; bounded strategic empathy, 71, 83–84; broadcast strategic empathy, 71–72, 84–85; and character identification, 70, 72, 73–76, 90n39, 91n48, 91n50; dynamics of, 79–85; and empathic inaccuracy, 80–82; as nourishment, 37–57; prototypes, 240–242, 249n16; and readers, 80–84; and situation, 73, 76–78; strategic empathizing, 83–84; techniques of, 72–78, 79–85; theory of, 70–78; unanswered questions on, 85–86

narrator, 85; concept of, 121–122; of dreams, 153–154; first-person narrator, 72, 76, 77–78, 85, 89–90nn38–39; third-person narration, 77–78, 85, 90n39. *See also* narrative

negative capability, 216

Neisser, Ulric, 58n20

nervous system. *See* neuroscience

neuroscience, 7, 13, 66, 236: and ambivalence, 230–234; and amygdala, 13, 15, 34n2, 67; and arts, 13–34; brain lesions in prefrontal cortex, 14–15; and conflict monitoring,

232, 247n3, 248n5; and drama, 13–16; and empathy, 13, 14–15, 61–62, 65–67, 86n3, 87n8, 88n21, 88–89nn23–25, 89n29; and face recognition, 30, 32–33; and films, 15, 30; and music, 19–22, 28–31; and narrative as nourishment, 39–44; and painting, 16–18, 26–28; and sublimity, 19–22; visual system, 17–18, 26–28. See also brain: imaging of; mirror neurons
New Criticism, 65
Ngai, Sianne, 62
Nichols, John G., 6
Nietzsche, Friedrich, 116n1
Nowak, Martin, 108
Nussbaum, Martha C., 87n8, 90n42

Oatley, Keith, 73, 239, 254–255
Oedipus Tyrannus (Sophocles), 16, 22
omniscient narrator, 85
O'Neill, Eugene, 14
Op Art, 26–27, 30
orbitofrontal cortex, 88n23, 117n12, 247–248n4, 248n7
Ortony, Andrew, 251, 252, 253, 257, 260, 262, 264–267
Osiel, Mark, 241–242
otherness, 168, 173–176
out-groups. See in-groups/out-groups

painting, 24, 32, 187–188, 195–199; absorptive, 179–183, 191–202; abstract, 26–27; and color-blindness, 31–32; colors in, 16–18, 22, 26, 27–28; depth perception in, 17–18; and embodied transparency, 194–200; and face recognition, 32–33; and Modernism, 26–28, 30; and prosopagnosia, 32–33; research on effects of, 16–18, 26–28
Palmer, Alan, 165, 167, 168
Palmer, Stephen, 110
Pamela (Richardson), 200
Pander Maat, H., 78, 92n72
Panksepp, Jaak, 20, 21, 87n8, 249n15, 254, 278

paralimbic area, 29
Paster, Gail Kern, 255
Patañjali, 244
Peanuts, 58n16
Peirce, C. S., 145
perception, 119; and concept of "narrator," 121–122; different grades of autonomy of voice and, 125–129; dream percepts, 146–147; and focalization, 119–120, 122–125, 133–134, 138n1, 138–139nn4–5; function of, in literary narrative, 119–129; as input for information gathering system, 129–135; in nonverbal narrative, 135–138; and voice, 119–138. See also voice and perception
Peretz, Isabelle, 6, 28–29, 35n13
performance/competence distinction, 109–111
performativity, 186
Perry, Tricia, 162
Persson, Per, 6
PET (positron emission tomography), 6, 20–21
Peterson, Candida C., 162
Peyron, Pierre, 182, 193
Phelan, Jim, 6
Picasso, Pablo, 26, 110
Piete filiale, La (Greuze), 196, 199, 200, 201
Pinker, Steven, 23, 117n13, 205, 225n1
Plantinga, Carl, 98, 249n14
Plays on the Passions (Baillie), 188–189
Plutarch, 255
Poetics (Aristotle), 13–16, 22, 24
poetry, 19, 24–25. See also specific poets
Pointillism, 27, 35n11
Popper, Karl, 39
Porton, Richard, 226n10
Pound, Ezra, 25
Practical Illustrations of Rhetorical Gesture and Action (Siddons), 189–191, 194
"practice of everyday life," 186
Pratt, Mary Louise, 46–47; display texts, 50

preference-rules, 218
prefrontal cortex, 13, 14–15, 232, 248n8, 249n16
presentism, 255–256
Preston, Stephanie D., 66–67, 247n1, 248n5
Priborkin, Klarina, 4, 161–178, 314
Prince, Gerald, 142
Prinz, Jesse J., 249n11
prosopagnosia, 32–33
prosopopoeia, 262
Proust, Marcel, 120
psychology: and Benjamin's Theory of Intersubjectivity, 168–169; of dreaming, 143–146; of empathy, 65–70. *See also* emotions; Theory of Mind
psychonarration, 77

Rabinowitz, Peter J., 82, 83, 92n76
Radford, Michael, 54
Raggatt, Peter T. F., 169–170
Ramachandran, V. S., 7
Rapson, Richard L., 63, 249n14
reality effects, 125, 136
regulative principles, 231, 237
representational hunger: cognitive and neurosciences regarding narrative and, 39–44; cognitive epistemology and hermeneutic theory on, 44–47; narrative as tool to feed, 37–57; and relationship of literature to fundamental human questions, 52–53; in Shakespeare's *Merchant of Venice*, 54–56; and themes expressing culturally complicated problems, 49–57; and Theory of Mind, 47–49, 51
representational systems, 110–111
representation function of language, 132
representations, embodied, 44–45
Richardson, Alan, 188
Richardson, Brian, 83, 84, 92n76, 224
Richardson, Samuel, 200
Ricoeur, Paul, 152
Riley, Bridget, 26–27
Rizzolatti, Giacomo, 15, 34n4, 184

Roach, Joseph, 186
Robbe-Grillet, Alain, 223–224
Robinson Crusoe (Defoe), 53
Rococo art, 192
Rolls, Edmund T., 88–89n24
Roston, Murray, 60n36
Rousseau, Jean-Jacques, 182–183
Russian Constructivism, 26
Ryan, Marie-Laure, 142, 150, 153

Sacre du printemps, Le (Stravinsky), 28
Sarris, Andrew, 97, 116n2
Saussure, Ferdinand de, 99
Saver, Jeffrey L., 249n16
Scarry, Elaine, 250n21
Schacter, Daniel L., 41, 42, 43
Schatz, Sara, 107, 109, 110
schemata, 42–43
Schleiermacher, Friedrich Daniel Ernst, 19
Schmidt, Neil, 209
Schneider, Ralf, 90n40
Schoenberg, Arnold, 3, 28, 29, 35n12
Schopenhauer, Arthur, 276
Schroeder, Timothy, 6
Schultz, Charles, 58n16
sciences, and the humanities, 1–9. *See also specific scientific fields*
"Secret Life of Walter Mitty, The" (Thurber), 217
self, 41, 58n14, 58n20; in dreaming, 147–148; formal self versus actual self, 264–265;
sentimentalism, 199–202
Seurat, Georges, 27
Shakespeare, William: *Julius Caesar* by, 41; *King Lear* by, 46, 227–230, 234, 236–238, 245–247; *Macbeth* by, 51, 251–280; *Merchant of Venice* by, 54–56, 60n36, 60n38; metaphors in plays by, 276–277; and presentism, 255–256
Shelley, Percy Bysshe, 24–25
Sherman-Stotland empathy scale, 88n20
Shore, Bradd, 57n8
Shyalaman, M. Night, 217
Siddons, Henry, 189–191, 194

Silas Marner (Eliot), 53
Singer, Tania, 15, 65–67, 88n22
situational meaning, law of, 258
Sixth Sense, The, 217–218, 225–226n7
Smith, Adam, 69
Smith, Greg, 6, 98
Smith, Murray, 121, 136
Snow, C. P., 1
Soap Bubble, The (Chardin), 179, 180, 191, 194, 198–199, 202n1
Social Cognition, 89n29
social exclusion, 247n2
sociopaths, 15
Solso, Robert, 110
Sophocles, 16, 22
source tags, 172, 178n6,
spectatorial competence, 104–105, 109
speech versus language, 99
Sperber, Dan, 23, 152
Spielberg, Steven, 101
Spinoza, Baruch, 254
Spivak, Gayatri Chakravorty, 83
Spolsky, Ellen, 3, 35n5, 37–60, 186, 202nn7–8, 314–315
Stallybrass, Peter, 258
Stam, Robert, 116n3
Stanford University, 19–20
Stanzel, Franz K., 77–78, 119, 124, 139n6
Stein, Gertrude, 24, 208–211, 212, 216
Steiner, Wendy, 209
Stenning, Keith, 43–44
Storey, Robert, 6
storytelling: Abbott on, 137–138; Herman on, 39; in Kingston's *Woman Warrior*, 170, 175–177; learning from, 63–64, 87–88n15; in prehistoric times, 133, 134, 137–138; purposes of, 41; universality of, 37, 38. *See also* narrative
Stravinsky, Igor, 3, 28
subject-position theory, 96
sublimity, 18–22, 25, 29
Sugiyama, Michelle Scalise, 131, 134, 138
superior temporal gyri, 29

"Susie Asado" (Stein), 210
sympathy, 62–63, 87n9, 89n32. *See also* empathy
syntagma, 101–102, 104–105, 112, 117n14,

Takahashi, Hidehiko, 236, 249n12
Talmud, 46
Tan, Ed, 253–254
Tangney, June Price, 236
Taylor, Charles, 174–175, 178n7
Taylor, Marjorie, 79
temporal cortex, and empathy, 88n23
Tender Buttons (Stein), 208–211
Terminator II, 101
theater arts. *See* drama; tragedy; *and specific plays*
theatricality: and performance, 186; of representation, 182–183
theories, adequacy of, 103–104
Theory of Intersubjectivity, 168–169
Theory of Mind, 162, 183, 202n5, 236; and absorptive paintings, 191–202; and body language, 185–186; definition of, 47, 162, 183; and embodied transparency, 187–191, 194–201, 203n11; and Fried's *Absorption and Theatricality*, 191–202; and mirror neurons, 47–48, 59n26, 184; and negotiation between psychology and history, 191–195; and nourishment, 47–49, 51; and sentimentalism, 199–202; underlying assumptions of, 183–187
"thrills" and "chills" phenomenon, 19–21, 25
Thurber, James, 217
Titchener, E. B., 64, 89n30
Tooby, John, 68, 129–130, 132–133
torture, 228, 240–244, 250n19; in Algeria, 244, 250n19; and Amnesty International, 250n22; in Argentina, 241–242; Bandura on, 244–245; and ethics, 227–228, 240–244; in Iraq, 229, 243–244; justifications for, 228, 242–245, 250n21; in *King Lear*, 227–230,